# SUSTAINABLE AGRICULTURAL PRODUCTION SYSTEMS

## *Traditional and Modern Perspectives*

# SUSTAINABLE AGRICULTURAL PRODUCTION SYSTEMS

## *Traditional and Modern Perspectives*

*Edited by*

**Romeo Rojas, PhD**

**Guillermo Cristian Guadalupe Martínez-Ávila, PhD**

**Julia Mariana Márquez-Reyes, PhD**

**Ma del Carmen Ojeda-Zacarías, PhD**

AAP | APPLE ACADEMIC PRESS

First edition published 2025

**Apple Academic Press Inc.**
1265 Goldenrod Circle, NE,
Palm Bay, FL 32905 USA

760 Laurentian Drive, Unit 19,
Burlington, ON L7N 0A4, CANADA

**CRC Press**
2385 NW Executive Center Drive,
Suite 320, Boca Raton FL 33431

4 Park Square, Milton Park,
Abingdon, Oxon, OX14 4RN UK



---

**Library and Archives Canada Cataloguing in Publication**

Title: Sustainable agricultural production systems : traditional and modern perspectives / edited by Romeo Rojas, PhD, Guillermo Cristian Guadalupe Martínez-Ávila, PhD, Julia Mariana Márquez-Reyes, PhD, Ma del Carmen Ojeda-Zacarías, PhD.
Names: Rojas, Romeo, editor. | Martínez-Ávila, Guillermo Cristian Guadalupe, editor. | Márquez-Reyes, Julia Mariana, editor. | Ojeda-Zacarías, Ma. del Carmen, 1966- editor.
Description: First edition. | Includes bibliographical references and index.
Identifiers: Canadiana (print) 2024040050X | Canadiana (ebook) 20240400518 | ISBN 9781774915547 (hardcover) | ISBN 9781774915554 (softcover) | ISBN 9781032684239 (ebook)
Subjects: LCSH: Sustainable agriculture. | LCSH: Agricultural systems. | LCSH: Agricultural innovations.
Classification: LCC S494.5.S86 S946 2025 | DDC 630.2/086—dc23

**Library of Congress Cataloging-in-Publication Data**

CIP data on file with US Library of Congress

---

ISBN: 978-1-77491-554-7 (hbk)
ISBN: 978-1-77491-555-4 (pbk)
ISBN: 978-1-03268-423-9 (ebk)

# About the Editors

**Romeo Rojas, PhD**
*Scientist, School of Agronomy, Universidad Autonoma de Nuevo Leon, Nuevo León, México*

Romeo Rojas, PhD, is a Scientist and Professor at the School of Agronomy at the Autonomous University of Nuevo León, Mexico, specializing in foods, biotechnology, and agricultural sciences. He is a member of the National Researches System (SNI) and the Mexican Society of Biotechnology and Bioengineering. He has been a thesis director of one PhD thesis, two MSc theses, and four BSc theses. Dr. Rojas is currently developing two research projects funded by the Agricultural Secretariat of Mexico, National Forestry Commission, and the National Mexican Research Council (SAGARPA-CONACYT; CONAFOR-CONACYT) and two collaborations on projects funded by the National Forestry Commission and National Mexican Research Council (CONAFOR-CONACYT). He graduated as a Food Science and Technology Engineer from the Autonomous Agrarian University Antonio Narro, where he specialized in agricultural sciences. His PhD degree in Food Science and Technology was earned at the Autonomous University of Coahuila, Mexico. He had academic stays in Portugal (Univeridade do Minho) and in Queretaro, México (Universidad Nacional Autónoma de Mexico). Dr. Rojas has published over 37 original research papers in indexed journals and 28 book chapters and has participated in and contributed to over 50 scientific meetings.

**Guillermo Cristian Guadalupe Martínez-Ávila, PhD**
*Scientist, School of Agronomy, Universidad Autonoma de Nuevo Leon, Nuevo León, México*

Guillermo Cristian Guadalupe Martínez-Ávila, PhD, is a Scientist and Professor at the School of Agronomy of the Autonomous University of Nuevo León, Mexico, specializing in foods, biotechnology, and agricultural sciences. He has been listed as an outstanding reviewer for journals such as *Food Research International, Heliyon, LWT-Food Science and Technology,* etc. He is a member of the National Researches System (SNI), Mexican Society of Biotechnology and Bioengineering, and Mexican Association of Food Chemistry. He has been a thesis director of five MSc theses, and one PhD theses. Dr. Martínez Ávila is currently developing two research projects funded by Agricultural Secretariat of Mexico, National Forestry Commission, and the National Mexican Research Council (SAGARPA-CONACYT; CONAFOR-CONACYT) and two collaborations on projects funded by the National Forestry Commission and National Mexican Research Council (CONAFOR-CONACYT). He graduated as Chemical pharmacobiologist from the Autonomous University of Coahuila, specializing in food sciences. His PhD degree in Biotechnology was awarded by the Autonomous University of Coahuila. He has had two academic stays, in Ireland (University College Cork) and in Fortaleza, Brazil (Universidade de Fortaleza). Dr. Martínez has published over 41 original research papers in indexed journals (JCR) and 10 book chapters and has participated in and contributed to over 50 scientific meetings.

**Julia Mariana Márquez-Reyes, PhD**
*Scientist, School of Agronomy, Universidad Autonoma de Nuevo Leon, Nuevo León, México*

Julia Mariana Márquez-Reyes, PhD, is a Scientist and Professor at the Universidad Autonoma de Nuevo Leon, School of Agronomy, Nuevo León, México, specializing in foods and biotechnology sciences. She is a member of the National Researches System (SNI). She has been a thesis director of several MSc and BSc theses. Dr. Reyes is currently developing four research lines: 1) Bioreactors with anaerobic activity for the removal of contaminants; 2) Phytoremediation for the control of heavy metals in water and soil; 3) Enzymatic activity and antioxidant capacity of plant organisms used in environmental biotechnology, and 4) Development of sustainable technologies. She graduated as an Environmental Engineer from the Universidad del Mar, Puerto Ángel, Oaxaca, México. Her MSc in Environmental Sciences was earned at Autonomous University of San Luis, Potosí, México, and her PhD degree in Biotechnology was awarded by Autonomous University of Nuevo León, México. Dr. Marquez has published eight original research papers in indexed journals and 12 original research in peer-reviewed and book chapters and has participated in and contributed to several scientific meetings.

**Ma del Carmen Ojeda-Zacarías, PhD**
*Scientist, Department of Food Research, School of Chemistry, Universidad Autónoma de Coahuila, Coahuila, Mexico*

Ma del Carmen Ojeda-Zacarías, PhD, is a Scientist, Universidad Autonoma de Nuevo Leon, School of Agronomy, General Escobedo, Nuevo León, México, specializing in engineering, biotechnology, and agricultural sciences. She is a reviewer for journals such has Plant Cell Biotechnology and Molecular Biology and Mexican Plant Breeding (Mexican Phytotechnics). She is a member of the National Investigation System (SNI), Mexican Society of Plant Micropropagation and Vegetable Automation, Mexican Society of Phytogenetics, and Mexican College of Agronomists, North Division. She has been a thesis director of two PhD theses, seven MSc theses, and two BSc theses. She graduated as an agronomist from the School of Agronomy, Autonomous University of Nuevo Leon (UANL), specializing in Biotechnology. Her PhD degree in Agricultural Sciences was awarded by the same institution. She has participated in three academic stays, at the Texas A&M University, USA; Center for Research and Advanced Studies of the IPN (CINVESTAV), the National Institute of Forestry; and Agricultural and Livestock Research (INIFAP), Mexico. Dr. Ojeda has published 30 original research papers in indexed journals and nine book chapters and participated and contributed to over 50 scientific meetings.

# Contents

# Contributors

**Pedro Aguilar-Zarate**
Tecnológico Nacional de México Campus Ciudad Valles, Departamento de Ingenierías, Carretera al Ingenio Plan de Ayala Km, Ciudad Valles, México

**Omar Guadalupe Alvarado-Gómez**
Autonomous University of Nuevo León, School of Agronomy, General Escobedo, Nuevo León, Mexico

**Nestor Aranda-Ledezma**
Universidad Autonoma de Nuevo Leon, School of Agronomy, Francisco I. Madero S/N, Ex Hacienda el Cañada, General Escobedo, Nuevo León, Mexico

**Sigifredo Arévalo-Gallegos**
Laboratorio de Biotecnología I, Facultad de Ciencias Químicas, Universidad Autónoma de Chihuahua, Circuito Universitario s/n Nuevo Campus Universitario, Chihuahua, México

**Julio C. Tafolla Arellano**
Laboratorio de Biotecnología y Biología Molecular, Departamento de Ciencias Básicas, Universidad Autónoma Agraria Antonio Narro, Calzada Antonio Narro, Buenavista, Saltillo, Coahuila, Mexico

**Víctor Ávila-Akerberg**
Instituto de Ciencias Agropecuarias y Rurales, UAEMéx, México

**Víctor Hugo Avendaño-Abarca**
Centro de Bachillerato Tecnológico Agropecuario, Madera, Chihuahua, México

**Israel Bautista-Hernández**
Universidad Autónoma de Nuevo León, School of Agronomy, General Escobedo, Nuevo León, México

**Vania Noemí Hernández Cárcamo**
Universidad Autónoma de Nuevo León, Facultad de Agronomía, Laboratorio de Biotecnología Microbiana,, General Mariano Escobedo, Nuevo León, México

**Adriana Tovar Delgado**
Laboratory of Molecular Biology and Biotechnology, Natural Resources Research Center (CIRENA), Chihuahua, México

**Edward Alexander Espinoza-Sánchez**
Laboratorio de Biotecnología I, Facultad de Ciencias Químicas, Universidad Autónoma de Chihuahua, Circuito Universitario s/n Nuevo Campus Universitario, Chihuahua, México

**Héctor Silos Espino**
Laboratorio de Biotecnología Aplicada, Instituto Tecnológico El Llano, Aguascalientes, Km, San Luis Potosí, México

**Celestino García-Gómez**
Universidad Autónoma de Nuevo León, Facultad de Agronomía, General Escobedo, Nuevo León, México

**Carmen Daniela González-Barriga**
Laboratorio de Cultivo de Tejidos, División de Ingeniería y Ciencias, Tecnológico de Monterrey, , Nombre de Dios, Chihuahua, México

**María Dolores González-Hernández**
Universidad Autónoma de Nuevo León. School of Agronomy, General Escobedo, Nuevo León, México

**Dulce Concepción González-Sandoval**
Centro de Bachillerato Tecnológico Agropecuario, Madera, Chihuahua, México

**Varela Gonzalez**
Michoacana University of San Nicolas of Hidalgo, Faculty of Chemical Pharmacobiology , Izintzuntzan, Matamoros. Morelia, Michoacan, Mexico

**Felix Varela Gonzalez**
Autonomous University of Tamaulipas-Mante Multidisciplinary Academic Unit. E. Cárdenas González, Poniente, Jardin, Ciudad Mante, Tamaulipas, Mexico

**Katty Shcarlen Hernández-Gaytán**
Universidad Autónoma de Nuevo León, Facultad de Agronomía, Ex-Hacienda El Canadá, General Escobedo, Nuevo León, México

**Melva Guadalupe Herrera-Godina**
Salud Pública, Universidad de Guadalajara, Guadalajara, Jalisco, México

**Silvana Escalante Hidalgo**
Licenciatura en ciencias de la Tierra, Facultad de Ciencias, UNAM, Mexico

**Blanca Flor Iglesias-Figueroa**
Laboratorio de Biotecnología I, Facultad de Ciencias Químicas, Universidad Autónoma de Chihuahua, Circuito Universitario s/n Nuevo Campus Universitario, Chihuahua, México

**Brenda Julián-Chávez**
Laboratorio de Biotecnología I, Facultad de Ciencias Químicas, Universidad Autónoma de Chihuahua, Circuito Universitario s/n Nuevo Campus Universitario, Chihuahua, México

**H. Lozoya-Saldaña**
Autonomous University of Chapingo, Phytotechnics Department, Mexico-Texcoco highway Km, Chapingo, Mexico

**Brenda Luna-Sosa**
Universidad Autonoma de Nuevo Leon, School of Agronomy, Francisco I. General Escobedo, Nuevo León, México

**Julia Mariana Márquez-Reyes**
Universidad Autónoma de Nuevo León, Facultad de Agronomía, General Escobedo, Nuevo León, México

**Guillermo Cristian Guadalupe Martínez-Ávila**
Universidad Autónoma de Nuevo León, School of Agronomy, General Escobedo, Nuevo León, México

**Nancy Medina-Herrera**
Universidad Autonoma de Nuevo Leon, School of Agronomy, General Escobedo 66050, Nuevo León, México

**Karina Moreno-León**
Departamento de Botánica, Universidad Autónoma Agraria Antonio Narro, Saltillo, Coahuila. México

**Guillermo Cristian Guadalupe Martínez-Ávila**
Universidad Autónoma de Nuevo León, School of Agronomy, General Escobedo, Nuevo León, México

**Jesús Arturo Muñiz-Jauregui**
Escuela Nacional de Estudios Superiores, Universidad Nacional Autónoma de México, Juriquilla, Querétaro, México

**Ma del Carmen Ojeda-Zacarías**
Universidad Autonoma de Nuevo Leon, School of Agronomy, General Escobedo, Nuevo León, México

**Emilio Olivares-Sáenz**
Universidad Autonoma de Nuevo Leon, School of Agronomy, General Escobedo, Nuevo León, México

**Gilberto Rodriguez Perez**
National Technology of Mexico-Valle del Yaqui Campus, Valledel Yaqui, Bácum, Ciudad Obregón, Sonora, Mexico

**Mohan Héctor Adrián García Ramírez**
Laboratory of Molecular Biology and Biotechnology, Natural Resources Research Center (CIRENA), Chihuahua, México

**Quintín Rascón-Cruz**
Laboratorio de Biotecnología I, Facultad de Ciencias Químicas, Universidad Autónoma de Chihuahua, Circuito Universitario s/n Nuevo Campus Universitario, Chihuahua, México

**Cristian Reyna-Morales**
Autonomous University of Nuevo León, Faculty of Agronomy, General Escobedo, Nuevo León, Mexico

**Romeo Rojas**
Universidad Autónoma de Nuevo León, School of Agronomy, General Escobedo, Nuevo León, México

**Rosa Elena Pérez Sánchez**
Michoacana University of San Nicolas of Hidalgo, Faculty of Chemical Pharmacobiology, Izintzuntzan, Matamoros, Morelia, Michoacan, Mexico

**Sugey Ramona Sinagawa-García**
Laboratorio de Biotecnología, Facultad de Agronomía, Universidad Autónoma de Nuevo León, General Escobedo, Nuevo León, México

**Tania Samanta Siqueiros-Cendón**
Laboratorio de Biotecnología I, Facultad de Ciencias Químicas, Universidad Autónoma de Chihuahua, Circuito Universitario s/n Nuevo Campus Universitario, Chihuahua, México

**Stephanie Solano-Ornelas**
Laboratorio de Biotecnología I, Facultad de Ciencias Químicas, Universidad Autónoma de Chihuahua, Circuito Universitario s/n Nuevo Campus Universitario, Chihuahua, México

**Julio C. Tafolla Arellano**
Laboratorio de Biotecnología y Biología Molecular, Departamento de Ciencias Básicas, Universidad Autónoma Agraria Antonio Narro, Calzada Antonio Narro, Buenavista, Saltillo, Coahuila, Mexico

**Jorge Ariel Torres-Castillo**
Universidad Autónoma de Tamaulipas, Instituto de Ecología Aplicada, Ciudad Victoria, Tamaulipas, México

**Darytza Mabel Treviño-Gómez**
Universidad Autónoma de Nuevo León. School of Agronomy, General Escobedo, Nuevo León, México

**Rigoberto E. Vázquez-Alvarado**
Universidad Autonoma de Nuevo Leon, School of Agronomy, General Escobedo, Nuevo León, México

**Juan Antonio Vidales-Contreras**
Universidad Autonoma de Nuevo Leon, School of Agronomy, General Escobedo 66050, Nuevo León, México.

**William Zárate-Martínez**
Campo Experimental Saltillo, Instituto Nacional de Investigaciones Forestales, Agrícolas y Pecuarias (INIFAP). Carretera Saltillo-Zacatecas km 342+119 No. 9515, Col. Hacienda de Buenavista, CP. 25315, Saltillo, Coahuila de Zaragoza, México

# Abbreviations

| | |
|---|---|
| ABA | abscisic acid |
| APX | ascorbate peroxidase |
| ATP | adenosine triphosphate |
| CAP | Center for Protected Agriculture |
| CCA | canonical correspondence analysis |
| CE | conventional extraction |
| COS | chito-oligosaccharides |
| Cu | copper |
| DAPG | diacetylphloroglucinol |
| DAS | days after sowing |
| DHAR | dehydroascorbate reductase |
| DMAPP | dimethylallyl diphosphate |
| EC | electrical conductivities |
| EOs | essential oils |
| GB | glycine betaine |
| GDP | gross domestic product |
| GPX | glutathione peroxidase |
| GR | glutathione reductase |
| GST | glutathione-S-transferase |
| HP | hydraulic pump |
| HR | hydrological regions |
| IAA | indole-3-acetic acid |
| IC | color index |
| IDW | inverse distance weighted |
| IPP | isopentenyl diphosphate |
| ISR | induced systemic resistance |
| LD | lethal dose |
| MAE | microwave-assisted extraction |
| MDFFS | model for the detection of Fipronil sulfide formation |
| MDHAR | monodehydroascorbate reductase |
| MLE | *Moringa* leaf extracts |

| | |
|---|---|
| MRDEI | model for areas prone to rapid degradation by environmental interactions |
| MS | Murashige and Skoog |
| MSDEI | model for areas prone to slow degradation by environmental interactions |
| NADPH | nicotinamide adenine dinucleotide phosphate |
| NCBI | National Center for Biotechnology Information |
| NDVI | normalized difference vegetation index |
| NOM | natural organic matter |
| OC | organic carbon |
| OECD | Organization for Economic Cooperation and Development |
| OEMS | EO monoterpenes |
| OH | Ohmic heating |
| OM | organic matter |
| PAL | phenylalanine ammonium lyase |
| PCR | polymerase chain reaction |
| POD | peroxidase |
| *pqq* | pyrroloquinoline quinone synthesis |
| PVC | polyvinyl chloride |
| ROS | reactive oxygen species |
| SFE | supercritical fluid extraction |
| SOD | superoxide dismutase |
| SPSS | Statistical Package for the Social Sciences |
| TALENs | transcription Activator-Like Effector Nucleases |
| TSS | total soluble Soluble solids Solids |
| TYLCV | tomato yellow leaf curl virus |
| UAE | ultrasound-assisted extraction |
| UV | ultraviolet light |
| Vis | visible light |
| WAS | water alteration state |
| WHO | World Health Organization |
| Zn | zinc |

# Preface

Due to the growth of world population, food production has increased exponentially, and with it, the ways to do it. However, there are crops that are the basis of the diet of several cultures (corn, wheat, sorghum, rice, etc.), cities, and entire countries. Through the years, they have been modified in terms of shape and quantities. With this, traditional systems have been implemented and modified to reach protected agriculture, hydroponic systems, and plant factories. For this, it is necessary to make use of phytosanitary systems, mechanize and automate them, as well as the use of technologies such as plant tissue culture. All these systems have evolved over time by using herbicides, pesticides, and other chemicals to improve production. In recent years, these systems have revolved around the use of biopesticides, bioherbicides, reducing the use of glyphosate, and other chemicals that erode the soil and can cause health damage. For this reason, water and soil bioremediation issues are also included that allow the integral and sustainable production of crops. Finally, in recent years, the production of functional foods has started from primary production (in the field) promoting the generation of crops that produce a greater amount of bioactive compounds, generating extraction and purification strategies for their application in other areas, such as the food, pharmaceutical industry up to modern perspectives.

The topics covered in this book are part of the research work conducted at the field, laboratory, and classroom levels with the experience of almost 70 years of training in the School of Agronomy and the 40 years of postgraduate studies in agricultural sciences. Therefore, it is necessary to present the different methods of production of basic crops from their inception to the perspectives and areas of opportunity that will allow readers to know and understand the science and technology of basic crops.

## CHAPTER 1

# In Vitro Propagation of *Agave montium-sancticaroli* for Conservation Purposes

MA DEL CARMEN OJEDA-ZACARÍAS[1], CRISTIAN REYNA-MORALES[1], H. LOZOYA-SALDAÑA[2], EMILIO OLIVARES-SÁENZ[1], R. E. VÁZQUEZ-ALVARADO[1], RIGOBERTO E. VÁZQUEZ-ALVARADO[3], and FÉLIX VALERA GONZALEZ[4]

[1]*Autonomous University of Nuevo León, Faculty of Agronomy, Col. Ex Hacienda El Canada, Gral. Escobedo, Nuevo León, Mexico*

[2]*Autonomous University of Chapingo, Phytotechnics Department, Mexico-Texcoco highway Km 38.5Chapingo, Mexico*

[3]*Michoacana University of San Nicolas of Hidalgo, Faculty of Chemical Pharmacobiology, Matamoros, Morelia, Michoacan, Mexico*

[4]*Autonomous University of Tamaulipas-Mante Multidisciplinary Academic Unit. E. Cárdenas González No. Jardin, Ciudad Mante, Tamaulipas, Mexico*

## ABSTRACT

Jarcia (*Agave montium-sancticaroli*) is a species native to Mexico, and used for the production of mezcal in different regions of Tamaulipas. In order to propagate the species in vitro, a propagation protocol using direct organogenesis was developed as an alternative method for its conservation. Apical buds of in vitro germinated seedlings of four weeks old were established in the basic salts of Murashige and Skoog (MS). For shoot induction, apical buds were used and subcultured in the same MS salts; after eight weeks, the number of explants with shoots and shoots per explant was evaluated, obtaining a mean of 2.77 and 3.6, respectively. The multiplication and

Sustainable Agricultural Production Systems: Traditional and Modern Perspectives.
Romeo Rojas, Guillermo Cristian Guadalupe Martínez-Ávila, Julia Mariana Márquez-Reyes, and Ma del Carmen Ojeda-Zacarías (Eds.)

elongation of shoots was obtained by using the combination of an auxin and a cytokinin in both stages, where the length of shoots was greater in treatment three with an average of 7.38 cm. Shoot rooting was superior when ANA was added to the culture medium, obtaining a greater number of roots per shoot and a slight decrease in length was also shown when combined with cytokinin.

## 1.1 INTRODUCTION

The *agave* genus is endemic to the American continent. More than 200 species of the genus are reported and around 150 are found in Mexico; the rest are distributed from the south of the United States of America, to Colombia and Venezuela, including the Caribbean islands, of the Bahamas, Trinidad and Tobago, Aruba and Curacao.[1] Agaves are species of great importance in arid and semi-arid regions of Mexico, due to the components they offer to the ecosystem;[1] Agaves are used in different regions of America for various purposes, such as, medicinal, obtaining fibers, construction material, live fences, ornamental, and fuel, as a support in decertified lands.[2,3] It has also been part of the human diet in Mexico for more than 9000 years.[4] Traditionally, they are propagated by vegetative means; asexually through the basal offspring of the rhizome, later propagation by bulbils produced by the inflorescence was incorporated, but propagation by seeds is not very common due to its lower uniformity and genetic quality.[5]

The agave is part of 1.25% of the national agricultural gross domestic product (GDP) and 3.04% of the agro-industrial production in Mexico. In 2016, tequila exports represented a value of 1203 million dollars. International demand is increasing—26 countries have increased consumption since 2003, and these countries together consume 184 million litres annually. The cultivated area in Mexico in 2016 was 111,420 hectares planted.[7] One of the main purposes of agave cultivation is the production of its distillates. The names "mezcal", "bacanora," and "tequila" refer to denominations of origin of Mexican distilled alcoholic beverages, made from agave, obtained by distillation of fermented juices with spontaneous or cultivated microorganisms, extracted from ripe maguey or cooked agave heads.[6]

The irresponsible exploitation of agave to produce distillates represents a serious problem in Mexico, it is considered that in the coming years there will be a deficit of raw material due to the overexploitation of young specimens, for which, at this time, it is essential to accelerate reproduction through plant biotechnology;[7,8] for this reason, it is intended to find efficient protocols for the multiplication of the species in a controlled manner, since in many cases the species of agave have been found in the category of threat or in danger of extinction, mainly those wild species.[9] Some authors recommend in vitro propagation by tissue culture as an alternative way of propagation to obtain a greater number of plants in less time, and taken to the field in equal development conditions to contribute to homogeneous plantations of the species in order to conserve wild populations such as *Agave montium-sancticaroli* and ensure the raw material used to make the "San Carlos" wine-mezcal from Tamaulipas.[10]

The *Agave montium-sancticaroli*, which is commonly called Jarcia, is a species native to Mexico, which is used mainly for the production of mezcal in different regions of Tamaulipas. This species is monocotyledonous and belongs to the Asparagaceae family, Agavoideae subfamily, its genus Agave, and the species *A. montium-sancticaroli*.[11,12]

It is a perennial plant, 1.5–2 m tall, 2–2.5 m in diameter, with an average of 50–80 leaves per plant. The leaves are 100–120 cm long and 9–12 cm wide in the middle, lanceolate, erect, rigid, fibrous, concave in shape, green-yellow, have a toothed margin with spines 4–6 mm long. long, small teeth between the largest, terminal spine 2.5–3.5 cm long, and each plant can produce from two to eight suckers. This species is distributed throughout the state of Tamaulipas, in the central region, between the Sierra de San Carlos and the Plain of the Gulf of Mexico; in the municipalities of Burgos, Cruillas, Jiménez, Padilla, San Carlos and San Nicolás.[10]

Although the nutritional and nutraceutical values of this species are unknown, a large number of species of the Asparagaceae family have shown medicinal and beneficial health characteristics.[13] There are some species of the Agave genus, which are used for the production of honey, syrup, fibers, ethanol, enzymes, and inulin. Normally, the carbohydrate content in Agave is high in pineapple, and inulin or agavin is a fructooligosaccharide of interest since it is used as a sugar substitute that can be used by people with diabetes as it is not digested by the digestive tract, in addition to being an alternative source of potassium, magnesium, calcium, and vitamin C.[2,14]

Agave cultivation grows worldwide every day; it is used in countries like Australia for the production of bioethanol and use of fiber, in addition to the production of inulin.[15] In some other countries (Indonesia, India, South Africa, Brazil, China, Kenya, Tanzania, Madagascar, and Thailand, among others), they produce more than 300,000 tons of Agave fiber.[16–18]

The value of the price of Agave has increased especially in recent years, making it a high profit margin crop for farmers, the value has increased from $1.00 MXN kilo-1 to a historical value of $30.00 MXN kilo-1 at the end of the year 2019, this due to the high demand for Agave pineapples.[19]

Due to the aforementioned background, Agave montium-sancticaroli was used in this study in order to develop a protocol that allows the massive multiplication of the species, in order to conserve the species in its natural habitat, as well as to increase the plantations of this species. species of Agave in order to produce mezcal. It is important to highlight the high sugar content of this species, which is higher than the level contained in A. americana subsp. protamericana, so it is highly appreciated by mezcaleros, which makes it a species potentially under risk of extinction, in addition to the low number of suckers and their small wild population, therefore; this species should be considered for in vitro research and its propagation by seed.[10]

Traditional propagation, in addition to being conditioned by the number of suckers produced by the mother plant, is limited by the period of production of these suckers.[20] On the other hand, in vitro propagation offers a greater number of shoots in less time, being able to reach more than 15 shoots per explant, which is a great advantage over the traditional method.[19,21]

## 1.2 MATERIALS AND EXPERIMENTAL METHODS

### *1.2.1 PLANT MATERIAL COLLECTION STAGE*

This research was carried out in the Plant Tissue Culture Laboratory of the Agricultural Sciences Campus, in General Escobedo, N.L., Mexico; and in the Plant Biotechnology laboratory, Marín unit, of the Autonomous University of Nuevo León. The plant germplasm used was collected from specimens of wild Agaves of the agave montium-sancticaroli species, from the municipalities of San Carlos and Jiménez, Tamaulipas, during the month of May of the year 2018. The plant germplasm used consisted of mature seeds collected from adult wild Agave plants of the species.[1]

**FIGURE 1.1** Collection of Agave sp seeds in the municipality of San Carlos, Tamaulipas. (a) Separation of floral branches from the quiote; (b) Separation of fruits; (c) Agave seeds exposed in the fruit.

### *1.2.2 ASEPTIC SEED ESTABLISHMENT STAGE*

The establishment began using mature seeds as a source of explant, which were cleaned and then selected the ones that were not damaged. In order to eliminate surface contaminants, they were washed with a commercial detergent and rinsed with drinking water; subsequently they were immersed in an antifungal solution with 3 $g.L^{-1}$ of 50% Captan for 6 hours; continuing with a rinse with abundant drinking water to finally be taken to the disinfestation process, which was carried out under aseptic conditions, under the laminar flow hood. Later the seeds were placed in a 70% v/v ethyl alcohol solution for 60 seconds and rinsed with sterile water. Subsequently, they were immersed in a commercial bleach solution (Cloralex®) at 10% and 20% v/v plus 0.02% polysorbate (Tween™ 20) for 15 min, rinsed with sterile water three times. After the disinfestation, the seeds were sown in the basic salts of Murashige & Skoog (MS), (Table 1.1)[21] added with 1.0 and 2.0 $mg.L^{-1}$ of BAP, in addition to 30.0 $mg.L^{-1}$ of sucrose. The pH of the medium was adjusted to 5.8 ± 0.02 using KOH or HCl 1.0 N and 4.5 $g.L^{-1}$ of Phytagel™ were added as a solidifying agent; the media were dosed in a Magenta™ box, with a capacity of 360.0 mL with 40.0 mL of medium per box and sterilized in an autoclave at 1.2 $kg.cm^{-2}$ for 15 min.

**TABLE 1.1** Components of Basic Salt MS, Supplemented with Sucrose.[22]

| Components | Quantity ($mg.L^{-1}$) | Components | Quantity ($mg.L^{-1}$) |
|---|---|---|---|
| $NH_4NO_3$ | 1600.000 | $CuSO_4{\cdot}5H_2O$ | 0.025 |
| $KNO_3$ | 1500.000 | $CoCl_2{\cdot}6H_2O$ | 0.025 |
| $CaCl_2{\cdot}2H_2O$ | 440.000 | $Na_2{\cdot}EDTA$ | 37.300 |
| $MgSO_4{\cdot}7H_2O$ | 370.000 | $FeSO_4{\cdot}7H_2O$ | 25.000 |
| $KH_2PO_4$ | 170.000 | Myo-inositol | 150.000 |
| $H_3BO_3$ | 6.200 | Glycine | 2.000 |
| $MnSO_4{\cdot}4H_2O$ | 22.300 | Nicotinic acid | 0.500 |

*Source*: Modified by Reyna-Morales.

After sowing, the experimental units were grown under controlled photoperiod conditions of 16 hours of light and 8 hours of darkness at a temperature of 24°C ± 2°C, under a completely randomized design with A × B factorial arrangement; where A is the concentration of Cloralex® and B is the level of BAP (Table 1.2), giving a total of four treatments, plus a control with 10 repetitions per treatment. After four weeks of remaining under these photoperiod and temperature conditions, the following variables were evaluated: asepsis percentage and seed germination.

The results obtained were analyzed with an analysis of variance using the statistical package Statistical Package for the Social Sciences (SPSS) and the mean comparison test using a Tukey test.

**TABLE 1.2** Disinfestation Treatments for *A. montium-sancticaroli* Seeds.

| Treatments | Ethanol (% V/V*) | Immersion Time (min) | Cloralex® (% V/V*) | BAP $mg.L^{-1}$ |
|---|---|---|---|---|
| 0 | 70.0 | 15.0 | 0.0 | 0.0 |
| 1 | 70.0 | 15.0 | 12.5 | 1.0 |
| 2 | 70.0 | 15.0 | 12.5 | 2.0 |
| 3 | 70.0 | 15.0 | 25.0 | 1.0 |
| 4 | 70.0 | 15.0 | 25.0 | 2.0 |

*Volume on volume.

### *1.2.3 SHOOT INDUCTION STAGE*

The apical shoots of 1.0 cm in length from the previous stage were used. They were planted in the basic MS salts, plus the addition of hormonal treatments and a control (Table 1.3). The subculture was performed under aseptic conditions, under a laminar flow hood. Once the transfer was concluded, the experimental units were grown under controlled conditions of photoperiod and temperature, similar to the previous stage. The experiment was established under a completely randomized design with factorial arrangement A × B plus the control, where A were the growth regulators and factor B the concentration of both BAP and ANA regulators, giving a total of four treatments, a control and 10 repetitions for each. After eight weeks, explants with shoots and shoot length were evaluated.

The results obtained were analyzed with an analysis of variance using the SPSS and the mean comparison test using the Tukey test.

**TABLE 1.3** Phytohormonal Treatments Used for the Induction of Shoots of *A. montium-sancticaroli.*

| Treatments | BAP (ppm) | ANA (ppm) |
|---|---|---|
| 0 | - | - |
| 1 | 1.0 | - |
| 2 | 2.0 | - |
| 3 | 1.0 | 0.3 |
| 4 | 2.0 | 0.3 |

In the figure, a hyphen (-) indicates the deficiency of the hormone.

### *1.2.4 SHOOT MULTIPLICATION STAGE*

The explants were subcultured again to basal MS medium, changing the hormonal levels (Table 1.4), giving a total of four treatments, plus the control and 10 repetitions for each one, later they were incubated under a completely random design with a 2 × 2 factorial arrangement where factor A were the growth regulators and factor B the concentrations of the hormones, these were grown in controlled conditions of photoperiod and temperature similar to the stage of induction of shoots, the explants remained in these conditions for eight weeks to be able to evaluate the Variables: Explants with buds, Buds per explant and Bud length.

The results obtained were analyzed with an analysis of variance using the statistical package SPSS and the mean comparison test was evaluated by means of a Tukey test.

### 1.2.5 SHOOT ROOTING STAGE

Rooting consisted of using shoots of 3.0 cm in length from the previous stage. The roots were removed from these and they were planted in the basic MS salts with the treatments described above in the shoot multiplication stage (Table 1.4). The subculture was performed under aseptic conditions, inside the laminar flow hood. Once the transfer was finished, the experimental units were rooted under controlled conditions of photoperiod and temperature, similar to the previous stage. After six weeks, the number and length of roots were evaluated.

The results obtained were analyzed with an analysis of variance using the statistical package SPSS and the mean comparison test was evaluated by means of a Tukey test.

**TABLE 1.4** Treatments Used for the Multiplication and Rooting of *A. Montium-sancticaroli* Shoots.

| Treatments | BAP (ppm) | ANA (ppm) |
|---|---|---|
| Control | - | - |
| 1 | 3.0 | 0.1 |
| 2 | 4.0 | 0.1 |
| 3 | 3.0 | 0.3 |
| 4 | 4.0 | 0.3 |

In the figure, a hyphen (-) indicates the deficiency of the hormone.

## 1.3 RESULTS AND DISCUSSION

### 1.3.1 ASEPTIC SEED ESTABLISHMENT STAGE

At this stage, the asepsis and germination of the seeds were evaluated after four weeks of being established in the culture medium. Microbial contamination reduces the efficiency in the establishment and propagation of *A. montium-sancticaroli*; for this reason, it is essential to start with a disinfestation process to be able to take the seeds to an in vitro environment.[26]

#### 1.3.1.1 ASEPSIS

Results were obtained during the aseptic establishment of *A. montium-sancticaroli* seeds, where T4 showed the best results, with an average of 100% asepsis (Table 1.5), this coincides with different Agave establishments by means of seeds (23.7, 24, 25).

**TABLE 1.5** Comparison of Estimated Means by Treatment. Dependent Variable: Asepsis.

| Treatments | | | Mean*** | Standard error | 95% confidence interval | |
|---|---|---|---|---|---|---|
| No. | Cl* | Min** | | | Lower limit | Upper limit |
| 0 | 0 | 0.0 | 0.00**e** | 0.033 | -0.066 | 0.066 |
| 1 | 12.5 | 7.5 | 0.20**d** | 0.033 | 0.134 | 0.266 |
| 2 | 12.5 | 15.0 | 0.56**c** | 0.033 | 0.494 | 0.626 |
| 3 | 25.0 | 7.5 | 0.82**b** | 0.033 | 0.754 | 0.886 |
| 4 | 25.0 | 15.0 | 1.00**a** | 0.033 | 0.934 | 1.066 |

*Bleach concentration in percent; **minutes under agitation; ***percentage of asepsis (1.00 = 100%). The letters (a–d) represent different homogeneous subsets.

### *1.3.1.2 GERMINATION*

While, germination was favored in the 25% treatments, being the T4 treatment the one that also achieved the best results, with 94% germination (Table 1.6 and Fig. 1.2); This response can be supported by what was mentioned by Domínguez-Rosales et al. and Abreu et al.,[23,26] who mention that less microbial contamination results in better germination.

**TABLE 1.6** Comparison of Estimated Means by Treatment. Dependent Variable: Germination.

| Treatments | | | Mean*** | Standard error | 95% confidence interval | |
|---|---|---|---|---|---|---|
| No. | Cl* | Min** | | | Lower limit | Upper limit |
| 0 | 0.0 | 0.0 | 0.070 abc | 0.039 | –.007 | .147 |
| 1 | 12.5 | 7.5 | 0.330 abc | 0.039 | .253 | .407 |
| 2 | 12.5 | 15.0 | 0.470 abc | 0.039 | .393 | .547 |
| 3 | 25.0 | 7.5 | 0.790 abc | 0.039 | .713 | .867 |
| 4 | 25.0 | 15.0 | 0.940 abc | 0.039 | .863 | 1.017 |

*Bleach concentration in percent; **minutes under agitation; ***percentage of asepsis (1.00 = 100%). The letters (a–d) represent different homogeneous subsets.

**FIGURE 1.2** Seeds of *A. montium-sancticaroli*: (a) 10 days after establishment; (b) with 20 days of establishment; (c) with 30 days of establishment.

### 1.3.2 SHOOT INDUCTION STAGE

Six weeks after subculture in MS medium, plus the addition of ANA, ANA helped shoot elongation, while BAP stimulated cell division and vegetative shoot cell formation in *A. montium-sancticaroli* explants. The variables under study were evaluated, which coincides with the results obtained in this stage by Ref. 27.

#### 1.3.2.1 FREQUENCY OF EXPLANTS WITH SHOOTS

Regarding the number of explants with shoots, it was found that treatment 4 obtained an average of 2.77 explants with shoots (Table 1.7), which coincides with that indicated by different authors, in *A. potatorum* Zucc, *Agave marmorata* Roezl and *Asparagus officinalis* L.,[9,27] where shoot induction is observed in treatments with combinations of auxins and cytokinins.

**TABLE 1.7** Comparison of Estimated Means by Treatment. Dependent Variable: Explants with Buds.

| Treatments | | | Mean*** | Standard error | 95% confidence interval | |
|---|---|---|---|---|---|---|
| No. | BAP* | ANA** | | | Lower limit | Upper limit |
| 0 | 0.0 | 0.0 | 0.350 **abc** | .178 | .000 | .700 |
| 1 | 1.0 | 0.0 | 1.650 **abc** | .178 | 1.300 | 2.000 |
| 2 | 2.0 | 0.0 | 2.350 **abc** | .178 | 2.000 | 2.700 |
| 3 | 1.0 | 0.3 | 2.433 **abc** | .178 | 2.084 | 2.783 |
| 4 | 2.0 | 0.3 | 2.767 **abc** | .178 | 2.417 | 3.116 |

*BAP concentration in $mg.L^{-1}$; **ANA concentration in $mg.L^{-1}$; ***number of explants with shoots. The letters (a–c) represent different homogeneous subsets.

#### 1.3.2.2 SHOOT FREQUENCY PER EXPLANT

The number of shoots per explant resulted in an average of 3.6 shoots per treatment 4 (Table 1.8), this is lower than the results obtained in *A. marmorata Roezl, A. salmiana*, and *Agave* spp., which could indicate that the concentrations and/or phytohormones used are not optimal.[9,25,28]

**TABLE 1.8** Comparison of Estimated Means by Treatment. Dependent Variable: Shoots per Explant.

| Treatments | | | Mean*** | Standard error | 95% confidence interval | |
|---|---|---|---|---|---|---|
| No. | BAP* | ANA** | | | Lower limit | Upper limit |
| 0 | 0.0 | 0.0 | 0.350 abc | 0.197 | -0.037 | 0.737 |
| 1 | 1.0 | 0.0 | 2.283 abc | 0.197 | 1.896 | 2.670 |
| 2 | 2.0 | 0.0 | 3.033 abc | 0.197 | 2.646 | 3.420 |
| 3 | 1.0 | 0.3 | 3.567 abc | 0.197 | 3.180 | 3.954 |
| 4 | 2.0 | 0.3 | 3.633 abc | 0.197 | 3.246 | 4.020 |

*BAP concentration in $mg.L^{-1}$; **ANA concentration in $mg.L^{-1}$; ***number of explants with shoots. The letters (a–c) represent different homogeneous subsets.

### *1.3.3 SHOOT MULTIPLICATION STAGE*

At this stage, the seedlings were subcultured and after eight weeks the number of explants with shoots, the shoots per explant, as well as the length of the shoots, were evaluated.

#### *1.3.3.1 BUDDING EXPLANTS*

The results obtained showed that all the treatments favored a greater number of explants with sprouts with respect to the control (Table 1.9 and Fig. 1.3). This coincides with that reported by,[23] who analyzes BAP and other phytohormones.

**FIGURE 1.3** Shoots of *A. montium-sancticaroli* from the mother plant after eight weeks of establishment.

**TABLE 1.9** Comparison of Estimated Means by Treatment. Dependent Variable: Explants with Buds.

| Treatments | | | Mean*** | Standard error | 95% confidence interval | |
|---|---|---|---|---|---|---|
| No. | BAP* | ANA** | | | Lower limit | Upper limit |
| 0 | 0.0 | 0.0 | 0.283 **ab** | 0.039 | 0.206 | 0.361 |
| 1 | 3.0 | 0.1 | 0.850 **ab** | 0.039 | 0.773 | 0.927 |
| 2 | 4.0 | 0.1 | 0.917 **ab** | 0.039 | 0.839 | 0.994 |
| 3 | 3.0 | 0.3 | 0.983 **ab** | 0.039 | 0.906 | 1.061 |
| 4 | 4.0 | 0.3 | 0.967 **ab** | 0.039 | 0.889 | 1.044 |

*BAP concentration in $mg.L^{-1}$; **ANA concentration in $mg.L^{-1}$; ***number of explants with shoots. The letters (a-c) represent different homogeneous subsets.

### *1.3.3.2 SHOOTS PER EXPLANT*

Likewise, it was found that treatments with 2, 3, and 4 were the best for the multiplication of shoots, where treatment 4 reached an average of 3.5 shoots (Table 1.10), this coincides with the observations made in *A. angustifolia*,[29] where it was found that BAP concentrations have a high significance in the number of shoots per explant.

**TABLE 1.10** Comparison of estimated means by treatment. Dependent variable: Shoots per explant.

| Treatments | | | Mean*** | Standard error | 95% confidence interval | |
|---|---|---|---|---|---|---|
| No. | BAP* | ANA** | | | Lower limit | Upper limit |
| 0 | 0.0 | 0.0 | 0.283 abc | 0.176 | -0.063 | 0.629 |
| 1 | 3.0 | 0.1 | 1.933 abc | 0.176 | 1.587 | 2.279 |
| 2 | 4.0 | 0.1 | 2.933 abc | 0.176 | 2.587 | 3.279 |
| 3 | 3.0 | 0.3 | 2.917 abc | 0.176 | 2.571 | 3.263 |
| 4 | 4.0 | 0.3 | 3.483 abc | 0.176 | 3.137 | 3.829 |

*BAP concentration in $mg.L^{-1}$; **ANA concentration in $mg.L^{-1}$; ***number of shoots per explant. The letters (a-c) represent different homogeneous subsets.

### *1.3.3.3 SHOOT LENGTH*

Regarding the length of shoots, T3 obtained an average of 7.4 cm (Table 1.11); this coincides with what was found in *A. angustifolia*,[29] where the relationship between auxins and cytokinins for a greater length of shoots is demonstrated.

**TABLE 1.11** Comparison of Estimated Means by Treatment. Dependent Variable: Shoot Length.

| Treatments | | | Mean*** | Standard error | 95% confidence interval | |
|---|---|---|---|---|---|---|
| No. | BAP* | ANA** | | | Lower limit | Upper limit |
| 0 | 0.0 | 0.0 | 0.633abc | 0.311 | 0.022 | 1.245 |
| 1 | 3.0 | 0.1 | 5.965abc | 0.311 | 5.353 | 6.577 |
| 2 | 4.0 | 0.1 | 7.047abc | 0.311 | 6.435 | 7.658 |
| 3 | 3.0 | 0.3 | 7.385abc | 0.311 | 6.773 | 7.997 |
| 4 | 4.0 | 0.3 | 6.160abc | 0.311 | 5.548 | 6.772 |

*BAP concentration in $mg.L^{-1}$; **ANA concentration in $mg.L^{-1}$; *** shoot length (cm). The letters (a-c) represent different homogeneous subsets.

## *1.3.4 SHOOT ROOTING STAGE*

The roots are an essential part for survival at the time of acclimatization. The length and number of roots per explant was evaluated after six weeks of subculture at this stage.

### *1.3.4.1 ROOTS PER EXPLANT*

The rooting capacity in the treatments was statistically higher than that obtained in the control, which lacked auxin (Table 1.12 and Fig. 1.4), this coincides with that reported in *A. angustifolia*,[30] where the use of indolebutyric acid (AIB) is recommended to induce rooting, contrary to what was reported in *A. vitoriae-reginae*,[31] where a high rooting rate has been obtained in treatments where auxins had not been used, and in *A. durangensis* where better results were obtained at low auxin concentrations of (0.1 $mg.L^{-1}$ AIB).[32]

**FIGURE 1.4** Rooting of *A. montium-sancticaroli* seedlings after six weeks of establishment in the medium.

**TABLE 1.12** Comparison of Estimated Means by Treatment. Dependent Variable: Roots per Explant.

| Treatments | | | Mean*** | Standard error | 95% confidence interval | |
|---|---|---|---|---|---|---|
| No. | BAP* | ANA** | | | Lower limit | Upper limit |
| 0 | 0.0 | 0.0 | 13.478 abc | 0.160 | 13.164 | 13.792 |
| 1 | 3.0 | 0.1 | 14.143 abc | 0.160 | 13.829 | 14.457 |
| 2 | 4.0 | 0.1 | 14.385 abc | 0.160 | 14.071 | 14.699 |
| 3 | 3.0 | 0.3 | 14.152 abc | 0.160 | 13.838 | 14.466 |
| 4 | 4.0 | 0.3 | 14.825 abc | 0.160 | 14.511 | 15.139 |

*BAP concentration in $mg.L^{-1}$; **ANA concentration in $mg.L^{-1}$; ***number of roots per explant. The letters (a–c) represent different homogeneous subsets.

### *1.3.4.2 ROOT LENGTH*

Root length was greater in treatment 3, with a mean of 14.36 cm (Table 1.13), these coincide with[9] about the concentrations of 3.0 and 10.0 $mg.L^{-1}$ used in *A. marmorata* Roezl.

**TABLE 1.13** Comparison of Estimated Means by Treatment. Dependent Variable: Length of Roots.

| Treatments | | | Mean*** | Standard error | 95% confidence interval | |
|---|---|---|---|---|---|---|
| No. | BAP* | ANA** | | | Lower limit | Upper limit |
| 0 | 0.0 | 0.0 | 12.592 abc | 0.217 | 12.164 | 13.020 |
| 1 | 3.0 | 0.1 | 13.485 abc | 0.217 | 13.057 | 13.913 |
| 2 | 4.0 | 0.1 | 13.908 abc | 0.217 | 13.480 | 14.336 |
| 3 | 3.0 | 0.3 | 14.360 abc | 0.217 | 13.932 | 14.788 |
| 4 | 4.0 | 0.3 | 13.318 abc | 0.217 | 12.890 | 13.746 |

*BAP concentration in $mg.L^{1}$; **ANA concentration in $mg.L^{1}$; ***root length (cm).

## 1.4 CONCLUSION

The in vitro propagation of *A. montium–sancticaroli* was successful through the use of growth regulators and the basic salts of Murashige and Skoog, which were optimal for all stages of in vitro development of the species. The disinfestation technique of the aseptic establishment of the seeds was adequate to eradicate the microbial load of these, achieving the in vitro establishment of the species. Shoot induction was achieved with the use of

6-N-Benzylaminopurine, causing a greater sprout induction by promoting cell differentiation. The multiplication and elongation of shoots was obtained by using the combination of an auxin and a cytokinin in these stages. Shoot rooting was greater when ANA was added to the culture medium, obtaining a greater number of roots per shoot.

## KEYWORDS

- **species**
- **endemic**
- **mezcal**
- **organogenesis**
- **regeneration**

## REFERENCES

1. García-Mendoza, A. J. Los Agaves de México. *Ciencias* **2007,** *87,* 14–23.
2. Bautista-D., N.; Arias A., G. C. Estudio Químico Bromatológico de Aguamiel de Agave Americana L. (Maguey). *Cienc. Invest.* **2008,** *11* (2), 46–51.
3. Espinosa-Barrera, L. A. Generalidades e Importancia de Los Agaves En México. *Herb. CICY* **2015,** *7* (10), 161–164.
4. Callen, E. O. Food Habits of Some Pre-Columbian Mexican Indians. *Econ. Bot.* **1965,** *19* (4), 335–343.
5. Guillot-Ortiz, D.; Van der Meer, P.; Laguna-Lumbreras, E.; Rosselló-Picornell, J. A. *El Género Agave L. En La Flora Alóctona Valenciana,* Vol. 3; Benito Alonso, J. L., Ed.; Revista Bouteloua, 2009.
6. Diario Oficial de la Federación. *Norma Oficial Mexicana NOM-070-SCFI-2016* [Online]. D.O.F.: México, 2016; pp 1–17.
7. Flores-Morales, A.; Castañeda-Hidalgo, E.; Sánchez-Pérez, F. J.; Romero-Aguilar, L.; Ruiz-Luna, J. Mecanismos de Conservación y Uso Del Maguey Pulquero Agave Salmiana En El Altiplano Mexicano. *Somas (Sociedad Mex. Agric. Sostenible)* **2009,** *6,* 1–10.
8. Cervantes-Ramírez, M. C. *Plantas de Importancia Económica En Zonas Áridas y Semiáridas de México*; 2005.
9. Aguilar-Jiménez, D.; Rodríguez-de la O, J. L.; Herrera-López, H. Obtención de Brotes in Vitro En Agave Potatorum Zucc. *Rev. Científica Multidiscip.* **2018,** *2* (29), 5–7.
10. García-Mendoza, A. J.; Jacques-Hernández, C.; Salazar-Bravo, Á.una Nueva Especie de Agave , SubgeneroLittae (Agavaceae) de Tamaulipas, México. *J. Bot. Res. Inst. Texas* **2007,** *1* (1), 79–84.

11. Chase, M. W.; Reveal, J. L.; Fay, M. F. A Subfamilial Classification for the Expanded Asparagalean Families Amaryllidaceae, Asparagaceae and Xanthorrhoeaceae. *Bot. J. Linn. Soc.* **2009,** *161* (2), 132–136.
12. Stevens, P. F. Angiosperm Phylogeny Website [Online], 2012. Recuperado de Versión. http://www.mobot.org/MOBOT/research/APweb/ (accessed 12, 2012).
13. Rai, P. S.; Bellampalli, R.; Dobriyal, R. M.; Agarwal, A.; Satyamoorthy, K.; Anantha Narayana, D. B. DNA Barcoding of Authentic and Substitute Samples of Herb of the Family Asparagaceae and Asclepiadaceae Based on the ITS2 Region. *J. Ayurveda Integr. Med.* **2012,** *3* (3), 136–140.
14. Ibañez-Vázquez, G.; Álvarez-Herrera, M.; Doria-Mendoza, F. J. *Cadena de Suministro: Inulina de Agave*; Universidad Autónoma de Tmaulipas, 2015, (834), pp 1–27. Recuperado de http://acacia.org.mx/busqueda/pdf/CADENA_DE_SUMINSTRO_ INULINA_DE_*AGAVE*.pdf
15. Azad, A. K. *Advances in Eco-Fuels for a Sustainable Environment.* In *Woodhead Publishing Series in Energy*; En, A. K. Azad, Ed.; 2018. https://doi.org/10.1016/ c2017-0-04211-8
16. Hulle, A.; Kadole, P.; Katkar, P. *Agave* Americana leaf fibers. *Fibers* **2015,** *3* (1), 64–75.
17. Santoso, B. Peluang Pengembangan *Agave* Sebagai Sumber Serat Alam. *Perspektif* **2015,** *8* (2), 84–95.
18. Smith, G. F.; Figueiredo, E. Notes on *Agave* Weberi J.F.Cels ex J. Poiss. (Agavaceae), a Largegrowing Species with Invasive Tendencies in Southern Africa. *Bradleya* **2015,** *33,* 161–170.
19. Sanschagrin, S. Las Principales Tendencias del Tequila a Tener en Cuenta en [Online], 2020. Recuperado el 8 de Julio de 2020, de TasteTequila https://tastetequila.com/ es/2020/the-top-tequila-trends-to-watch-for-in-2020/#:~:text= Aunque los precios del *Agave*,kilo para fin de año
20. Garibay-Kuri, S. V. *Enraizamiento de Hijuelos en Viveros de "Agave Azul Tequilero" Agave tequilana weber. Utilizando hormonas,* 1988.
21. López-Acevedo, L.; Merino-Pérez, Y. E.; Enríquez-Del-Valle, J. R.; Rodríguez-Ortíz, G.; Lagunas-Sánchez, Z. C. Organogénesis *In Vitro* en Tejidos de Tallo de *Agave* Marmorata y *Agave* Angustifolia. *Rev. Mexicana Agroecosist.* **2018,** *5* (2), 98–105.
22. Murashige, T.; Skoog, F. A Revised Medium for Rapid Growth and Bio Assays with Tobacco Tissue Cultures. *Physiol. Plant.* **1962,** *15,* 473–497.
23. Domínguez-Rosales, M. S.; Alpuche-Solís, Á. G.; Vasco-Méndez, N. L.; Molphe-Balch, E. P. Efecto de Citocininas En La Propagación in Vitro de Agaves Mexicanos; Revista Fitotecnia Mexicana. 2008, pp 317–322.
24. Mejía-franco, R.; Arzate-Fernández, A. Capacida Embriogénica de Callos Inducidos en Ejes Embrionarios Cigoticos de *Agave* angustifolia Haw. *Revista Fitotecnia Mexicana* **2011,** *34* (2), 101–106.
25. Silos-Espino, H.; Lourdes Tovar-Robles, C.; González-Cortés, N.; Méndez-Gallegos, S. J.; Rossel-Kipping, D.; Aguascalientes-San Luis Potosí El Llano, C.; México, A. Estudio Integral del Maguey (*Agave* salmiana): Propagación y Valor Nutricional. *Revista Salud Pública Nutrición* **2011,** (5), 75–82.
26. Abreu, E.; Sosa-del-Castillo, M.; Ascunce-del-Sol, G.; González, G. Efecto de Antibióticos en la Propagación *in vitro* de *Agave* Fourcroydes Lem. *Biotecnol. Veg.* **2016,** *16* (1), 31–36.

27. Millán-Soto, G.; Robert, M. L.; Tiznado-Hernández, M. E.; Gutiérrez, A.; Esqueda, M. Organogénesis de Rizoma de Espárrago (Asparagus Officinalis L.) por Combinación de Auxinas y Citocininas. *Agrociencia* **2019,** (53), 549–561.
28. Molphe-Balch, E. P.; Esparza-Araiza, M. J.; Pérez-Reyes, M. E. Conservación *In Vitro* de Germoplasma de *Agave* spp. Bajo Condiciones de Crecimiento Retardado. *Rev. Fitotec. Mex.* **2012,** *35* (4), 279–287.
29. Ríos-Ramírez, S. C.; Enríquez-del Valle, J. R.; Rodríguez-Ortíz, G.; Ruíz-Luna, J. Benzylaminopurine and Indol-3-Acetic Acid Concentrations in in Vitro Proliferation of Agave Angustifolia Adventitious Shoots. *Cienc. Investig. Agrar.* **2017,** *44* (3), 285–294.
30. Enríquez-del-Valle, J. R.; Carrillo-Castañeda, G.; Rodríguez-de-la-O, J. L. Sales Inorgánicas y Ácido Indolbutírico en el Enraizado *In Vitro* de Brotes de *Agave Angustifolia. Rev. Fitotec. Mex.* **2005,** 28, Correo.
31. Martínez-Palacios, A.; Ortega-Larrocea, M. P.; Chávez, V. M.; Bye, R. Somatic Embryogenesis and Organogenesis of *Agave* Victoriae—Reginae: Considerations for its conservation. *Plant Cell Tissue Organ Cult.* **2003,** *74,* 135–142.
32. Godoy-Díaz, C. G.; Núñez-Palenius, H. G. *Micropropagación de Agave Durangensis en un Sistema de Inmersión Temporal (SIT).* Jóvenes en la ciencia, 2015; pp 60–65.

# CHAPTER 2

# Effects of Elicitors on Productivity, Fruit Quality, and Health of Tomato (*Solanum lycopersicum* L.) under Greenhouse Conditions

A. K. CHAM[1], MA DEL C. OJEDA-ZACARÍAS[1], H. LOZOYA-SALDAÑA[2], E. OLIVARES-SÁENZ[1], OMAR GUADALUPE ALVARADO-GÓMEZ[1], R. E. VÁZQUEZ-ALVARADO[1], and GILBERTO RODRIGUEZ PEREZ[3]

[1]*Autonomous University of Nuevo León, Faculty of Agronomy, Street Francisco Villa S/N, Col. Ex Hacienda El Canada, Gral. Escobedo, Nuevo León, Mexico*

[2]*Autonomous University of Chapingo, Phytotechnics Department, Mexico-Texcoco highway Km 38.5, Chapingo, Mexico*

[3]*National Technology of Mexico-Valle del Yaqui Campus, Valledel Yaqui, Bácum, Ciudad Obregón, Sonora, Mexicol*

## ABSTRACT

In Mexico, the tomato (*Solanum lycopersicum* L.) occupies the largest greenhouse surface area at the national level, which is why there is the demand for alternative products, such as organic products to control diseases and pests, which are not harmful to humans and the environment. There is not only increase in the demand for alternative products such as organic products for the control of diseases and pests, but also the use of inducing molecules, also called elicitors. The objective of this investigation is to know the effect of three elicitors—Activane® (1.8 g $L^{-1}$), Micobiol® (3 m $L^{-1}$), and Stemicol® (2.5 g $L^{-1}$) on the incidence of diseases, vigor, yield, and fruit quality of the

Sustainable Agricultural Production Systems: Traditional and Modern Perspectives.
Romeo Rojas, Guillermo Cristian Guadalupe Martínez-Ávila, Julia Mariana Márquez-Reyes, and Ma del Carmen Ojeda-Zacarías (Eds.)

two tomato cultivars of bola and saladette produce under greenhouse conditions. Elicitors decreased the incidence and severity of presented diseases and increased plant survival. On the other hand, the elicitors Micobiol® and Activane® were able to increase plant height, stem diameter, weight, and fruit diameter. In general, elicitors were able to reduce disease, increase plant survival and also significantly improved production. These results provide positive directions for the potential use of these elicitors and induce resistance in plant protection.

## 2.1 INTRODUCTION

In Mexico, tomato (*Solanum lycopersicum* L.) crop is considered as one of the most important vegetables, since it occupies a prominent place within the horticultural crops that are produced under protected agriculture conditions. However, the greenhouse environment is not problem free. The higher humidity, temperature, and lush green foliage create conditions that enable certain diseases and insects to thrive.[1,2] The crop is of greater importance nationally and internationally, due to its wide consumption, harvested area, and the economic value of the production. In addition to its economic and social importance in the world's production systems, it has increasing nutritional relevance because it is a significant source of antioxidant substances (lycopene, beta-carotene) and vitamins.[3–5]

China is the main tomato producing country in the world followed by the United States, India, Turkey, Egypt, Italy, Iran, Spain, Brazil, Mexico, Uzbekistan, Russia, among others.[6] While Mexico's production of an approximately about 43,903 hectares is coming from shade and greenhouse conditions as well as open field cultivation, with six states producing approximately 53 percent of the national production in in 2020. Moreover, it permits a large domestic and international market during most of the year.[2]

Tomato is prone to various diseases; these diseases cause low yield and quality and even total loss. Due to their aggressiveness, only preventive treatments show a certain degree of effectiveness, but once the symptoms appear they no longer have an effect, which is why many applications of synthetic chemical pesticides are made.[7] Most pesticides are costly and harmful to humans and the environment, so their use should be limited to emergencies. However, it is recommended to use combined strategies of pesticides (with emphasis on biological control), organic fertilizers, and associated culture management and ecological procedures.[8,9] In that case, the demand for alternatives such as organic products for disease and pest management and the use of inducing molecules also called elicitors increases.[10,11] Elicitors are natural or inorganic substances that, once applied to plants in a preventive way, help to reduce or avoid damage caused by diseases, pests, or adverse abiotic factors.[12] They are presented as the most promising method among the pathogen control alternatives that ensure better tissue vigor, better post-harvest quality,[13,14] production profitability, and cost reduction due to reduced use of pesticides and healthy products.[15] For this reason, the main objective of this work was to study the potentiality of selected elicitors to confer broad spectrum disease resistance and production of the two tomato cultivars under greenhouse conditions.

## 2.2 EXPERIMENTAL METHODS AND MATERIALS

### *2.2.1 STUDY AREA AND PLANT MATERIAL*

The cultivation of this vegetable is very important for the economy of many countries in the world, including Mexico, which is why it has been chosen as a model plant in this study. We choose two important cultivars, "Bola" and "Saladette," produced under greenhouse conditions. Therefore, above-mentioned cultivars were cultivated under the greenhouse "Israelita" during the summer–spring of 2019–2020 at the Center for Protected Agriculture (CAP) of the Fac.of Agronomy of the Autonomous University of Nuevo León, Nuevo León, Mexico.

### *2.2.2 EXPERIMENTAL DESIGN AND TREATMENTS*

The experiment consisted of four treatments—T1: Control, T2: Activane®, T3: Micobiol®, and T4: Stemicol® with four repetitions giving a total of 256 plants; these were established under a completely randomized design, where the foliar application (eight application at different stages of the crop) was to evaluate the effect of the elicitors on tomato productivity, fruit quality, and health. The recommended dose and product information were obtained from the producing company, "Lida de Mexico" (Table 2.1).

**TABLE 2.1** Doses of the Elicitors Applied to the Leaves of Both Tomato Cultivars.

| Elicitor | Elicitor source | Elicitor properties | Dosis applied |
|---|---|---|---|
| Activane® | Crustaceans | Chito-oligosaccharides (COS) | $1.8\ g\ L^{-1}$ |
| | Vegetable origin | Phenols | |
| Micobiol® | Yeast | β-glucan | $3\ mL\ L^{-1}$ |
| Stemicol® | Crustaceans | Chito-oligosaccharides (COS) | $2.5\ g\ L^{-1}$ |

### 2.2.3 DATA ANALYSIS

Treatment comparisons according to the findings were performed using the Tukey ($P \leq 0.05$) test to assess the association between the variables with statistically significant differences between the treatments. The statistical analyses were done with SPSS software ver. 21.0 (IBM 2015) program for Windows.

### 2.2.4 PARAMETERS OF STUDY

#### 2.2.4.1 VEGETATIVE GROWTH

The data of the variables of interest were taken at seven day intervals at the same time for greater precision. The following agronomic variables of growth and yield were evaluated according to the methodology of 16 and 17. A measuring tape was used to measure plant height (cm) and stem diameter was measured with a Digital Vernier Caliper in millimetre (mm). The SPAD units was evaluated with a SPAD brand Minolta model 502 Plus, and the weight of the fruits from several harvest was evaluated by using an electronic scale. Meanwhile, the fruit diameter was also measured with a Digital Vernier Caliper in millimetres (mm).

#### 2.2.4.2 FRUIT QUALITY VARIABLES

The fruit quality evaluation was carried out with four fruits per plant, when they presented an intense and uniform red surface. Meanwhile, the fruit diameter was also measured with a Digital Vernier Caliper in millimetres (mm). The pH values and total soluble solids were evaluated for these variables using a potentiometer brand "HANNA" model 211 and a digital refractometer, Atago; the quantification of the titratable acid content was carried out by extracting 1.0 g of fruit material crushed in a tube with 9 mL of purified water. The mixture was titrated with 0.1N sodium hydroxide (NaOH), using 0.1% phenolphthalein as indicator, meanwhile the results were expressed as a percentage of citric acid present, The fruit color was objectively evaluated where L*, a*, and b* values were measured with a Chin SPEC model colorimeter previously calibrated.[18,19]

#### *2.2.4.3 DISEASE EVALUATED*

The incidence of the disease was estimated considering the presence of healthy and diseased plants at 50, 100, and 150 days after transplant; the diseases present during the development of the crop were evaluated according to the scale proposed by Moreno et al.,[20] which is described as follows—0 = no symptoms (0%): 1 = necrotic spots (20%); 2 = necrotic lesions (40%); 3 = necrotic lesions spots on the surface of the leaf (60%); 4 = necrotic lesions distributed (80%); 5 = fully affected (wilting) or dead (100%).

#### *2.2.4.4 MOLECULAR IDENTIFICATION OF DISEASES*

Root, stem, leaf, and fruit samples were also collected from the two tomato cultivars in the greenhouse. For fungi and bacteria, the samples consisted of plant roots, leaves, and fruits with symptoms of pathogen attack. While for the virus the sample consisted of 2 g of the newest leaf located in the aerial apex with symptoms indicating the presence such as interveinal chlorosis, mottling, and leaf deformation. The samples were taken to the Phytopathology Laboratory, where the corresponding analysis was subsequently carried out for each one. (1) Isolation of bacterial and fungi strains from diseased tomato plants, (2) obtaining of bacterial DNA and amplification with polymerase chain reaction (PCR), and (3) RNA extraction and virus detection by generic PCR Begomovirus.

### ***2.2.5 SEQUENCING AND DATA ANALYSIS***

The procedure for the corroboration of the identification of bacteria, fungi, and viruses by comparing sequences 16S for bacteria and 18S/5.8S/28S for fungi was revised. The amplified products were sequenced and their sequences were compared with GenBank of National Center for Biotechnology Information (NCBI) (www.ncbi.nlm.nih.gov) of the quantitative values observed; only those with the greatest identity were considered.

## 2.3 RESULTS AND DISCUSSION

### *2.3.1 VEGETATIVE GROWTH*

The production of the these two tomato cultivars under greenhouse conditions through elicitation showed a significant effect on vegetative growth. The highest weekly plant height in saladette cultivar was in treatments with Micobiol® (21.5 cm) followed by Stemicol® (20.6) and Activane® (19.4). Meanwhile, for stem diameter, the highest records were registered for Stemicol® (10.0 mm). Also, the variables of plant height evaluated in the Bola cultivar showed significant value (P = 0.05) for these variables, registering the highest average in stem diameter with Stemicol® (10.0 mm). Leaf SPAD value showed a significant difference (*P* = 0.05) between the means of the Saladette cultivar, obtaining the highest value for the treatment Micobiol® (T3) with 46.99, followed by Control (T1), Stemicol® (T4), and Activane® (T2) with average values of 45.17, 4.14 and 44.33, respectively. Meanwhile, the Bola cultivar didn't show a significant value (*P* = 0.05) between the treatments of this variable (Table 2.2).

**TABLE 2.2** Comparison of Weekly Means for the Variables Plant Height, Stem Diameter, Leaf Length, and Number of Leaves for the Two Tomato Cultivars.

| Treatment | Weekly plant height (cm) | Stem diameter (mm) | Number of leaves | SPAD |
|---|---|---|---|---|
| **Saladette cultivar:** | | | | |
| T1—Control | 18.6875 c | 8.9013 c | 14.7812 a | 45.1750 b |
| T2—Activane® | 19.4375 b | 9.2125b c | 15.2343 a | 44.3312 c |
| T3—Micobiol® | 21.5000 a | 9.9894 b | 14.9218 a | 46.9984 a |
| T4—Stemicol® | 20.6250 b | 10.0788 a | 15.0156 a | 45.1406 b |
| **Bola cultivar:** | | | | |
| T1—Control | 28.8667 c | 11.6500 a | 18.4969 a | 50.6870 a |
| T2—Activane® | 30.7500 a | 11.9233 a | 18.7188 a | 50.0235 a |
| T3—Micobiol® | 30.2000 b | 12.0193 a | 19.1240 a | 50.6258 a |
| T4—Stemicol® | 29.7500 b | 11.7816 a | 18.6563 a | 49.7156 a |

**Means with the same letter within each column do not differ statistically (*P* = 0.05).

The application of the elicitors in the plants of both cultivars promoted the increase of the weekly height of the plant and leaf length for the treated plants in comparison with the height of the plants used as control, being the treatment T3—Micobiol® that most influences these variables. Results obtained by García et al.[10] reported a positive effect of the application of elicitors of natural origin; it was found that their application promoted the accumulation of total dry biomass, the increase in plant height compared to plants that were not treated.

## 2.3.2 *FRUIT WEIGHT AND DIAMETER*

The statistical analysis for the parameters under study presented a significant difference ($P = 0.05$) for the variables weight and equatorial diameter of the fruits in the saladette cultivar. Meanwhile, a significant difference was registered for the variables of weight and fruit polar diameter in the Bola cultivar. For both cultivars, the highest value of fruit weight and polar and equatorial diameter were recorded in the treatment Micobiol® with (128.0 g, 64.1 mm and 52.4 mm). The variables evaluated above allow knowing the balance between the weight and the diameter of the fruit in the harvest of the two cultivars of tomato (Table 2.3).

**TABLE 2.3** Fresh Fruit Weight (Grams), Polar, and Equatorial Diameter (mm) of the Two Tomato Cultivars as Affected by Elicitors.

| Treatment | Fresh weight of fruit (g) | Fruit diameter (mm) | |
|---|---|---|---|
| | | Polar | Equatorial |
| Saladette cultivar: | | | |
| T1—Control | 107.13 b | 60.12 a | 50.69 ab |
| T2—Activane® | 91.46 c | 60.02 a | 47.23 b |
| T3—Micobiol® | 128.07 a | 64.10 a | 52.45 a |
| T4—Stemicol® | 111.45 b | 63.33 a | 51.11 ab |
| **Bola cultivar:** | | | |
| T1—Control | 169.03 b | 49.83 b | 69.38 a |
| T2—Activane® | 131.01 c | 52.24 ab | 68.13 a |
| T3—Micobiol® | 176.21 a | 53.16 ab | 69.46 a |
| T4—Stemicol® | 175.20 a | 54.66 a | 69.46 a |

**Means with the same letter within each column do not differ statistically ($P = 0.05$).

The results of this investigation were found between the range of fruit weight and fruit diameter for both tomato varieties. According to Grijalva et al.,[21] who reported that in Saladette tomato, the weight of a tomato fruit ranges between 20 and 135 g, and the diameter between 25 and 59 mm.

While in ball tomato the weight of the fruit ranges between 20 and 220 g, and the diameter is between 25 and 59 mm, respectively.

### 2.3.3 *YIELD*

The highest fruits yield was obtained from the plants treated with elicitors in comparison with the untreated ones. The statistical analysis did not show any significant difference ($P = 0.05$) for the Bola cultivar. Nevertheless, in the Saladette cultivar, a significant difference ($P = 0.05$) was recorded with the highest average in the treatments Activane® (T2) with 70 fruits per square meter and 7.2 kgs per square meter. The treatments with elicitors Control (T1) presented the lowest average for the number and weight of fruits (Fig. 2.1).

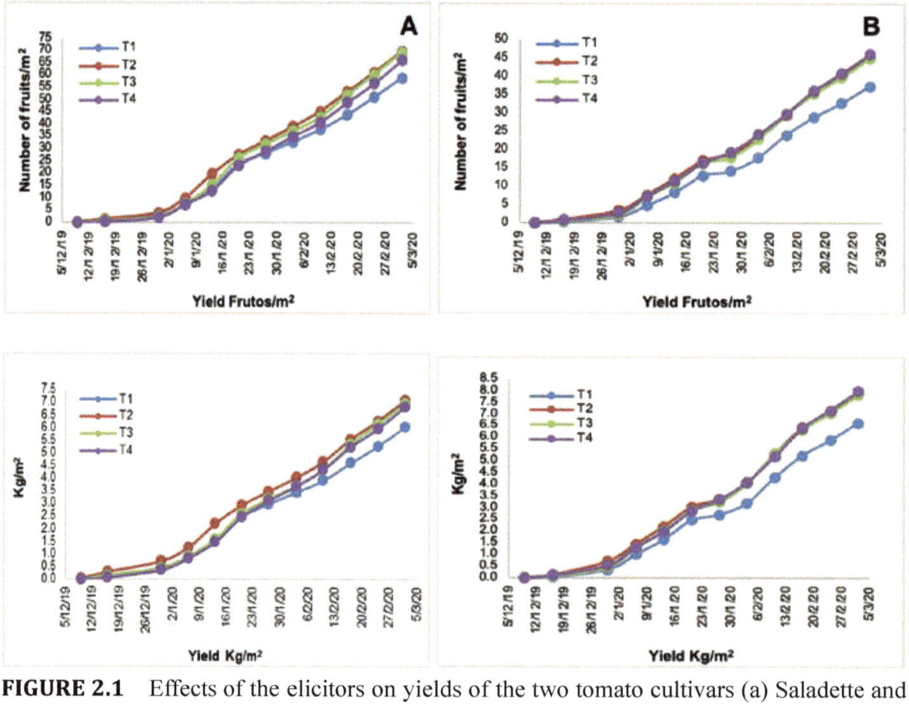


**FIGURE 2.1** Effects of the elicitors on yields of the two tomato cultivars (a) Saladette and (b) Bola under greenhouse conditions.

The results obtained in relation to the total production in the two varieties harvested during the 2020 summer-spring cycle did not show a significant difference ($P \leq 0.05$) between the treatments evaluated for the Bola variety.

The elicitors increased the fruit number of the treated plants and avoided the reduction in the total yield. These results coincided with García et al.,[10] who, when evaluating the effect of two elicitors on the tomato crop, found that the use of elicitors from plant extracts had a positive result on the height of the treated plants, where the accumulation of biomass prevented the reduction in performance.

### 2.3.4 *FRUIT QUALITY*

#### 2.3.4.1 *FRUIT COLOR*

The colors of the ripe fruits were deep and uniform red. Visual perception was not different between the treatments for the two cultivars evaluated during the period of the evaluations.

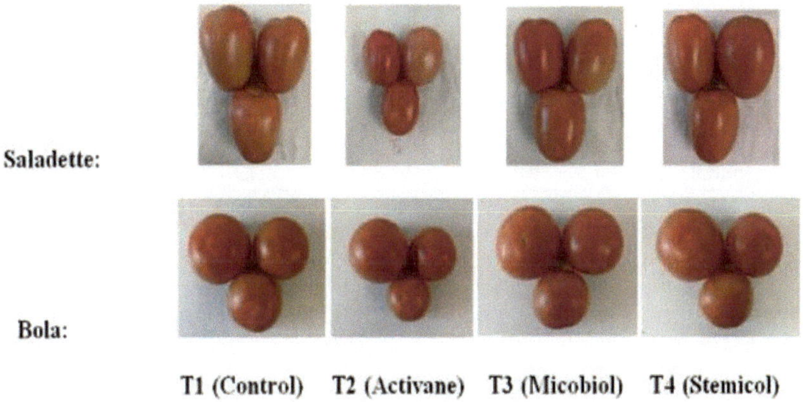


**FIGURE 2.2** Ripe fruits of intense and uniform red color in both tomato cultivars were selected and evaluated for greater precision of this variable.

##### 2.3.4.1.1 *Coordinates (a *: b *: C *)*

The statistical analysis of the parameter a*, under the conditions of the study, did not show a significant difference ($P \leq 0.05$) for the two varieties. While for the parameters b* and C*, they showed a significant difference ($P \leq 0.05$) between the means corresponding to the Saladette variety, the treatment T4 (Stemicol®) being the one that most influences these coordinates (Table 2.4). According to Hornero-Méndez et al.,[22] the carotenoid lycopene is the dye that gives the tomato its red color. High levels of beta-carotenoids give it an orange color. It is important that the tomatoes are bright and uniform in color, without immature green patches or dots. On the other hand, López and Gómez[23] reported that, for tomato fruits, the C* coordinate is not an adequate indicator of maturity, since it is essentially an expression of purity for a specific color.

**TABLE 2.4** Comparison of Means of the Fruits Color Coordinates of the Two Tomato Cultivars as Affected by Elicitors.

| Treatments | Luminosity (L*) | Hue angle (h*) | Red contribution (a*) | Yellow contribution (b*) | Chroma (C*) | Color index (IC) |
|---|---|---|---|---|---|---|
| | | | **Saladette cultivar:** | | | |
| T1—Control | 36.57 a | 46.18 a | 20.70 a | 21.50 b | 29.84 b | 27.06 a |
| T2—Activane® | 37.11 a | 50.38 a | 19.96 a | 23.76 ab | 31.50 ab | 23.52 a |
| T3—Micobiol® | 34.97 a | 46.19 a | 21.51 a | 22.41 ab | 31.12 ab | 27.06 a |
| T4—Stemicol® | 35.78 a | 47.36 a | 21.87 a | 23.93 a | 32.51 a | 26.08 a |
| | | | **Bola cultivar:** | | | |
| T1—Control | 38.76 a | 48.55 c | 18.10 a | 20.53 a | 27.44 a | 23.10 a |
| T2—Activane® | 39.16 a | 54.17 a | 16.13 a | 22.45 a | 27.79 a | 18.80 b |
| T3—Micobiol® | 38.86 a | 51.45 b | 18.05 a | 22.36 a | 28.86 a | 21.19 ab |
| T4—Stemicol® | 38.26 a | 50.45 b | 17.82 a | 21.61 a | 28.175 a | 22.05 ab |

**Means with the same letter within each column do not differ statistically ($P = 0.05$).

The statistical analysis of the color values expressed as hue angle (h*), showed a significant difference ($P \leq 0.05$) for the Bola variety, while for Saladette it did not show a significant difference between the variables. The values of this cylindrical coordinate showed a high value for the treatment (Activane®) T2 in the two varieties (54.17 and 50.38), the T1 being the one that presented the lowest values of hue angle (h*) (Table 2.4), López and Gómez[23] reported h* values of 50.3 for fruits at consumption maturity (red color). Similarly, Hornero-Méndez et al.,[22] reported h* values of 55 when the fruit reached maturity for consumption.

The statistical analysis of the values obtained for the color index presented significant difference ($P \leq 0.05$) for the Bola variety, while no significant difference between the variables for the Saladette variety (Table 2.4), the lowest values were in the T2 (Activane®) in both varieties with (18.80 and 23.52), followed by treatments T4 (Stemicol®), T3 (Micobiol®), and T1 (Control). We can compare these results with color indices obtained by López and Gómez,[23] who found a direct correlation between this expression and the degree of maturation, with CI values around 21.8 for fruits at consumption maturity in tomato.

### 2.3.4.2 TOTAL SOLUBLE SOLIDS (TSS), PH, AND TITRATABLE ACIDITY OF FRUITS

In this study, the total soluble solids (Brix) and antioxidant capacity (DPPH) in fruits has registered a significant difference ($P = 0.05$) for the Saladette cultivar, obtaining the lowest value for the treatment Micobiol® (T3) with 5.26, followed by Activane® (T2), Control (T1), and Stemicol® (T4) with values of 5.37, 5.43, and 5.52, respectively. While, for the Bola cultivar, it did not show significant difference ($P = 0.05$) between the variables. The reported pH values are within the range (4.23–4.59) obtained for the four treatments of both tomato cultivars. The statistical analysis did not present a significant difference for the variables of the Bola cultivar. On the other hand, analysis regarding the acidity of the fruits, showed a significant difference for the two cultivars with the highest being treatment Activane® (T2) with 0.52% of citric acid, followed by the other three treatments with an average of 0.35% (Fig. 2.3).

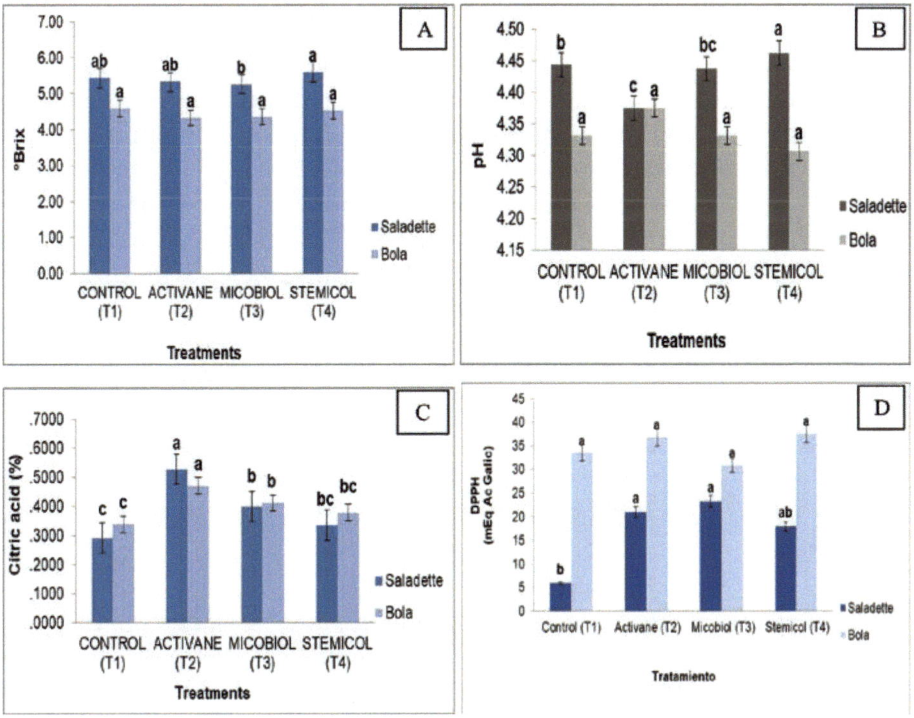


**FIGURE 2.3** Elicitation effects on: (a) Total soluble solids (°Brix), (b) pH, (c) Titratable acid, and (d) antioxidant capacity in fruits of the two tomato cultivars.

The Brix degree results coincide with that of Ramírez-Godina et al.,[24] who indicated a range from 5.95 to 6.63. Brix, with a significant difference ($P = 0.05$) between the two different tomato cultivars, was evaluated. On the other hand, our data regarding fruit pH value were similar to those reported by Jiménez et al.,[25] who indicated a range of pH 3.51 to 4.51, between three genotypes and tomato cultivars with a significant difference ($P = 0.05$) due to the effect of the cut-off date.

In this study, a significant difference ($P \leq 0.05$) was found for pH in the Saladette variety, while no significant differences were found in the Bola variety. The reported pH values are within the range (4.23–4.59) obtained for the four treatments of both varieties. The statistical analysis did not present a significant difference ($P \leq 0.05$) for the variables of the Bola variety (Fig. 2.3). The results were similar to those reported by Jiménez et al.,[25] who indicated a range from 3.51 to 4.51, between three tomato genotypes and varieties with a significant difference ($P \leq 0.05$) due to the effect of the cut-off date.

Los resultados de ácido cítrico presente muestran que el tratamiento T2 Activane® es el que más influye sobre esta variable. Estos resultados coinciden con[26] quien reportó que la cantidad típica de acidez en tomate es el ácido cítrico y oscila entre 0.2–0.6%. The production of the greenhouse tomato crop through the applications of the elicitors in the present study significantly affected the activity of the 2-2-diphenyl-1-picrylhydrazyl (DPPH) radicals, with significant differences ($P \leq 0.05$) for the DPPH antioxidant activity on the Saladette variety. While the Bola variety did not show significant differences ($P \leq 0.05$). The treatments with these elicitors studied showed their capacity to increase the antioxidant activity in fruits of the two tomato cultivars. Both elicitors Activane® and Micobiol® showed to be the best suitable in the obtention of a high antioxidant activity in fruits.

### *2.3.5 INCIDENCE AND SEVERITY OF DISEASES AND PLANT SURVIVAL*

The incidence of diseases was observed mainly in plants of treatment Control (T1). However, the result of the treatments (T2, T3, and T4) showed an incidence of (3–17%) for *Enterobacter cloacae*, (6–45%) for *Oidiopsis Taurica*, and (3%) for Tomato Yellow leaf Curl Virus (TYLCV) in the Bola cultivar and (6–7%) for *Oidiopsis taurica* and (3%) for TYLCV in saladette cultivar (Table 2.5).

**TABLE 2.5** Incidence and Severity of Diseases in the Two Tomato Cultivars Growing in Greenhouse Conditions as Affected by Elicitors.

| | Bacteria (Enterobacter cloacae) | | | | Fungi (Oidiopsis taurica) | | | | Virus (TYLCV) | | | |
|---|---|---|---|---|---|---|---|---|---|---|---|---|
| | Incidence (%) Days after transplant | | | Severity (scale) | Incidence (%) Days after transplant | | | Severity (scale) | Incidence (%) Days after transplant | | | Severity (scale) |
| | 50 | 100 | 150 | | 50 | 100 | 150 | | 50 | 100 | 150 | |
| **Bola cultivar:** | | | | | | | | | | | | |
| T1 | 0 | 0 | 0 | 0 | 0 | 7 | 21 | 1 | 0 | 0 | 3 | 1 |
| T2 | 0 | 0 | 0 | 0 | 0 | 6 | 19 | 0 | 0 | 0 | 0 | 0 |
| T3 | 0 | 0 | 0 | 0 | 0 | 6 | 13 | 0 | 0 | 0 | 0 | 0 |
| T4 | 0 | 0 | 0 | 0 | 0 | 6 | 16 | 0 | 0 | 0 | 0 | 0 |
| **Saladette cultivar:** | | | | | | | | | | | | |
| T1 | 6 | 10 | 17 | 4 | 0 | 10 | 45 | 2 | 0 | 0 | 3 | 1 |
| T2 | 0 | 3 | 3 | 1 | 0 | 7 | 20 | 1 | 0 | 0 | 0 | 0 |
| T3 | 6 | 6 | 6 | 2 | 0 | 6 | 22 | 1 | 0 | 0 | 0 | 0 |
| T4 | 6 | 6 | 6 | 2 | 0 | 6 | 25 | 1 | 0 | 0 | 3 | 1 |

The application of the elicitors decreased the presence of the symptoms of the diseases during the course of the cycle for the treated plants. Treatments T3 and T4 had a greater positive effect on the incidence of diseases. According to García et al.[10] found that the application of the elicitors not only decreased the incidence of diseases, but also promoted the accumulation of total dry biomass, the increase in height and a lower yield loss per plant compared to plants that were not treated. As shown in both experiments, where the incidence of the diseases and the severity of their symptoms during the evaluations were low due to the use of the elicitors, this might be due to the fact that the applied active compounds of these elicitors activated some mechanism related to the defense of the plant, to face these conditions.

For plant survival, a significant difference ($P = 0.05$) was recorded for the Bola cultivar, while for the Saladette type cultivar did not present a significant difference ($P = 0.05$). Saladette showed the highest averages of plants in the treatments Micobiol® (T3) with (94%) and Activane® (T2) with (91%). Finally, the lowest percentages were presented in the Bola cultivar in the treatment Control (T1) with (47%) (Fig. 2.4).

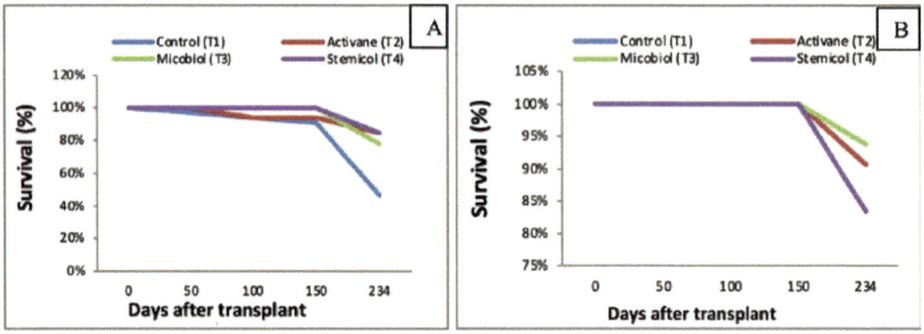


**FIGURE 2.4** The plant survival trend shown in the two tomato cultivars: (a) Saladette and (b) Bola under greenhouse conditions as affected by elicitors.

Mortality in the treatments was influenced by the incidence and severity of the diseases. The Bola and Saladette type varieties presented the highest survival rates at 234 days after transplantation (ddt) in the plants treated with the elicitors. A probable reason for this trend is that the elicitors promoted the induction of resistance in the treated plants, thus reducing the incidence and severity of diseases and improving survival. These results coincided with García et al.[10] who, when evaluating the effect of two elicitors on the produccion de tomato crop, found that the incidence of the disease caused by *F. oxysporum* and the severity of its symptoms during the first evaluation, were reduced by the application of the elicitors of origin natural, which could activate some mechanism related to the defense of the plant, to face biotic stress conditions.

### *2.3.6 MORPHOLOGICAL AND MOLECULAR IDENTIFICATION OF DISEASES*

Out of the two tomato cultivars sampled and with symptoms of necrosis on roots, stems, and leaves, bacterial, fungal and viral isolations were obtained, most of which were characterized by having a white or creamy color. After sequencing them (Fig. 2.5), the bacteria (*Enterobacter Cloacae*) with an identity of 97.8%, fungi (*Oidiopsis taurica*), and Tomato Yellow Leaf Curl Virus with an identity of 99.6% were identified.

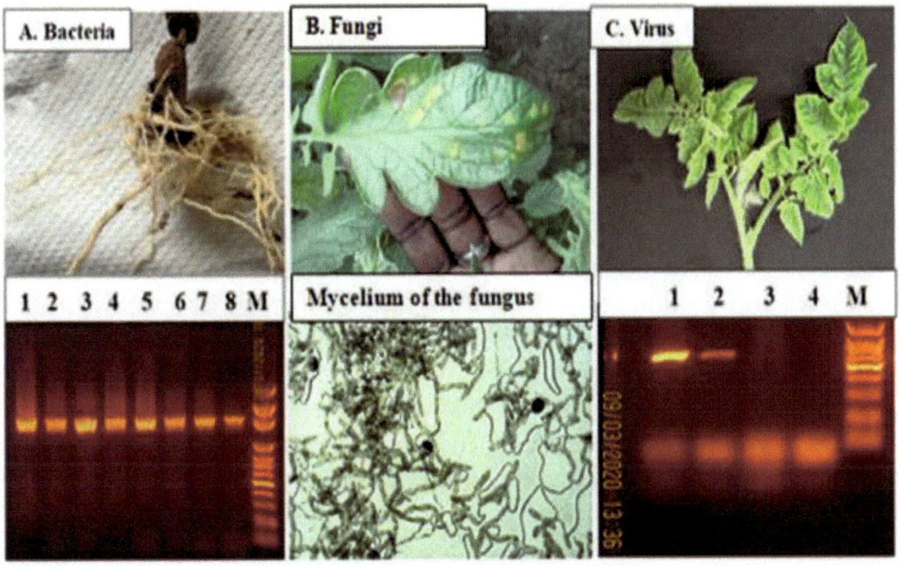


**FIGURE 2.5** Bacterial, fungal, and viral diseases found during the experiment. (A) PCR amplification of the phytopathogenic bacteria in lane 3. (B) Morphological identification of the phytopathogenic fungus powdery mildew with microscopic photographs of the fungus mycelium. (C) PCR amplification of virus DNA (TYLCV) in lanes 1–2, greenhouse tomato samples from FAUANL. (M) Molecular weight marker.

In this study, we describe a new bacterial disease in tomato crops caused by *Enterobacter Cloacae* bacteria with symptoms of small irregular spots on the leaves and brown necrosis was observed at the tips of the margins, subsequent followed by defoliation and total drying of the plant. Morphological and molecular analyzes were performed for bacterial isolation and re-isolated bacteria on an artificial infection in tomato and chili seedlings, thus fulfilling Koch's postulates. However, sequencing of the 16S RNA gene information was inconclusive.[27] Found various genera of enterobacteria present in soil and tomatoes from three fertigation greenhouses in Mexico. On the other hand,[10] reported this bacterial disease affecting up to 4% of pepper seedlings (*Capsicum annuum* L.) under greenhouses in Chihuahua, Mexico.

The virus (TYLCV) belongs to the *Begomovirus* genus, Geminiviridae family, which stands out enormously for its incidence and severity in tomato crops. It is a DNA virus carried by the *Bemisia tabaci*. The states of Baja California, Nayarit, San Luis Potosí, Sinaloa, among others have reported this endemic disease. The virus that causes the disease has also been detected in desert regions with low humidity, as is the case in some coastal areas of

Sonora, Mexico, and more recently in the agricultural area of Caborca under shade houses and greenhouses.[28]

## 2.4 CONCLUSIONS

The growth and production of these two tomato cultivars were favored by the application of elicitors. The ones that most influenced in terms of plant height, stem diameter, weight, and fruit diameter were Micobiol® and Activane®, while for the variables like fruit color, Brix and pH it was Stemicol®, for its part Activane® had significant effect on titratable acid of both cultivars. Moreover, the incidence and severity of diseases decreased and survival increased in plants treated with elicitors. The elicitors Activane®, Micobiol®, and Stemicol® may represent an important potential in greenhouse tomato production to counteract the negative effects of increased incidence and severity of diseases and to prevent unfavourable conditions in the development of tomato crops.

## KEYWORDS


- **cultivars**
- **disease**
- **incidence**
- **quality**
- **deverity**


## REFERENCES


1. FIRA. *Panorama Agroalimentario Tomate Rojo* [Online]. 2019. diees@fira.gob.mx
2. Servicio de Información Agroalimentaria y Pesquera (SIAP) [Online], 2019. https://www.gob.mx/siap/acciones-y-programas/produccionagricola-33119
3. Marín, L. L. *Manual Técnico del Cultivo de Tomate (Solanum Lycopersicum) Costa Rica.* Instituto Nacional de Innovación y Transferencia en Tecnología Agropecuaria. San José, C.R. *INTA* **2017,** (126): 978–9968.
4. Souri, M. K.; Dehnavard, S. Tomato Plant Growth, Leaf Nutrient Concentrations and Fruit Quality Under Nitrogen Foliar Applications. *Adv. Hort. Sci.* **2018,** *32* (1), 41–47.

5. Souri, M. K.; Tohidloo, G. Effectiveness of Different Methods of Salicylic Acid Application on Growth Characteristics of Tomato Seedlings Under Salinity. *Chem Biol. Technol. Agric.* **2019,** *6* (1): 26.
6. FAOSTAT. *Statistics Division Food and Agriculture Organization of the United Nations Viale delle Terme di Caracalla* [Online]. Rome, Italy. 2018, http://www.fao.org/faostat/en/#home
7. Bustamante, J. D.; Vázquez, A.; Trujillo, C.; Reye, R.; Escalona, F. Manual Para el Cultivo del Jitomate en Bioespacio e invernadero. Secretaria de Agricultura, Ganadería, Desarrollo Rural, Pesca y Alimentación, Instituto Nacional de Investigaciones Forestales, Agrícolas y Pecuarias, Centro de Investigación Regional Pacífico Sur, Campo Experimental Zacatepec. *Zacatepec, Morelos. Libro Técnico* **2013,** *11,* 190.
8. Souri, M. K.; Naiji, M.; Kianmehr, M. H. Nitrogen Release Dynamics of a Slow Release Urea Pellet and Its Effect on Growth, Yield, and Nutrient Uptake of Sweet Basil (*Ocimum basilicum* L.). *J. Plant Nut.* **2019,** *42* (6): 604–614.
9. Ebrahimi, M.; Mousavi, A.; Souri, M. K.; Sahebani, N. Can Vermicompost and Biochar Control Meloidogyne Javanica on Eggplant? *Nematology* **2021,** (1), 1–12.
10. García, E.; Robledo, A.; Mendoza, B.; Solís, G.; González, S. M. Efecto de Elicitores de Origen Natural Sobre Plantas de Tomate Sometidas a Estrés Biótico. *Rev. Mex. Cienc. Agric.* **2018,** Esp. (20).
11. Simmons, E.; *Biostimulants Working Group. USDA. AAPFCO Admin Seminar*; Palm Springs: CA. 2018; pp 1–11.
12. INTAGRI. La Inducción de Defensa en las Plantas a través de Elicitores. Serie Fitosanidad Núm. 92. *Artículos Técnicos de INTAGRI. México* 2017; p 6.
13. Souri, M. K.; Hatamian, M. Aminochelates in Plant Nutrition; A Review. *J Plant Nut.* **2019,** *42* (1), 67–78.
14. Souri, M. K.; Bakhtiarizade, M. Biostimulation Effects of Rosemary Essential Oil on Growth and Nutrient Uptake of Tomato Seedlings. *Sci. Hort.* **2019,** (243), 472476.
15. Mota, J. B. *Las Fitovacunas y Su Efecto Vigorizante y Protector en Los Cultivos*. V Congreso Nacional de Fertirrigación y Nutrición en Hortalizas. Lida de México, S.A. de C.V. 2019; pp 1–83.
16. Muñoz. *Condiciones Agroclimática de Mexico y la Horticultura Protegida. Manual de Producción de Tomate en Invernadero*. Intagri, Castellano (Ed), *Gto. Mexico.* **2009,** 44–60.
17. Urrieta-Velázquez, J. A.; Rodríguez-Mendoza, M.; Ramírez-Vallejo, P.; Baca-Castillo, G. A.; Ruiz-Posada, L.; Cueto-Wong, J. A.; Variables de Producción y Calidad de Tres Selecciones de Jitomate de Costilla (*Solanum lycopersicum* L.). *Rev. Chap. Serie Hort.* **2012,** *18* (3), 371–381.
18. Ruiz, M. Á.; Rodríguez, M. S. Parámetros de Calidad Interna de Hortalizas y Frutas en la Industria Agroalimentaria. Negocio Agroalimentario y Cooperativo. *Fichas de Referencia Número* (005) **2014,** 1–18.
19. Benito-Bautista, P.; Arellanes-Juárez, N.; Pérez-Flores M. E. Color y Estado de Madurez del Fruto de Tomate de Cáscara. *Agron. Mesoam.* **2016,** *27* (1), 115–130.
20. Moreno, L. P.; Mendoza, G. H.; Celedón, B. M.; Galván, M. F.; Hernández, L. R.; Franco, A. C. Incidencia, Severidad y Detección de Virus Fitopatógenos en Lechuga, en el Estado de Querétaro, México. Ciencias Biológicas, Agropecuarias y del Medio Ambiente. *Acta. Univ.* **2016,** *26* (2), 3–11.

21. Grijalva, C. L.; Robles, C. F.; Macías, D. R. Híbridos de Tomate Para la Producción en Invernadero en el Noroeste de Sonora. Instituto Nacional de Investigaciones Forestales y Agropecuarias, Centro de Investigaciones Regional del Noroeste, Campo Experimental Caborca. Caborca, Sonora, México. *Folleto Cientifico* **2009,** (1), 1–34.
22. Hornero-Méndez, D.; Mínguez-Mosquera, M. I. Chlorophyll Disappearance and Chlorophyllase Activity during Ripening of *Capsicum Annuum* L Fruits. *J. Sci. Food Agric.* **2002,** (82), 1564–1570.
23. López, A. F.; Gómez, P. A. Comparison of Color Indexes for Tomato Ripening. *Hortic. Bras.* **2004,** (22), 534–537.
24. Ramírez-Godina, F.; Robles-Torres, R.; Foroughbakhch-Pournabav, A.; Benavides-Mendoza, J. L.; Hérnández–Piñero, M. H.; Reyes-Váldez, M.; Alvarado-Vázquez, M. A.; Yield and Fruit Quality Evaluation in Husk Tomato Autotetraploids (*Physalis ixocarpa*) and Diploids. *Aust. J. Crop Sci.* **2013,** (7), 933–940.
25. Jiménez, E.; Robledo, A.; Benavides, T. A.; Ramírez, F.; Ramírez, H.; De la Cruz, E. Calidad de Fruto de Genotipos Tetraploides de Tomate de Cáscara (*Physalis ixocarpa* Brot.). *Uni. Cien.* **2012,** *28* (2), 153–161.
26. Nielsen, S. *Food Analysis*. Purdue University. Editorial *Springer*. **2009,** (4), 587–657.
27. García, G. T.; Sáenz, H. H.; Silva, R.; Morales, N. C.; Vancheva, T.; Koebnik, R.; Ávila, Q. G. *Enterobacter Cloacae*, an Emerging Plant-Pathogenic Bacterium Affecting Chili Pepper Seedlings. *Plant Pathol. J.* **2018,** *34* (1), 1–10.
28. Guerrero, J. C.; Fernández, E.; Zamora, E.; Virus del Rizado Amarillo de la Hoja del Tomate (TYLCV) o Virus de la Cuchara Bajo Casas Sombra. Enfermedades de Las Plantas (Cultivos Protegidos). *ENFPL*-CP-(001) **2016,** 1–4.

# CHAPTER 3

# Evaluation of Growth in Three Cultivars *Lilium* Asiatic Under Greenhouse Hydroponic Conditions

DULCE CONCEPCIÓN GONZÁLEZ-SANDOVAL[1], ADRIANA TOVAR DELGADO[2], HÉCTOR ADRIÁN GARCÍA RAMÍREZ[2], KARINA MORENO-LEÓN[3], WILLIAM ZÁRATE-MARTÍNEZ[4], and VÍCTOR HUGO AVENDAÑO-ABARCA[1]

[1]*Centro de Bachillerato Tecnológico Agropecuario #124. Calle Ahumada, Col. Obrera, Madera, Chihuahua, México*

[2]*Laboratory of Molecular Biology and Biotechnology, Natural Resources Research Center (CIRENA), Chihuahua, México*

[3]*Departamento de Botánica, Universidad Autónoma Agraria Antonio Narro, Calz, Antonio Narro No. 1923, Col. Buenavista, Saltillo, Coahuila. México*

[4]*Campo Experimental Valles Centrales de Oaxaca, Instituto Nacional de Investigaciones Forestales, Agrícolas y Pecuarias (INIFAP), Melchor Ocampo No. 7, Col. Santo Domingo Barrio Bajo, Etla, Oaxaca, México*

## ABSTRACT

In Mexico, there is great interest in the production of cut ornamentals and one of the crops to which more interest is being given is the cultivation of *Lilium* spp. because it is one of the most economically important species in the flower industry. The commercialization of flowers depends on quality parameters and to comply with these, good management in the production system must be in place. The cultivation of *Lilium* in optimal environmental

Sustainable Agricultural Production Systems: Traditional and Modern Perspectives.
Romeo Rojas, Guillermo Cristian Guadalupe Martínez-Ávila, Julia Mariana Márquez-Reyes, and Ma del Carmen Ojeda-Zacarías (Eds.)

and nutritional conditions will have increases in its quality; same that is accentuated with the use of hydroponic technology where there is an efficient absorption of nutrients, availability of water, and reduction in the use of pesticides, allowing more profitable production. The objective of this research was: to assess growth in *Lilium* Asiatic of the cultivars Pavia (A), Dinamix (B), and Tresor (C) in order to see the variability of growth between them. The average results of the evaluated variables include plant length: A = 99.76, B = 110.02, and C = 101.74 cm; fresh biomass: A = 190.31, B = 216.20, and C = 203.97 g; dry biomass: A = 24.08, B = 29.56, and C = 25.60 g and number of buttons: A = 7.50, B = 6.50, and C = 5.50. The results indicated that the genetic variability of the cultivars had a different behavior in each of the evaluated variables. The hydroponic system allowed to meet the quality parameters for the commercialization of this Lillium species due to the availability of water and nutrients, an advantage of this type of production.

## 3.1 INTRODUCTION

Floriculture in Mexico is an agricultural activity that benefits from the wide and varied range of climates present in the country, as well as the fertility of its soils, which allow the production of a wide variety of flowers, which are mainly marketed in the internal market.[1] Mexico has approximately 22,700 ha of land devoted to ornamental crops, the market value of cut flowers at the consumption level is estimated at 14,000 million pesos.[2] Floriculture generates a production value of 5445 million pesos, which triggers around 188,000 permanent jobs, 50,000 temporary jobs and one million indirect jobs in the ornamental market.[3] About 90% of the national production is destined for national markets and the rest is sent to international markets. More than 70 varieties of cut flowers and foliage are grown in the country, concentrating 90% of the surface in five states of the country, State of Mexico, Morelos, Puebla, Veracruz, and Michoacan. The main cultivated species are roses, gladiolas, carnations, chrysanthemums, and *Lilium* spp. it is estimated that they represent 50% of the cultivated area with a value of almost 80% of the value of production.[4]

*Lilium* spp is one of the species of economic importance in the production of cut flowers, occupying the fifth place in terms of the value of production in the floricultural branch, with an economic benefit of 310 million pesos, occupying a production area of 124 ha. with a generation of 438,040 stems.[5] The *Lilium* plant developed under optimal environmental and nutritional conditions will have larger bulbs, increasing the length and quality of the floral stems.[6] The implementation of hydroponic systems is an alternative to achieve the efficient use of inputs due to the efficient absorption of nutrients,

### 3.3.4 NUMBER OF BUDS

The average number of buttons was A = 7.50, B = 6.50 and C = 5.50 (Fig. 3.4) finding significant differences between treatments, according to the agricultural production manual[21,22] for this species of *Lilium* was higher than average which is 2–4 buttons for the 14/16 bulb caliber; in addition to complying with the Extra quality standard (4 or more flowers) in accordance with regulation (CEE)[13] 32 n° 316/68. These results were possibly presented by the application of the nutrient solution which has the optimal nutritional requirements for the crop; in addition to the non-competition for nutrients and water that the hydroponic system provides.[6] For *Lilium* the recommendations should be for each type or class due to its remarkable genetic variability.[22,23]

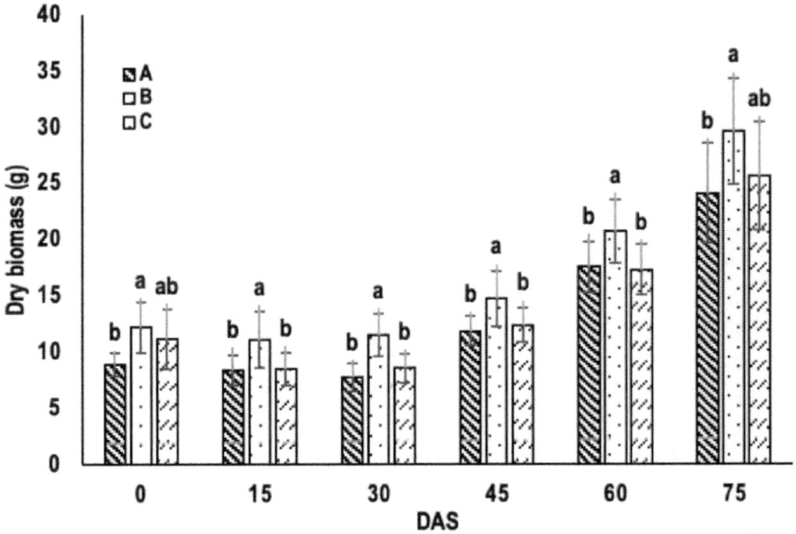


**FIGURE 3.3** Dry biomass (g) during the agricultural cycle of *Lilium* Asiatic cultivars: A = Pavia; B = Dynamix, and C = Tresor; DAS = Days after sowing. Values with different letters are statistically different according to the Tukey test ($p \leq 0.05$). Values with a letter in common are not statistically different according to the Tukey test ($p \leq 0.05$).

### 3.3.5 TOTAL IRRIGATION VOLUME

During the production cycle, 19 crop irrigations were carried out, representing a total volume of 2575.44 L water applied, a total of water used per plant is 5.11 L.

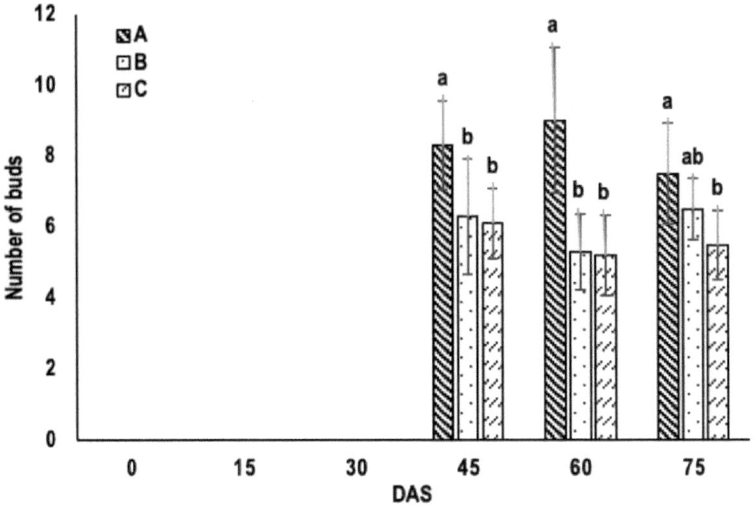


**FIGURE 3.4** Number of buds during the agricultural cycle of *Lilium* Asiatic cultivars: A = Pavia; B = Dynamix, and C = Tresor; DAS = Days after sowing. Values with different letters are statistically different according to the Tukey test ($p \leq 0.05$). Values with a letter in common are not statistically different according to the Tukey test ($p \leq 0.05$).

## 3.4 CONCLUSION

The differences between the evaluated variables of each of the *Lilium* Asiatic cultivars provide a guideline to continue conducting research and to be able to define the behavior of each cultivar, since each one presents a specific genetic variability.

## KEYWORDS

- **cutting ornamentals**
- **hydroponics**
- **quality**
- **commercialization**
- **genetic variability**

## REFERENCES


1. Tecnológico de Monterrey (TecM). .Estado de México Hacia una Identidad de Conocimiento, el Desarrollo Tecnológico y la Innovación, Con Ventajas Comparativas y Competitivas Para las Empresas, los Negocios y el Bienestar Social. Estudio Sectorial Flores y Horticultura Ornamental. México, 2011.
2. Asociación de Consumidores Orgánicos (ACO). México, Tercer Lugar en Cultivo de Flores [Online], 2018. https://consumidoresorganicos.org/2018/10/10/mexico-tercer-lugar-en-cultivo-flores/ (accessed Feb 23, 2022).
3. Torres, L. R. Floricultura Mexicana: Gran industria en el Anonimato [Online], 2014. http://tierrafertil.com.mx/floricultura-mexicana-gran-industria-en-el-anonimato/ (accessed Feb 25, 2022).
4. Secretaría de Agricultura y Desarrollo Rural (SADER). Floricultura Actividad Económica más Importante del Sector Agropecuario en el Edomex [Online], 2015. https://www.gob.mx/agricultura%7Cedomex/articulos/floricultura-actividad-economica-mas-importante-del-sector-agropecuario-en-el-edomex (accessed Feb 20, 2022).
5. Ramírez, H.; Benavides, A. Horticultural Science and Industry in Mexico an Overview. *Chron. Hortic.* **2003,** *43* (3), 20–25.
6. Alberto, V. A. Efecto del Suministro de Calcio en el Desarrollo de la Planta y Calidad de la Flor *Lilium* spp. Tipo Asiático, Cultivada en hidroponía. Tesis de licenciatura. Universidad Autónoma Agraria Antonio Narro: Saltillo, Coahuila, México, 2012.
7. Asociación Hidropónica Mexicana (AHM). Ventajas de la Hidroponía [Online], 2012. http://hidroponia.org.mx/cultivo-hidroponico/ventajas-de-la-hidroponia/ (accessed Feb 2, 2022).
8. Barrantes-Infante, B. L.; Bertsch Hernández, F. Curvas de Absorción de Nutrimentos Para Tres Variedades de Lirios (*Lilium* sp.) y afinamiento del programa de fertilización en una finca comercial en Heredia, Costa Rica. *Agron. Costarricense* **2013,** *36* (2).
9. Van Os, E. A. Closed Soilless Growing Systems: A Sustainable Solution for Dutch Greenhouse Horticulture. *Water Sci. Technol.* **1999,** *39* (5), 105–112.
10. Cruz-Crespo, E.; Arévalo-Galarza, L.; Cano-Medrano, R.; Gaytán-Acuña, E. A. Soluciones Pulso en la Calidad Poscosecha de Lisianthus *(Eustoma grandiflorum* Raf.) cv. ‘echo blue’. *Agric. Técnica en México* **2006,** *32* (2), 191–200.
11. Giri, T. K.; Beura, S.; Behera, S.; Acharjee, S. Response of Asiatic *Lilium* Hybrid cv. Tresor to Foliar Application of Different Group of Nutrients. *Int. J. Curr. Microbiol. Appl. Sci.* **2017,** *6* (9), 3280–3286.
12. Seyedi, N.; Mohammadi, T. A.; Allahyari, M. S. Investigating of the Effects of Calcium Concentration Under Hydroponic Conditions on Quantitative and Qualitative Growth of *Lilium* Tresor. *J. Ornam. Hortic. Plants* **2013,** *3* (1) 19–24.
13. Reglamento (CEE) nº 316/68 del Consejo, de 12 de marzo de 1968, por el que se Establecen Normas de Calidad de las Flores Cortadas Frescas y Los Follajes Frescos [Online]. https://publications.europa.eu/en/publication-detail/-/publication/f3b379a0-0843-4f50-9de4-08c6f4da2f28/language-es (accessed Feb 21, 2022).
14. Herreros Delgado, L. M. Cultivo del *Lilium* (azucena híbrida) [Online], 1983. Hojas Divulgadoras. Ministerio de Agricultura, pesca y alimentación.
15. International Flower Bulb Centre (IBC) [Online], 2010. http://flowers.flowers-expo.com/flowers2010/en/358.html_(accessed Feb 10, 2022).

16. Di Benedetto, A.; Tognetti, J. Técnicas de Análisis de Crecimiento de Plantas: su Aplicación a Cultivos Intensivos. *RIA. Revista de Investigaciones Agropecuarias* **2016,** *42* (3), 258–282.
17. Betancourt-Olvera, M.; Rodríguez-Mendoza, M. N.; Sandoval-Villa, M.; Gaytán-Acuña, E. A. Fertilización Foliar una Herramienta en el Desarrollo del Cultivo de *Lilium* cv. Stargazer. *Revista Chapingo. Serie Horticultura* **2005,** *11* (2), 371–378
18. Toit, E. S.; Robbertse, P. J.; Niederwieser, J. G. Plant Carbohydrate Partitioning of *Lachenalia* cv. Ronina During Bulb Production. *Sci. Hortic.* **2004,** *102* (4), 433–440.
19. Artacho-Vargas, P.; Pinochet-Tejos, D. Producción de Materia Seca y Absorción de Nitrógeno del Cultivo del Tulipán (*Tulipa gesneriana* L.). *Agrociencia* **2008,** *42* (1), 37–45.
20. Correa, M. Caracterización de los Estados de Desarrollo y Determinación de las Curvas de Absorción de Nutrientes en *Lilium* sp. *Cvs.* Miami, Navona y Fangio para flores de corte. Tesis Mag. Sc. Santiago, Chile. Pontificia Universidad Católica de Chile, 2003.
21. Herreros, D. L. M. Cultivo de *Lilium.* Servicio de Extensión Agraria. Ministerio de Agricultura Pesca y Alimentación: Madrid, España, 1983.
22. Ortiz, L. Manual Técnico Para la Producción de *Lilium* [Online], 2013. http://es.scribd.com/doc/137797645/Manual-parala-produccion-de-Lilium-spp#scribd (accessed Feb 11, 2022).
23. Torres Miranda, E.; Morales Rosales, E. J.; Pérez López, D. D. J. Vida en Florero de *Lilium* "Brindisi" y "Menorca" Fertilizados con Nitrato y Oxido de Calcio. *Cienc. Agrícolas Infor.* **2009,** *18,* 4–12.

## CHAPTER 4

# Importance of Secondary Metabolites in Plant Defense Mechanisms

WILLIAM ZÁRATE-MARTÍNEZ[1], KARINA MORENO-LEÓN[2], DULCE CONCEPCIÓN GONZÁLEZ-SANDOVAL[3], VÍCTOR HUGO AVENDAÑO-ABARCA[3], ADRIANA TOVAR DELGADO[4], and HÉCTOR ADRIÁN GARCÍA RAMÍREZ[4]

[1]*Campo Experimental Saltillo, Instituto Nacional de Investigaciones Forestales, Agrícolas y Pecuarias (INIFAP). Carretera Saltillo-Zacatecas km 342+119 No. 9515, Col. Hacienda de Buenavista, CP. 25315, Saltillo, Coahuila de Zaragoza, México*

[2]*Departamento de Botánica, Universidad Autónoma Agraria Antonio Narro. Calz. Antonio Narro, Col. Buenavista, CP, Saltillo, Coahuila. México*

[3]*Centro de Bachillerato Tecnológico Agropecuario #124, Calle Ahumada, Col. Obrera, CP, Madera, Chihuahua, México*

[4]*Laboratory of Molecular Biology and Biotechnology, Natural Resources Research Center (CIRENA), Chihuahua, México*

### ABSTRACT

Plants synthesize a wide variety of organic molecules, including secondary metabolites, which do not directly participate in photosynthetic or respiratory processes, nutrient assimilation, solute transport, or protein, carbohydrate, or lipid synthesis. Secondary metabolites are synthesized in small quantities, in certain genera, families or species of plants. Secondary metabolites are grouped into four classes: terpenes, phenolic compounds, glycosics, and alkaloids. The molecular structure of phenolic compounds is characterized by the fact that an aromatic ring is attached to a hydroxyl functional group. According

Sustainable Agricultural Production Systems: Traditional and Modern Perspectives.
Romeo Rojas, Guillermo Cristian Guadalupe Martínez-Ávila, Julia Mariana Márquez-Reyes, and Ma del Carmen Ojeda-Zacarías (Eds.)

to their chemical structure, they are a very diverse group ranging from simple molecules (phenolic acids) to complex polymers (tannins and lignin). Flavonoid pigments are also found in the group. In plants, phenolic compounds can be biosynthesized in different pathways, but there are basically two pathways involved: the shikimic acid pathway, where most phenols are synthesized in plants, and the malonic acid pathway, a less used pathway in higher plants.

## 4.1 INTRODUCTION

An important characteristic of plants is that they carry out two types of metabolism: primary and secondary. The first takes place in the cells of all living things and the substances produced are called primary metabolites. The latter are directly involved in the growth, development and reproduction of plants. Examples of these compounds are sugars, proteins, amino acids, and nucleic acids.[1] However, plants also have a secondary metabolism, which allows them to produce and accumulate compounds of a diverse chemical nature.[2] Although they are not essential for life like the primaries, they do play a very important role in adapting to environmental stress and defending against potential predators and pathogens.[3] The compounds that result from this process are called secondary or specialized metabolites, which mediate plant–environment interactions (and allow the production and accumulation of compounds of a diverse chemical nature).[4]

Primary metabolism is important for cell growth, development, and reproduction.[5] Secondary metabolism plays an important role in the evolutionary adaptation of plants and is therefore involved in processes specific to the characteristics of the plant ecology.[6] Plants have great metabolic diversity and a great genetic capacity to rapidly generate new compounds due to gene duplication and divergence.[7] Therefore, plants are able to potentially adapt to new environments at a faster rate than that explained by the mutation process.[8] Therefore, plants constantly synthesize a large number of primary and secondary metabolites.[9] So too, similar patterns of metabolite production may emerge between closely related and distant taxa that have evolved under similar environmental conditions.[10]

## 4.2 PRIMARY METABOLITES

Primary metabolites are molecules that are directly related to vital functions of the plant, necessary for its growth and development.[9,11] Some of the primary metabolites produced by plants include carbohydrates, protein amino acids, proteins, lipids, fatty acids, some carboxylic acids, etc.12 Plants contain numerous organic compounds called "primary metabolites," among which are sugars or carbohydrates, which are produced in the process of photosynthesis.[13] Hormones are signaling molecules that control the growth and development of plants, and because they are present at very low levels, convert them into atypical primary metabolites.[14] It is important to mention that some primary metabolites, such as proteins and polysaccharides, have antiviral characteristics.[12]

## 4.3 SECONDARY METABOLITES

Secondary metabolites are synthesized in small amounts in plants and have no direct role in processes, such as photosynthetic, respiration, solute transport, nutrient assimilation, solute transport, protein, carbohydrate, or lipid synthesis.[15,16] Secondary metabolites are related to the plant's defense system, since they help it to overcome biotic and abiotic stress conditions and participate in the attraction of pollinators.[17] More than 40,000 different molecules of secondary metabolites are known, which can be found in plants.[2] Because secondary metabolites participate in plant defense against abiotic and biotic stress, they are considered important drivers of plant diversification and evolution.[18,19] On the other hand, secondary metabolites are synthesized in different concentrations, depending on their function in the plant.20 It is important to note that many functions of secondary metabolites are still unknown.[17]

The study of the biological and structural functions of secondary metabolites is of great importance since, based on this knowledge, they have been given uses and applications in different industries.[21] In general, secondary metabolites are responsible for the smell, taste, and color of the plant, and also for its medicinal properties.[22]

Biotic or abiotic stress conditions trigger a cascade of biochemical signals, which induce the expression of genes related to the biosynthesis of secondary metabolites; Likewise, the discovery of new compounds and their elicitors may allow the manipulation of their biosynthetic pathways, increasing their yield in tissue and cell cultures.[23]

## 4.4 CLASSIFICATION OF SECONDARY METABOLITES

Various criteria have been considered for the classification of secondary metabolites: chemical structure (presence of rings or sugars), composition (containing or not nitrogen), solubility in inorganic solvents or water, and biosynthetic pathway.[2,24] Classifies secondary metabolites into four main classes: terpenes, phenolic compounds, alkaloids, and glycosides. The most used criterion to classify secondary metabolites in plants has been the biosynthetic pathway, finding three large groups: terpenes, phenolic compounds, and alkaloids.[1,2,25] Secondary metabolites are classified into three large chemical classes: terpenes (aromatic and volatile compounds that provide aroma and flavor to plants), phenolic compounds (compounds with a protective function, as well as generators of attractive colors for plant pollination and seed dispersal) and alkaloids (natural compounds to repel herbivorous organisms).

### *4.4.1 TERPENES*

Terpenes, isoprenoids or terpenes constitute the largest group of secondary metabolites.[26] They are a class of organic compounds, insoluble in water, derived from isoprene, a hydrocarbon with five carbon atoms.[2] In flowers and fruits of various plants, we can find terpenes as mixtures of volatile compounds with specific odors, some examples of these plants are lemon, mint, eucalyptus, ginger, and basil.[27]

#### *4.4.1.1 BIOSYNTHESIS OF TERPENES*

Terpenes are synthesized from primary metabolites, which can be carried out by at least two different pathways: the mevalonate pathway (which binds three Acetyl-CoA molecules) and the non-mevalonate pathway (which produces isopentenyl diphosphate [IPP] and dimethylallyl diphosphate [DMAPP]). This shows the importance of primary metabolism in plants since they are the source of terpenes and other secondary metabolites. In this case, the presence of Acetyl-CoA in the IPP pathway and Pyruvate plus Glyceraldehyde 3-Phosphate, in the DMAPP pathway, is essential for the synthesis of terpenoids in plants.[27]

#### *4.4.1.2 CLASSIFICATION OF TERPENES*

Terpenes are classified according to the number of isoprene they have (5°C): terpenes of two isoprenes are called monoterpenes (10°C); Terpenes of three isoprenes are called sesquiterpenes (15°C), terpenes of four isoprenes are called diterpenes (20°C), terpenes of six isoprenes are called triterpenes

(30°C), terpenes of eight isoprenes are called tetraterpenes (40°C) and when the terpenes contain more than eight isoprene, we speak of polyterpenes.[2]

#### *4.4.1.3 FUNCTION OF TERPENES*

Terpenes participate in the primary and secondary metabolism of plants, as well as intervene in various biological functions.[28] In primary metabolism, terpenes can be found as photosynthetic pigments (carotenes), electron carriers (ubiquinone and plastiquinone), as well as growth regulators (gibberellin), can be part of cell membranes and participate in protein glycosylation.[29] In the secondary metabolism, they participate as defense molecules, insecticides, toxic compounds, and dissuasive food for insects, likewise, in some plants, it is responsible for attracting pollinators.[30]

### ***4.4.2 PHENOLIC COMPOUNDS***

Phenolic compounds have a hydroxyl group attached directly to an aromatic hydrocarbon.[31] Also, phenolic compounds are known as a type of multipurpose bioactive, widely distributed throughout the plant kingdom.[32] Chemically, phenolic compounds are a very diverse group of secondary metabolites, from simple elementary molecules to highly polymerized compounds.[33]

#### *4.4.2.1 BIOSYNTHESIS OF PHENOLIC COMPOUNDS*

Phenolic compounds can be sintered by (a) the shikimic acid pathway, (b) the malonate/acetate pathway or (c) by both pathways, an example of the latter being flavonoids). The shikimic acid pathway allows the synthesis of simple phenols, phenolic acids, coumarins, lignans and all derivatives of phenylpropane.[34] Polyacetate pathway provides quinones and xanthones.[35] The mixed pathways combine precursors of the shikimic acid pathway and the polyacetate pathway, an example of a secondary metabolite generated from the mixed pathways are flavonoids.[36]

#### *4.4.2.2 CLASSIFICATION OF PHENOLIC COMPOUNDS*

The classification of phenolic compounds is carried out according to the number of carbons present in the molecule, within the main phenolic compounds, we have simple phenols, phenolic acids, acetophenones, phenylacetic acids, hydroxycinnamic acids, coumarins, and flavonoids.[37] Lignans, tannins, and phlobaphens are also phenolic compounds; phlobaphenes have a more complex structure.[38]

#### 4.4.2.3 *ROLE OF PHENOLIC COMPOUNDS*

In plants, phenolic compounds can perform various functions: with molecules that oxidize rapidly and act as antioxidants.[39] Phenolic compounds also act as plant growth inhibitors.[40] The seeds accumulate significant amounts of phenolic compounds, which act as a filter so that oxygen does not reach the embryo and thereby inhibit its germination.[41] In addition, phenolic compounds accumulate on the surface of the leaves, blocking up to 90% of ultraviolet radiation.[42] Phenolic compounds are responsible for giving color and aroma to the fruits, making them palatable to herbivores, which favors the dispersal of seeds through feces.[43] Participate in the process of allelopathy.[44] Likewise, phenolic compounds participate in the synthesis of phytoalexins, which are toxic substances for microorganisms.45 Phenolic compounds generate bitter flavors or unpleasant textures in plants, thus protecting them from herbivores.[46]

### ***4.4.3 ALKALOIDS***

Alkaloids contain a nitrogen atom in their structure. More than 27,000 alkaloids are known, which are found mainly in the flowers of plants.[47] The type and quantity of alkaloids varies within the same plant, to the point that some structures may not have.[48] Alkaloids, in addition to having basic nitrogen atoms, may also contain some neutral or weakly acidic compounds.[49] Additionally, alkaloids are low molecular weight structures, and a total of 12,000 alkaloids have been isolated from different plant species, which represents approximately 20% of the secondary metabolites of plant origin.[50] Physically, most alkaloids are solid. They are mainly found in the families, such as Leguminoceae, Rutaceae, and Solanaceae.[51]

#### 4.4.3.1 *BIOSYNTHESIS OF ALKALOIDS*

The biosynthetic routes of alkaloids are one of the greatest challenges that chemistry has faced from the XX century to the present due to their structural diversity and the diversity of plants that produce them. Advances in enzymatic studies have meant a great evolution for the elucidation of biosynthetic routes for the formation of alkaloids in nature; however, not all of them have been elucidated.[52]

The basic biosynthesis pathway for most alkaloids is defined by the following reactions:

a) condensation of an aldehyde or ketone with an amine to form a Schiff base.
b) the closure of a ring coupled to an aromatic system.
c) oxidation of the hydroxyl group of a phenolic compound to form a highly reactive intermediate radical.[52]

#### *4.4.3.2 CLASSIFICATION OF ALKALOIDS*

The alkaloids present in plants belong mostly to the group of true alkaloids, which are characterized by having an intracyclic nitrogen, additionally they are compounds that predominate as salts in plants.[53] There is no uniform classification for alkaloids, so several criteria are used: biosynthetic origin, according to their pharmacological properties and according to their distribution by plant families.[54] Alkaloids are classified into various categories, depending on the heterocyclic rings and their biosynthetic precursor.[50]

The classification according to biogenesis considers three types of alkaloids: (a) true alkaloids, (b) protoalkaloids and (c) pseudoalkaloids.[51]

a) **True alkaloids.** This type of alkaloids share a highly reactive nitrogenous heterocyclic ring with powerful biological activity. Many of them conjugate with acid and form a salt. Almost all true alkaloids are solid with a bitter taste.56 In the plant, it can be found in three forms: free, as N oxide or as a salt.[55]

b) **Protoalkaloids.** These alkaloids are synthesized from amino acids, they also contain a nitrogen atom.[57]

c) **Pseudoalkaloids.** These are not derived directly from amino acids.55 They can be synthesized from phenylalanine or acetate. Caffeine is the most common example of pseudoalkaloids.[51]

Classification according to the structure of the ring, this classification is the most complete.

a) **Tropane alkaloid.** These alkaloids have a tropane ring (C4N backbone), these alkaloids are abundantly found in the family Solanaceae.[19]

b) **Pyrrolizidine alkaloids.** The pyrrolizidine nucleus is distinctive to this group of alkaloids, found mostly in plants of the asteraceae and fabaceae families.[19] La mayoría de los alcaloides de pirrolizidina se encuentran en las plantas como N-óxidos.58 These alkaloids are not edible, senecionine is the popular alkaloid of this type.[59]

c) **Piperidine alkaloids.** The piperidine nucleus is the basic ring system of this group of alkaloids. The presence of odor is the common characteristic of piperidine alkaloids. Lobeline is one of the important alkaloids of this group.[60]

d) **Quinoline alkaloid.** This type of alkaloid is obtained exclusively from the bark of plants of the cinchona genus. The main alkaloids of this specific group are cinchonine, cinchonidine, quinine, and quinidine.[60]

e) **Isoquinoline alkaloids.** Isoquinoline alkaloids are found primarily in higher plants.[60] The isoquinoline nucleus is the basic structural feature. These groups of alkaloids have properties, such as antiviral, antifungal, antioxidant, among others.61 This group of alkaloids is also classified as follows: simple isoquinoline alkaloids, benzylisoquinoline alkaloids, and bisbenzylisoquinoline alkaloids.[62]

f) **Indole alkaloids.** These alkaloids are derivatives of tryptophan.[63,64] Some alkaloids in this group include simple tryptamine derivatives, and others in which integration has occurred of regular monoterpenoid or diterpenoid units.[65,66] These alkaloids are classified as follows: simple indole alkaloids (aplisinopsin, gramine), bisindoles (indirubine, 6,6′-dibromoindigotine), simple tryptamine alkaloids (tryptamine), and cyclotryptamine alkaloids.[67,68]

g) **Steroidal alkaloids.** The phenanthrene 1,2-cyclopentane rings is the characteristic of this type of alkaloids.[69] They originate from higher plants of the Liliaceae and Solanaceae families mainly. These alkaloids are divided into several subtypes, aminopregnanes being the simplest type.[70] The other types of steroidal alkaloids are salamander-type (cycloneosamandione), jerveratrum-type (jervina), spirosolane-type (soladulcidin), solanidine-type (rubijervina), cerveratrum-type (3,6-cevanediol), conanine-type (didimeline),

Buxus-type ( cyclobuxin), pregnane-type (20α-dimethylamino-3β-senecioylamino-16β-hydroxy-pregn-5-ene), cephalostatins/ritterazines (ritterazines a), and miscellaneous steroidal alkaloids (bufotoxin).[71]

h) **Imidazole alkaloid.** The imidazole ring structure is characteristic of this type of alkaloid. This class of alkaloids are structurally different and show a wide range of biological activities.[72]

i) **Purine alkaloids.** These alkaloids consisting of a purine ring and pentose sugar together with another pyrimidine base.[74] Caffeine is the most common purine alkaloid, quite popular to find in plants.[73]

j) **Pyrrolidine alkaloids.** These compounds are mainly found in plants of the Leguminosae family in different concentrations, although it is more common to find this compound in flowers.[75]

#### *4.4.3.3 ROLE OF ALKALOIDS*

Although alkaloids are not essential for plant development, there is sufficient evidence to indicate their importance.[76] Previously, alkaloids were considered as growth regulators, it is now known that alkaloids defend plants from insects and herbivores.[77]

## 4.5 IMPORTANCE OF SECONDARY METABOLITES IN AGRICULTURE

Secondary metabolites are related to the ecological interactions between plants and the environment, especially in the defense of the plant against predators and pathogens, they act as attractants or repellents of animals.78 Most secondary metabolites are synthesized from intermediate metabolites of primary metabolism, mainly from the Krebs cycle, glycolysis, and the shikimate pathway.[79] Plants are constantly changing and adapting to the environment, among the main adaptations are the formation of various organic molecules,[19] which do not have a direct function with primary metabolic processes, such as photosynthesis, respiration, assimilation, and transport of nutrients, and are known as "secondary metabolites."[2] Various works have been reported, in which they have made applications of molecules that act as elicitors in plants, with the aim of activating the defense mechanisms of plants. The application of phenolic acids to a tomato crop activated the enzymatic defense mechanisms of the plant, since greater activity of enzymes such as phenylalanine ammonium lyase (PAL) and catalase (CAT) was quantified.[80]

## 4.6 CONCLUSION

Secondary metabolites are organic compounds that are synthesized in plants in small quantities. These compounds are related to the defense system of plants to deal with predators, pathogens, or as allelopathic agents of survival and reproduction of the same plants. There is a large quantity and diversity of secondary metabolites and phenolic compounds are a very diverse group of secondary metabolites that are mostly found in plants. Due to the fact that there is interest in understanding the biosynthesis of these compounds, they will continue to be studied and finally generate actions that stimulate and activate the defense mechanisms against biotic and abiotic stress conditions.

## KEYWORDS

- **secondary metabolites**
- **phenolic compounds**
- **terpenes**

## REFERENCES

1. Adeyemi, M. A Review of Secondary Metabolites from Plant Materials for Post Harvest Storage. *Int. J. Pure Appl. Sci. Technol.* **2011,** *6* (2), 94–102.
2. Ávalos, G. A. y Pérez-Urria, C. E. Metabolismo Secundario de Plantas. Reduca (Biología). *Serie Fisiología Vegetal.* **2009,** *2*, 119–145.
3. Lustre, S. H. Los Superpoderes de las Plantas: Los Metabolitos Secundarios en su Adaptación y Defensa. *Revista Digital Universitaria* **2022,** *23* (2). 1–8.
4. Erb, M.; Kliebenstein, J. D. Plant Secondary Metabolites as Defenses, Regulators, and Primary Metabolites: The Blurred Functional Trichotomy. *Plant Physiol.* **2020,** *184* (1), 39–52.

5. Rojas C. M.; Senthil-Kumar M.; Tzin V.; Mysore K. S. Regulation of Primary Plant Metabolism During Plant–Pathogen Interactions and Its Contribution to Plant Defense. *Front. Plant Sci.* **2014,** *5,* 17.
6. Nishida, R. Chemical Ecology of Insect–Plant Interactions: Ecological Significance of Plant Secondary Metabolites. *Biosci. Biotechnol. Biochem.* **2014,** *78,* 1–13.
7. Pichersky, E.; Gang, D. R. Genetics and Biochemistry of Secondary Metabolites in Plants: An Evolutionary Perspective. *Trends Plant Sci.* **2000,** *5,* 439–445.
8. Moore, B. D.; Andrew, R. L.; Külheim, C.; Foley, W. J. Explaining Intraspecific Diversity in Plant Secondary Metabolites in an Ecological Context. *New Phytol.* **2014,** *201,* 733–750.
9. Martín, G. D. A. Los Compuestos Fenólicos, Un Acercamiento A Su Biosíntesis, Síntesis Y Actividad Biológica. *Revista De Investigación Agraria Y Ambiental.* **2018,** *9* (1), 81–104.
10. Hatcher, C. R.; Ryves, D. B.; Millett, J. The Function of Second Metabolites in Plant Carnivory, *Ann. Bot.* **2020,** *125* (3), 399–411.
11. Fernie, A. R.; Pichersky, E. Focus Issue on Metabolism: Metabolites, Metabolites Everywhere. *Plant Physiol.* **2015,** *169,* 1421–1423.
12. Wang, W.; Wang, S. X.; Guan, H. S. The Antiviral Activities and Mechanisms of Marine Polysaccharides: An Overview. *Marine Drugs* **2012,** *10.* 2795–2816.
13. Paumier, M.; Verdecia, D. M.; Ramírez, J. L.; Herrera, R. S.; Leonard, I.; Santana, A.; Méndez, Y. El Contenido de Metabolitos Primarios de Gliricidia Sepium en una Zona del Valle del Cauto, Cuba. *REDVET Revista Electrónica de Veterinaria.* **2018,** *19* (4), 1–8.
14. Ahanger, M. A.; Ashraf, M.; Bajguz, A.; Ahmad, P., Brassinosteroids Regulate Growt in Plants Under Stressful Environments and Crosstalk with Other Potential Phytohormones. *J. Plant Growth Regul.* **2018,** *37,* 1007–1024.
15. Kroymann J. Natural Diversity and Adaptation in Plant Secondary Metabolism. *Curr. Opin Plant Biol.* **2011,** *14,* 246–251.
16. Ramakrishna, A.; Ravishankar, G. A. Influences of Abiotic Stress Signals on Secondary Metabolites in Plants. *Plant Signal. Behav.* **2011,** *6,* 1720–1731.
17. Berini, J. L.; Brockman, S. A.; Hegeman, A. D.; Reich, P. B.; Muthukrishnan, R.; Montgomery, R. A.; Forester, J. D. Combinations of Abiotic Factors Differentially Alter Production of Plant Secondary Metabolites in Five Woody Plant Species in the Boreal-Temperate Transition Zone. *Front Plant Sci.* **2018,** *9,* 1257–1273.
18. Stevenson, P. C.; Nicolson, S. W.; Wright, G. A. Plant Secondary Metabolites in Nectar: Impacts on Pollinators and Ecological Functions. *Func. Ecol.* **2017,** *31,* 65–75.
19. Hartmann, T. From Waste Products to Ecochemicals: Fifty Years Research of Plant Secondary Metabolism. *Phytochemistry* **2007,** *68,* 2831–2846.
20. Pinard, D.; Fierro, A. C.; Marchal, K.; Myburg, A. A.; Mizrachi, E. Organellar Carbon Metabolism is Coordinated with Distinct Developmental Phases of Secondary Xylem. *New Phytol.* **2019,** *222,* 1832–1845.
21. Zillich, O. V.; Schweiggert-Weisz, U.; Eisner, P.; Kerscher, M. Polyphenols as Active Ingredients for Cosmetic Products. *Int. J. Cosmet. Sci.* **2015,** *37,* 455–464.
22. Castillo, G. E.; Martínez, S. I. *Manual de Fitoterapia,*3ra ed; Elsevier, 2021; p 596.
23. Kim, H. J.; Ono, E.; Morimoto, K.; Yamagaki, T.; Okazawa, A.; Kobayashi, A.; Satake, H. Metabolic Engineering of Lignan Biosynthesis in Forsythia sp. Cell Culture. *Plant Cell Physiol.* **2009,** *50* (12), 2200–2209.

24. Kozikowski, A.; Tückmantel, W.; Böttcher, G.; Romanczyk, L. Studies in Polyphenol Chemistry and Bioactivity. 4. Synthesis of Trimeric, Tetrameric, Pentameric, and Higher Oligomeric Epicatechin-Derived Procyanidins Having All-4β, 8-Interflavan Connectivity and Their Inhibition of Cancer Cell Growth through Cell Cycle Arrest. *J. Org. Chem.* **2003,** *68* (5), 1641–1658.
25. Hopkins, W. G. *Physiologie Vegetale. De Boeck y Larcier*, 2da ed; 2003; pp 1–495.
26. Vranová E.; Coman, D.; Gruissem, W. Structure and Dynamics of the Isoprenoid Pathway Network. *Mol. Plan.* **2012,** *5,* 318–333.
27. Olivoto, T.; Nardino, M.; Carvalho, I. R.; Follmann, D. N.; Szareski, V. J.; Ferrari, M.; Pelegrin, A. J.; Souza, V. Q. Plant Secondary Metabolites and Its Dynamical Systems of Induction in Response to Environmental Factors: A Review. *Af. J. Agric. Res.* **2017,** *12,* 71–84.
28. Veitch, G. E.; Boyer, A.; Ley, S. V. The Azadirachtin Story. *Angew. Chem. (International ed. in English).* **2008,** *47* (49), 9402–9429.
29. Loreto, F.; Dicke, M.; Schnitzler, J. P.; Turlings, T. C. J. Plant Volatiles and the Environment. *Plant Cell Environ.* **2014,** *37,* 1905–1908.
30. Rodrigues, M.; Festucci-Buselli, R. A.; Campos, S. L.; Campos, O. W. Azadirachtin Biosynthesis Induction in Azadirachta indica A. Juss Cotyledonary Calli with Elicitor Agents. *Braz. Arch. Biol. Technol.* **2014,** *57* (2): 155–162.
31. Velderrain-Rodríguez, G. R.; Palafox-Carlos, H.; Wall-Medrano, A.; Ayala-Zavala, J. F.; Chen, C. Y. O.; Robles-Sánchez, M.; Astiazaran-García, H.; Alvarez-Parrilla, E.; González-Aguilar, G. A. Phenolic Compounds: Their Journey After Intake. *Food Func.* **2014,** *5,* 189–197.
32. Xu, G.; Ye, X.; Liu, D.; Ma, Y.; Chen, J. Composition and Distribution of Phenolic Acids in Ponkan (Citrus poonensis Hort. ex Tanaka) and Huyou (Citrus paradisi Macf. Changshanhuyou) During Maturity. *J. Food Compos. Anal.* **2008,** *21* (5): 382–389.
33. Lattanzio, V.; Kroon, P. A.; Quideau, S.; Treutter, D. Plant Phenolics—Secondary Metabolites with Diverse Functions. In: *Recent Advances in Polyphenol Research,* Vol 1; Daayf, F.; Lattanzio, V.; Eds.; Wiley-Blackwell: Oxford, 2008; pp 1–35.
34. Vattem, D. A.; Randhir, R.; Shetty, K. Cranberry Phenolics-Mediated Elicitation of Antioxidant Enzyme Response in Fava Bean (Vicia faba) Sprouts. *J. Food Biochem.* **2005,** *29,* 41–70.
35. Lin, D. R.; Hu, L. J.; You, H.; Sarkar, D.; Xing, B. S.; Shetty, K. Initial Screening Studies on Potential of High Phenolic-Linked Plantclonal Systems for Nitrate Removal in Cold Latitudes. *J. Soils Sediment.* **2010,** *10,* 923–932.
36. Cheynier, V.; Comte, G.; Davies, K. M.; Lattanzio, V.; Martens, S. Plant Phenolics: Recent Advances on their Biosynthesis, Genetics, and Ecophysiology. *Plant Physiol. Biochem.* **2013,** *72,* 1–20.
37. Seabra, R. M.; Andrade, P. B.; Valentão, P.; Fernandes, E.; Carvalho, F.; Bastos, M. L. In *Biomaterials from Aquatic and Terrestrial organisms*; Fingerman, M., Nagabhushanam, R., Eds.; Science Publishers: Enfield, NH, USA. 2006; pp 115–174.
38. Dai, J.; Mumper, R. J. Plant Phenolics: Extraction, Analysis and their Antioxidant and Anticancer Properties. *Molecules (Basel, Switzerland)* **2010,** *15* (10), 7313–7352.
39. Sumczynski, D.; Kotásková, E.; Družbíková, H.; Mlček J. Determination of Contents and Antioxidant Activity of Free and Bound Phenolics Compounds and In Vitro Digestibility of Commercial Black and Red Rice (Oryza sativa L.) Varieties. *Food Chem.* **2016,** *211,* 339–346.

40. Sudha, R. C.; Sudhakar, C. Effect of Plant Growth Retardants on Growth, Yield and Economics of Kharif Pigeonpea (Cajanus cajan. L). *Int. J. Chem. Stud.* **2018,** *6,* 1495–1498.
41. Weidner, S.; Chrzanowski, S.; Karamać, M.; Król, A.; Badowiec, A.; Mostek, A.; Amarowicz, R. Analysis of Phenolic Compounds and Antioxidant Abilities of Extracts from Germinating Vitis californica Seeds Submitted to Cold Stress Conditions and Recovery after the Stress. *Int. J. Mol. Sci.* **2014,** *15* (9): 16211–16225.
42. Verdaguer, D.; Jansen, M. A. K.; Llorens, L.; Morales, L. O.; Neugart, S. UV-A Radiation Effects on Higher Plants: Exploring the Known Unknown. *Plant Sci.* **2017,** *255,* 72–81.
43. Valenta, K.; Burke, R. J.; Styler, S. A.; Jackson, D. A.; Melin, A. D.; Lehman, S. M. Colour and Odour Drive Fruit Selection and Seed Dispersal by Mouse Lemurs. *Sci. Rep.* **2013,** *3,* 2424.
44. John, J.; Sarada, S. Role of Phenolics in Allelopathic Interactions. *Allelopath. J.* **2012,** *29* (2): 215–230.
45. Ahuja, I.; Kissen, R.; Bones, A. M. Phytoalexins in Defense Against Pathogens. *Trends Plant Sci.* **2012,** *17,* 73–90.
46. Marsh, K. J.; Wallis, I. R.; Kulheim, C.; Clark, R.; Nicolle, D.; Foley, W. J.; Salminen, J.-P. New Approaches to Tannin Analysis of Leaves can be Used to Explain in Vitro Biological Activities Associated with Herbivore Defence. *New Phytol.* **2019,** *225,* 488–498.
47. Babbar, N. An Introduction to Alkaloids and their Application in Pharmaceutical Industry. *Pharm. Innov. J.* **2015,** *4* (10), 74–75.
48. Ng, Y. P.; Cho, T.; Or, T.; Ip, N. Y. Plant Alkaloids as Drug Leads for Alzheimer's Disease. *Neurochem. Int.* **2015,** *89,* 260–270.
49. Manske, R. H. F.; Holmes, H. L. *The Alkaloids: Chemistry and Physiology*; Academic Press, 2014; p 558.
50. Kaur, R.; Arora, S. Alkaloids-Important Therapeutic Secondary Metabolites of Plant Origin. *J. Crit. Rev.* **2015,** *2* (3), 1–8.
51. Dey, P.; Kundu, A.; Kumar, A.; Gupta, M.; Lee, B. M.; Bhakta, T.; Dash, S.; Kim, H. S. Analysis of Alkaloids (Indole Alkaloids, Isoquinoline Alkaloids, Tropane Alkaloids). *Recent Adv. Nat. Prod. Anal.* **2020,** 505–567. DOI: 10.1016/B978-0-12-816455-6.00015-9.
52. Cordell, G. A.; Choi, T. Alkaloids and Their Biosynthesis, Phytochemistry and Pharmacognosy. In *Encyclopedia of Life Support Systems (EOLSS)*; 2011, pp 1–17.
53. Roy A. A Review on the Alkaloids an Important Therapeutic Compound from Plants. *Int. J. Plant Biotechnol.* **2017,** *3* (2), 1–9.
54. Kumar, S. Alkaloidal Drugs: A Review. *Asian J. Pharm. Sci. Technol.* **2014,** *4,* 107–119.
55. Dewick, P. M.; John Wiley & Sons. *Medicinal Natural Products: A Biosynthetic Approach.* John Wiley & Sons, Ltd: Segunda edición. 2002. DOI: 10.1002/0470846275.
56. Aniszewski, T.; *Alkaloids—Secrets of Life. Alkaloid Chemistry, Biological Significance, Applications and Ecological Role*; Elsevier Science, 2007.
57. Chini, C.; Bilia, A. R.; Keita, A.; Morelli, I. Protoalkaloids from Boscia Angustifolia. *Planta Med.* **1992,** *58* (5):476.
58. Liddell, J. R. Pyrrolizidine Alkaloids. *Nat. Prod. Rep.* **2002,** *19,* 773–781.
59. Schardl, C. L.; Grossman, R. B.; Nagabhyru, P.; Faulkner, J. R.; Mallik, U. P. Loline Alkaloids: Currencies of Mutualism. *Phytochemistry* **2007,** *68,* 980–996.
60. Bentley, K. W. β-Phenylethylamines and the Isoquinoline Alkaloids. *Nat. Prod. Rep.* **2006,** *23,* 444–463.

61. Hu, J. F.; Hamann, M. T.; Hill, R.; Kelly, M. The Manzamine Alkaloids. The Alkaloids. *Chem. Biol.* **2003,** *60,* 207–285.
62. Thomas, R. Biogenetic Speculation and Biosynthetic Advances. *Nat. Prod. Rep.* **2004,** *21,* 224–248.
63. Fresneda, P. M.; Molina P. Application of Iminophosphorane-Based Methodologies for the Synthesis of Natural Products. *Synlett* **2004,** *1,* 1–17.
64. Knölker, H. J. Synthesis of Biologically Active Carbazole Alkaloids Using Selective Transition-Metal-Catalyzed Coupling Reactions. *Chem. Lett.* **2008,** *38,* 8–13.
65. Kam, T.; Choo, Y. Bisindole Alkaloids. *Alkaloids: Chem. Biol.* **2006,** *63,* 1–363.
66. Schardl, C. L.; Panaccione, D. G.; Tudzynski, P. Ergot Alkaloids—Biology and Molecular Biology. *Alkaloids: Chem. Biol.* **2006,** *63,* 45–86.
67. Somei, M.; Yamada, F. Simple Indole Alkaloids and those with a Nonrearranged Monoterpenoid Unit. *Nat. Prod. Rep.* **2004,** *21,* 278–311.
68. Somei, M.; Yamada, F. Simple Indole Alkaloids and those with a Non-Rearranged Monoterpenoid Unit. *Nat Prod Rep.* **2005,** (1), 73–103. DOI: 10.1039/b316241a.
69. Ata, A.; Andersh, B. J. Buxus Steroidal Alkaloids: Chemistry and Biology. *Alkaloids: Chem. Biol.* **2008,** *66,* 191–213.
70. Dey, P.; Kundu, A.; Chakraborty, H. J.; Kar, B.; Choi, W. S.; Lee, B. M.; Bhakta, T.; Atanasov, A. G.; Kim, H. S. Therapeutic Value of Steroidal Alkaloids in Cancer: Current Trends and Future Perspectives. *Int. J. Cancer* **2018,** *145,* 1731–1744.
71. Zhao, H.; Wang, X. Y.; Li, M. K.; Hou, Z.; Zhou, Y.; Chen, Z.; Meng, J. R.; Luo, X. X.; Tang, H. F.; Xue, X. Y. A Novel Pregnane-Type Alkaloid from Pachysandra Terminalis Inhibits Methicillin-Resistant Staphylococcus Aureus In Vitro and In Vivo. *Phytother. Res.* **2015,** *29,* 373–380.
72. Jin, Z. Imidazole, Oxazole and thiazole alkaloids. *Nat. Prod. Rep.* **2006,** *23,* 464–496.
73. Ashihara, H.; Yokota, T.; Crozier, A. Springer. Purine Alkaloids, Cytokinins, and Purine-Like Neurotoxin Alkaloids. *Nat. Prod.* **2013,** 953–975.
74. Lean, M. E.; Ashihara, H.; Clifford, M. N.; Crozier, A. Purine Alkaloids: A Focus on Caffeine and Related Compounds in Beverages. In *Teas, Cocoa and Coffee: Plant Secondary Metabolites and Health*; 2011; pp 25–44.
75. Robertson, J.; Stevens, K. Pyrrolizidine Alkaloids. *Nat. Prod. Rep.* **2014,** *31,* 1721–1788.
76. Kartsev. V. G. Natural Compounds in Drug Discovery. Biological Activity and New Trends in the Chemistry of Isoquinoline Alkaloids. *Med. Chem. Res.* **2004,** *13,* 325–336. DOI: 10.1007/s00044-004-0038-2.
77. Shakhnoz, S. A.; Yunusov, M. *Natural Compounds-Alkaloids*; Springer: Nueva York, Nueva York. 2013. DOI: 10.1007/978-1-4614-0560-3.
78. Sepúlveda-Jiménez, G.; Porta, H.; Rocha, M. La Participación de los Metabolitos Secundarios en la Defensa de las Plantas. *Rev. Mex. Fitopatol.* **2003,** *2,* 355–363.
79. Camacho-Escobar, M. A.; Ramos-Ramos, D. A.; Ávila-Serrano, N. Y.; Sánchez-Bernal, E. I.; López-Garrido, S. J. Las Defensas Físico-Químicas de las Plantas y su Efecto en la Alimentación de Los Rumiantes. *Terra Latinoam.* **2020,** *38* (2), 443–453.
80. Zárate-Martínez, W.; González-Morales, S.; Ramírez-Godina, F.; Robledo-Olivo, A.; Juárez-Maldonado, A. Efecto de los Ácidos Fenólicos en el Sistema Antioxidante de Plantas de Tomate Solanum Lycopersicum Mill.). *Agron. Mesoam.* **2021,** *32* (3), 854–868.

## CHAPTER 5

# Importance, Function, and Applications of the Microbiome in Agricultural Production Systems

HÉCTOR ADRIÁN GARCÍA RAMÍREZ[1], ADRIANA TOVAR DELGADO[1], DULCE CONCEPCIÓN GONZÁLEZ-SANDOVAL[2], VÍCTOR HUGO AVENDAÑO-ABARCA[2], KARINA MORENO-LEÓN[3], and WILLIAM ZÁRATE-MARTÍNEZ[4]

[1]*Laboratory of Molecular Biology and Biotechnology, Natural Resources Research Center (CIRENA), Chihuahua, México*

[2]*Centro de Bachillerato Tecnológico Agropecuario #124, Madera, Chihuahua, México*

[3]*Departamento de Botánica, Universidad Autónoma Agraria Antonio Narro, Saltillo, Coahuila, México*

[4]*Campo Experimental Valles Centrales de Oaxaca, Instituto Nacional de Investigaciones Forestales, Agrícolas y Pecuarias (INIFAP), Oaxaca, México*

### ABSTRACT

The microorganism–plant interaction is of utmost importance since it depends, to a large extent, on the optimal growth and development of plants. For this reason, some agricultural production systems adopt the application of microorganisms in their systems; however, the effect of these microorganisms is influenced by different biotic and abiotic factors, nevertheless, advances in studies and new technologies have made it possible to answer different questions about microorganism–plant interactions. Nowadays, the

Sustainable Agricultural Production Systems: Traditional and Modern Perspectives.
Romeo Rojas, Guillermo Cristian Guadalupe Martínez-Ávila, Julia Mariana Márquez-Reyes, and Ma del Carmen Ojeda-Zacarías (Eds.)

study of the plant microbiome has opened up a wide variety of biotechnological applications for new and future sustainable agricultural production systems, and these systems can be applied based on the knowledge of the importance, function, and mechanisms of the microbiome; in addition, this made it possible to develop strategies to solve problems associated with production systems, predicting the behavior of the crop and microbial diseases to such an extent that currently the selection and manipulation of microbiomes is sought to potentiate ecosystems in order to increase interactions. positive with the plant.

## 5.1 INTRODUCTION

Microorganisms are associated with plants, in such a way that they can develop endophytic and rhizospheric interaction networks, they are capable of developing beneficial interactions through different mechanisms which can directly or indirectly affect the development of the plant. When we talk about microorganisms with biotechnological applications, particularly applied in agricultural production systems, we usually think of isolated microorganisms as one or two unique individuals performing a particular function, we look for a microorganism that promotes growth, another who is a biocontrol agent, another that is capable of fixing nitrogen, solubilizing phosphorus, etc. These kinds of microorganisms have been widely studied and used in different important crops, such as vegetables, fruit trees, and ornamental plants.[1,2] For the use of microorganisms in the agricultural area, time is spent in isolation, in vitro tests and in interaction with the plant or with some phytopathogen to finally be scaled to some production system.[3] Nowadays, the new strategies for the use of microorganisms in agriculture are oriented toward the study of microbial communities, as well as the application of key microbiota to improve the growth, health and yield of plants, since better results have been demonstrated in the application of isolated strains of microorganisms, in such a way that the new production systems must be based on the application, care and management of microbiomes since these play a fundamental role in the development of the plant and at the same time help microorganisms to perform and potentiate their function, or help the ecological balance, playing a fundamental role in the control of phytopathogenic microorganisms and in the plant development.[4]

The study of microbial communities in agricultural production systems is not a new concept; however, and thanks to new sequencing technologies, it has allowed us to study and understand microbial communities in a production system, as well as innovation and improvements in the biotechnological application of microorganisms in new sustainable production systems. For this reason and thanks to new technologies, they have allowed us to study other microorganisms such as archaea, algae, nematodes and protists, which assume an important role in the health of plants and at the same time allow us to have a broader view of the interaction of the microbiome.[5] Likewise, recent works are aimed at exploring microbiomes and their beneficial effects on plants, as well as predicting outbreaks/control of diseases and plant performance, in such a way that the new strategies that have emerged in recent years involve the manipulation of microbiomes to improve the microorganism–plant interaction in different production systems. In this chapter, we will focus on describing the importance of the microbiome in the development and proper functioning of the plant, as well as the mechanisms of action and its present and future applications in agricultural production systems, which are based on the knowledge generated from new techniques.

## 5.2 PLANT MICROBIOME

The term microbiome has been very changing through the years and fields of study; however, it can be summarized as: the entire habitat, environmental conditions, microorganisms and the set of all their genomes,[6] on the other hand, the term microbiota refers to the set of microorganisms (without considering the entire genome) present in a given environment. Both terms can be confused, used indistinctly or even taken as synonyms; especially, when we talk about the microbiome in plants, we will be including microorganisms (including their genome) that develop and interact in a given environment (root, leaves, seeds, soil, and among others).[6–8] Although it is true that the microorganism–plant relationship leads to the diversification and co-evolution of both. Plants develop different types of strategies to recruit those useful microorganisms to form a microbial community with all the genetic, biochemical, and physiological tools that provide benefits; this microbial community depends on the communities present in the soil and on the host plant, since the relationship between the microbiota and the plant (holobiont) depends largely on the genes present in the plant, which determine its growth, development, and health.[8–10] However, the genes present in the plant are also determinants in the endophytic microbiome.[11] In such a way that the microorganism–plant relationship is determined by the genes of both parties, as well as the environment in which it develops. However, there are microbial communities that persist in most of the microbiota associated with a particular host, these microorganisms contain particular genes that provide key functions for the plant. The set of these microorganisms is called the core microbiota.[8,12,13]

## 5.3 BIODIVERSITY

The members of the microbiota are composed of both beneficial, neutral, or pathogenic interactions, the individuals of the community have certain beneficial traits; however, the effect depends on many factors, such as the conditions of the plant, the environment, and to the composition and diversity of the microbiota.[10] Given that the microorganism–plant interaction is designed to survive in stressful environmental conditions, there is a difference between the microbiome present in a crop, a plant in the wild and a plant in cultivation, in addition to the fact that biogeography also plays a fundamental role in the microbial composition.[14] The plant microbiome is made up of a great biodiversity in different compartments: soil, rhizosphere, phyllosphere, and endosphere. In these compartments, we can find protozoa, fungi, nematodes, bacteria, archaea, and microscopic algae; however, the practice of different intensive agricultural production systems has significantly altered the biodiversity of the ecosystem. The use of monocultures, pesticides, and the low use of biofertilizers are some of the factors that affect the development of communities and microbial diversity.[15,16] Currently, there are studies that identify the microbiome of wild and agricultural plants in order to identify microbial biodiversity, its function, and its possible applications.

It is known that the rhizosphere contains greater diversity, since the integration of the microorganisms present in the soil and the microorganisms of the plant root are concentrated here, bacteria are the most abundant microorganisms, followed by fungi, actinomycetes, protozoa, and algae. Bacteria are the most studied group of microorganisms, not only because of their wide diversity, but also because of their functions and applications in production systems. Different sources agree that Proteobacteria are the most abundant family in the rhizosphere and in the soil, followed by of Bacteriodates, Firmicutes, Actinobacteria and Planctomycetes, Chloroflexi, Verrucomicrobia, Acidobacteria, Gemmatimonadetes, and Parcubacteria,[10,17] it is worth mentioning that the structure and proportion of each bacterium varies depending on the depth of the soil and the characteristics of the plant. Similarly, the Proteobacteria, Bacteroidetes, Actinobacteria, Chloroflexi,

Elusimicrobia, and Firmicutes are the most abundant family in the phyllosphere,[10,18] while the most abundant families in the endosphere are Proteobacteria, Firmicutes, Bacteriodetes (these can vary considerably if they are root or leaf), and Actinobacteria.[10,19] However, it is known that bacterial OTUs can be found in different compartments (root and leaf endosphere or rhizosphere), these communities vary in proportion depending on the compartment and the crop variety.[14]

The most predominant family in the soil are the basidiomycetes and ascomycetes, despite the fact that arbuscular fungi and ectomycorrhizae are among the most studied and valued in some sustainable production systems, these groups are very little present in the plant microbiome.[10] However, the abundance of these families varies depending on the crop and type of soil.[20] Although fungi are the main pathogenic microorganisms in plants, there are numerous genera such as *Penicillium*, *Aspergillus*, *Cladosporium*, *Fusarium*, *Trichoderma*, *Pochonia*, *Chlamydospora*, *Rhizopus*, *Saccharomycetales*, *Glomus*, among others, that make up the plant microbiome and also play an extremely important and beneficial role for the plant.[17,21]

On the other hand, it has also been described that there are different communities of protists in the soil of forests and grasslands belonging to the *Rhizaria*, *Amoebozoa*, and *Alveolata* groups,[22] protists play an important role in controlling microbial communities since it shapes the structure of the microbiome, serving as a selector of functional and beneficial ranges for the plant,[23] changes in protist diversity lead to alterations in other microbial interaction networks, thus changing the ecological function of crop-associated biomes.[24] Within such a diverse and complex ecosystem, algae also help the good development of the plant since the metabolites produced by the algae induce a series of physiological events in the plant that help different functions such as adaptation to biotic and abiotic factors, as well as helps the functioning of the microbial community.[25] Members of the Cyanophyceae, Bacillariophyceae, and Euglenophyceae families can be identified within agricultural production systems,[26] *Trentepohliales*, *Watanabea*, and *Apatococcus* have also been found in the tropical plant phyllosphere.[27]

## 5.4 MECHANISMS

In experimental models, it has been defined that the microbiome of the rhizosphere depends to a large extent on the host plant, the selected microbiota can change depending on the stage of development since each stage is according to the needs of the plant.[28,29] However, this recruitment does not only depend on the plant, but also on other abiotic factors and especially on the local microbiota, which is why the holobiont is established based on the function of its members, that is, it may be that a plant has different taxonomic compositions in its microbiota; however, the functional structure of these microbial communities tends to be similar.[30,31] That is why when we talk about sustainable production systems, we should not only think about the application of growth-promoting microorganisms or biocontrol agents, here it is proposed to study a complex network, where not only the plant and its microbiome intervene, but not also genes associated with functions that lead to the sustenance of the holobiont.

### *5.4.1 GROWTH PROMOTION*

It is well known that in the soil, in the rhizosphere and endosphere there is a great diversity of microorganisms capable of providing benefits to the plant, of which we can highlight the promotion of growth through mechanisms, such as phosphate solubilization, nitrogen fixation, production of siderophores and phytohormones, or provides protection to the plant against the attack of pathogenic microorganisms through the production of secondary metabolites,[3,32–37] likewise, the microbiota present in the rhizosphere also has the ability to help the adaptation of the plant under different abiotic stress conditions such as salinity and drought.[38–42] Next, we will summarize the most common growth promotion mechanisms.

- **Phosphate solubilization:** in agriculture, the use of rhizospheric and endophytic bacteria of the genus *Azotobacter, Bacillus, Beijerinckia, Klebsiella, Micrococcus, Burkholderia, Enterobacter, Erwinia, Flavobacterium, Microbacterium, Pseudomonas, Rhizobium,* and *Serratia* among others are well known for solubilizing phosphorus on the ground.[43–48] Genes involved in pyrroloquinoline quinone synthesis (*pqq*) contribute to the solubilization of P.[49] Some works have shown the presence of these genes in phytopathogens, saprophytes, endophytes, symbionts microorganisms, for which it describes preferential associations that occur between certain genes that were established in different taxa.[49] In addition, different endophytic fungi of the genus *Acremonium, Aspergillus, Paecilomyces, Cryptococcus, Fusarium, Curvularia, Rhodotorula, Cladosporium, Alternaria, Phaeomoniella, Chaetomium, Colletotrichum,*

*Berkleasmium, Rhizoctonia, Geomyces, Leptospora, Phyllosticta, Microdochium, Neotyphodium, Ophiognomonia, Glomus, Penicillium, Rhizopus, Trichoderma, Xylaria,* and *Wallemia* are often used as a strategy to increase the availability of P in agriculture.[50] It is worth mentioning that these fungi regulate phosphorus solubilization through the *pqq* genes.[51]

- **Nitrogen fixation:** nitrogen fixation can occur by symbiotic bacteria of legumes (*Rhizobia*), endophytic or free-living microorganisms, such as Cyanobacteria (*Anabaena, Nostoc*), *Azospirillum, Azotobacter, Gluconoacetobacter diazotrophicus,* and *Azocaru,* among others.[34] The genes involved in nitrogen fixation (*nif* genes) in symbiotic and free-living bacteria are *nifH,D,K* that code for a structural subunit of dinitrogenase reductase and the two subunits of dinitrogenase; however, there are other genes involved in regulation (*nifE,N,X,Q,W,V,A,B,Z,S*).[52] However, the mechanism of recruitment by the plant to nitrogen-fixing bacteria is not well defined; some studies that mention that some strains of *Pseudomonas* secrete homoserine for the recruitment of *Rhizobium.*[53] Within the microbiome of the fixative nodules there are bacteria that despite not being nitrogen fixers, play a very important role in the nodulation and in the availability of nitrogen within sustainable production systems, that is why the study of the function of these microorganisms is also important since they could be used as inoculants to improve nitrogen fixation in new and improved production systems.[54]
- **Iron uptake:** iron is one of the most important nutrients for the plant, especially in the functioning of numerous enzymes,[55] the iron that is present in the soil is found in its $Fe^{+3}$ form, which is not assimilable,[56] one of the mechanisms to assimilate Fe is the production of siderophores, through the production of these metabolites not only does Fe become available, but also biological interactions are generated that lead to the selection and control of the microbiota,[57] derived from this, have been demonstrated that members of the microbiome associated with the natural rhizosphere secrete siderophores that alter the pathogen–plant relationship.[58] Through the use of sequencing and assembly of genomes, it allows the massive identification of microorganisms associated with agricultural production systems, as well as the different varieties of siderophores produced, which allows us to identify the microbial consortia that can be used

in different types of systems of agricultural production, this strategy was carried out in a vine production system, obtaining consortiums of the genus *Pseudomonas.*[59]

- **Production and regulation of phytohormones:** it is well known that the microorganisms belonging to the plant microbiota are capable of synthesizing or controlling the synthesis of phytohormones.[34] Indole-3-acetic acid (IAA) is the most commonly studied phytohormone produced by microorganisms,[59] especially in rhizospheric and endophytic bacteria of the genus *Bacillus, Pseudomonas, Brevibacillus, Bradyrhizobium, Variovorax, Sphingomonas, Mycobacterium, Rhodococcus, Azospirillum, Rhizobium*, among others.[43,60] Phytopathogenic and symbiotic fungi are also capable of producing phytohormones,[61] such is the case of *Fusarium, Trichoderma, Daldinia,* and *Macrophomina.*[62,63] In addition to auxins, gibberellins are produced by endophytic microorganisms of the genus *Azospirillum,*[64] fungi of the genus *Aspergillus, Emericella, Penicillium*, can produce gibberellins.[63,65] Auxins are produced by different pathways, where different types of genes that code for enzymes that participate in the degradation of tryptophan for the production of IAA participate. The *iaaM* and *iaaH* genes have been characterized in phytopathogenic microorganisms such as *A. tumefaciens* (the genes so located in the pTi plasmid), *P. savastanoi, Pseudomonas syringae, Pantoea agglomerans*, have also been described in symbiotic bacteria of the genus *Rhizobium* and *Bradyrhizobium*. These genes are located on plasmid pIAA1. On the other hand, *ipdC* genes have been identified in some beneficial bacteria.[66]

### *5.4.2 PROTECTION AGAINST MICROBIAL DISEASES*

Sustainable production systems have exerted a great demand for biocontrol agents, since this strategy provides a sustainable alternative, reduces the use of chemical pesticides, and is also safe for human health and the environment.[67] New strategies for plant microbiome studies have allowed the discovery and identification of new biocontrol agents, secondary metabolites, and the investigation of microbial communities and their effect on the control of phytopathogens.[68] The strategies for the study of the plant microbiome and its effect on phytopathogens can be divided into different aspects; the study of microorganisms that directly affect phytopathogens, the microbiota that indirectly affects the phytopathogen (microorganisms that help strengthen the plant or predatory microorganisms), the study of strains that produce secondary metabolites, among others, these study strategies will help to understand the role of the different microbial communities and their effect on phytopathogenic microorganism.[5] Likewise, it has been identified that the initial composition and function of the soil microbiome predetermines the response of the plant to the attack of phytopathogenic microorganisms, such as *Pseudomonas* and *Bacillus*, which have an influence on the control of diseases and the survival of plants.[69] However, and despite the fact that antimicrobial activity is usually associated with a few species, it is known that the abundance and diversity of species belonging to the Proteobacteria, Firmicutes, and Actinobacteria groups play a fundamental role in the suppression of fungal diseases.[70] However, there are works where the relationship of the microbiome in the rhizosphere are similar; Proteobacteria, Bacteroidetes, Acidobacteria, Actinobacteria, Firmicutes, among others; however, the difference between both microbiomes led to the detection of a flavobacterial genome that is much more abundant in the rhizosphere of the variety resistant to tomato wilt (*Ralstonia solanacearum*).[71] The information obtained from the study of the plant microbiome can facilitate the identification of beneficial microorganisms or the full use of microbiomes for disease control. Most of the time we overlook the ecological function of many microorganisms, especially the regulatory function of microbial communities, for this some protists are specialists, not only regulates and shapes the microbiome of the plant, they have the ability to reduce the pathogenic effect of some phytopathogens, in addition to improving the suppressive effect of pathogens of some microorganisms by improving the production of secondary metabolites with antimicrobial activity.[23,72] Algae and Cyanobacteria have also been shown to have an antifungal effect against phytopathogens, such as *Sclerotinia sclerotiorum* and several species of *Rhizoctonia*, *Phytophthora*, *Pythium*, *Fusarium*, *Verticillium*, *Botrytis cinerea*, and *Colletotrichum* spp. in different fruits and horticultural plants,[73–75] and some species of algae have been proven to significantly reduce the effect of nematodes in banana plants.[76]

The protective effect of phytopathogens is also related at a functional level, particularly the presence of genes that code for secondary metabolites, such as genes linked to nonribosomal peptide and polyketide synthases.[69,70] The production of antimicrobial compounds, such as 2,4-diacetylphloroglucinol (DAPG) (phlD+) or phenazine (phzC+) produced by bacteria of the genus *Pseudomonas* is linked to the presence of genes *phl* and *phz*C,D.[77–79] In some fungi, such as *Trichoderma*, they contain genes that regulate the secretion of crucial enzymes for the biocontrol process, the genes involved in the production of Chitinases (*ech42, chi33, nag1, chi*$^{18-13}$), Glucanases (*Bgn, lam, glu*), Proteases (*Tvsp, Prb1, TaAsp, SL41*).[80] As in the rhizosphere, the diversity of the microbiome of the phyllosphere has an effect on the development of phytopathogens that affect the surface of the plant (leaves, stems, and fruits).[81] However, there is disagreement if the biocontrol agents applied can affect the beneficial microbiota since some studies report that the plant microbiota can be affected by the biocontrol agent that is applied in the production system;[82] however, other works recommend the application of consortiums of biocontrol agents, such as *Bacillus subtilis* and *Trichoderma harzianum*, since it reduces diseases by increasing beneficial bacteria in the rhizosphere.[83] In addition to the study of genes associated with secondary metabolites that suppress phytopathogens in microorganisms, genes for resistance to pathogens present in the plant can also be studied for the sustainable management of crops,[84] since these genes trigger an immune response in the plant, the same response can be triggered or modulated thanks to the plant–microbiome interaction.[85] Such is the case of the *epl1* genes present in *T. atroviride*, which are related to systematic protection against *A. solani* and *B. cinerea* in tomato crops.[86] In the case of bacteria, *Pseudomonas* produce induced systemic resistance (ISR) in different types of plants by different mechanisms such as recognition of extracellular components (flagella and lipopolysaccharides), iron-regulated metabolites, and antibiotic production.[87] Likewise, the metabolites produced recruit a beneficial microbiota for the suppression of diseases.[88]

### *5.4.3 GROWTH UNDER ABIOTIC STRESS*

The plant is an organism that needs to adapt to different environmental changes such as drought and salinity, these abiotic stress factors have allowed the plant to develop adaptation mechanisms, where the plant microbiome plays a fundamental role in these mechanisms.[89,90] It is known that through intensive, unsustainable agricultural processes, the indiscriminate use of chemical products, and coupled with global drought, soil salinity problems are becoming more noticeable;[91] however, there are plant-associated microorganisms, particularly bacteria belonging to the phyla Proteobacteria, Actinobacteria, Firmicutes, and Archaea related to the phyla Crenarchaeota and Euryarchaeota able to tolerate high concentrations of salt and also have the ability to increase salt tolerance in the plant.[92] The soil bacteria that help induce salt tolerance are Actinobacteria, Firmicutes, Proteobacteria, while the most representative in nodules are those of the genus *Bradyrhizobium*,[93] in addition, the microbial communities adapt to different salinity conditions in such a way that the selected community by this factor can induce the growth and adaptation of the plant under these conditions.[94]

As previously mentioned, the plant determines the composition of the microbiota based on environmental factors, during drought, the plant is capable of recruiting microorganisms that help the plant adapt, the genus *Bacillus* and *Klebsiella* are the most abundant in the endosphere and rhizosphere, the isolated bacteria were tested in chili (*Capsicum annuum L.*) cultivation, which adapted to drought conditions.[95] In sorghum, drought retards the development of the root microbiome and causes a greater abundance of monodermal bacteria, which have a positive impact on plant growth; this also demonstrates that drought also plays a role in restoration and adaptation of the plant microbiomes,[96] since it has been correlated with the increase of endophytic Actinobacteria in grass roots.[97] Recent studies confirm that this change in the plant microbiome persists even when water stress stops.[98,99] The work carried out on rice shows that endophytic Actinobacteria are more predominant during drought and these promote plant growth during this period. Therefore, it is a key taxon for adaptation to drought conditions, particularly in rice crops.[98]

## 5.5 MICROBIOMES AS NEW STRATEGIES IN PRODUCTION SYSTEMS

In this chapter, the functioning of the plant has been described in association with microbial communities present in the soil, rhizosphere, and endosphere. These interactions form a complex network that benefits the development of the plant, as has been mentioned throughout the text, the new trend of the use of microorganisms in sustainable agricultural production systems is based on the study of the microbiome, through the application of new massive sequencing techniques.[100] The knowledge generated through these technologies and the study of the plant–microbiome interaction can be used to increase agricultural productivity in an efficient and sustainable way, as well as the search for new microbial inoculants and their secondary metabolites.[101]

One of the strategies to increase the supply of production is based on studying and selecting the beneficial bacteria that form interactions with the plant, increasing the supply of nitrogen through biological fixation, some studies propose the use of next-generation sequencing to evaluate production parameters microbial ecology in order to better identify bacterial inoculants in less time, such is the case of the identification of different species of *Bradyrhizobium*, which are found in bean nodules from different production systems, agroindustrial and organic agroecosystem.[102] Genomics has also helped in the sequencing of complete genomes of growth-promoting microorganisms that through their analysis, can identify genes involved in resistance to drought. This is the case of the identification of genes involved in the promotion of plant growth in a strain of *Pseudomonas*, through genomic analysis.[103]

The future of new sustainable production systems, in addition to being aimed at studying the functioning of microbial biodiversity, are focused on managing the microbiome in a precise and personalized way in order to obtain more sustainable, predictive, and inclusive production systems, for this, it is intended to develop strategies that based on the knowledge of the microbiomes, can carry out a specific management for each host plant, these strategies should be applied not only in traditional systems in the field, but also in controlled environments, such as greenhouses, hydroponics, among others.[104] The benefits of the microbiome is the protection against pathogens; therefore, by knowing the taxa present in the microbiome of different hosts, it is possible to predict the plant–pathogen interaction, the taxa associated with the survival of plants can be transferred by transplanting soil with it.[69] The study of the effect of phytopathogenic microorganisms, such as *Verticillium dahliae* on tomato microbiota demonstrates effectors with antimicrobial effect, which is recently proposed for the manipulation of soil microbiomes, representing a resource of new antibiotics that could be used beneficially.[105]

Nowadays, the new concepts for a sustainable agriculture, not only depend on the processes but also on the quality of the product, sustainable processes, and their ecological impact on microbial biodiversity, that is why new works in the study of microbial inoculants evaluate their effect in an integral and holistic way, that is to say that they coexist and adapt with the microbial communities of the place and that they have an effect on the control of diseases, nutritional value and their ecological impact on the microbial biodiversity in the soil and the rhizosphere of the plant.[106]

Metabolomics is another tool that helps to meet new challenges for the development of sustainable technologies and their application in agriculture. Through this tool, we can unravel the chemical integrations between the microorganism–plant and microorganism–microorganism, in order to molecularly detect signaling and/or communication, as well as the identification of antimicrobial compounds, phytohormones, bacteriocins, siderophores, among others, together with genomics, metabolomics also helps to identify new bioactive compounds.[107] It has been shown how phytopathogenic microorganisms such as *Zymoseptoria tritici* induce a change in the metabolome of wheat, the change in metabolism directly affects its microbiota, which makes it more vulnerable to diseases caused by this pathogen.[108] It has also been found that the soil microbiome of plants affects the leaf metabolome, in such a way that it affects the content of amino acids which affect the feeding of leaf insects; therefore, these types of strategies are not only focused on the control of microbial diseases or abiotic stress, but also on the sustainable control of agricultural pests.[109] The effect of the microbiome on the plant not only affects pests and diseases, but also the production of secondary metabolites that are highly relevant for agricultural production and at the same time for consumer health.[110]

The new trend for the sustainable development of agriculture is to alter the plant microbiome that is, to manipulate the plant holobiont by designing a strategic microbiome in such a way that it improves the growth of the crop, and at the same time protects it from pathogens and of abiotic stress. For this, there are different strategies, such as inoculation of seed or seedling consortia, use of specific and selective substrates and amendments, inoculation of probiotics, transplantation of microbiomes, and direct injection into tissues or wounds.[111,112] Same is the case of the identification of microbial communities in composts for the suppression of diseases.[113] Knowledge of the microbial composition of composts helps to modify the microbial communities with the purpose of the reduction of diseases caused by soil microorganisms, such as *Phytophthora capsici.*[114] Strategies, such as fumigation and application of organic amendments for the selection of soil microbiota have been effective in suppressing *Ralstonia solanacearum* in tomato.[115] Inoculation of bacteria such as *Agrobacterium* sp. affects the structure and richness of microbial communities, particularly Firmicutes, Actinobacteria, Bacteroidetes, and Proteobacteria, as well as different species of *Bacillus*; the structural changes of the microbiome allowed better growth and development of nodulation in bean crops.[116]

Newer methods that at the same time become more complex and questioned by different social sectors, are genetic and microbial manipulation techniques at a microbial consortium level, to add or eliminate functions that alter the plant ecosystem in a specific way.[117] The design of microbiomes to increase the productivity and yield of crops can be improved by means of new genomic engineering tools, metabolomics, computational, and functional genomics tools; however, for this to be carried out, strategies must be improved or developed that allow improving the biosafety, bioprotection, and biocontainment to modify and use modified microorganisms in sustainable agricultural production systems.[118]

## 5.6 CONCLUSION

Microorganisms are the basis of many sustainable agricultural production systems through the application of biofertilizers in the form of organic fertilizers, biofertilizers, and/or biological control agents; however, these practices contain many limitations such as the environmental conditions of each region, the type of cultivation, the phytopathogenic microorganism, and the biome of the region, due to these disadvantages, many producers choose to use chemical implements; however, these production systems cause secondary effects to health and the environment. Due to this, there are many works oriented to the study of the plant–microbiome interaction and the beneficial mechanisms of this holobiont. It should be noted that the new advances in genomic technologies have helped to investigate in depth the microbiome associated with plants and how it plays a fundamental role in the development of plants. These technologies help us unravel the mechanisms of action of microorganisms, as well as the genes involved in each mechanism and how each beneficial effect is thanks to the set of microbial communities that selectively associate with a particular host. The microbiome of the rhizosphere and endosphere of the plant is selected according to the type of host and environmental conditions; despite this, we can distinguish that the selection of the microbiome can be given according to the taxon or according to the specific and necessary functions for the plant. Once the microbiome is selected, the microbial communities produce metabolites that help the proper development of the plant, mechanisms, such as the production of siderophores, phytohormones, phosphate solubilization, nitrogen fixation, and the inhibition of pathogenic agents, they also help the adaptation of the plant under different conditions of abiotic stress such as salinity and drought. Thanks to all the knowledge generated by different studies with different crops or experimental models, they have made it possible to generate advances in microbial biotechnology in such a way that there are new perspectives in the application of microbiomes that lead to sustainable production systems and at the same time increase the crop yields without endangering human health and the environment.

## KEYWORDS


- **plant microbiome**
- **sustainable agriculture**
- **microorganism**
- **phytopathogen**
- **growth promoting**


## REFERENCES


1. Yadav, A. N.; Singh, J.; Singh, C.; Yadav, N. In *Current Trends in Microbial Biotechnology for Sustainable Agriculture*, 2021.
2. Rouphael, Y.; Colla, G. Synergistic Biostimulatory Action : Designing the Next Generation of Plant Biostimulants for Sustainable Agriculture. *Front. Plant Sci.* **2018**, *9* (1665), 1–7. https://doi.org/10.3389/fpls.2018.01655.
3. Review, M. Advances in Plant Growth-Promoting Bacterial Inoculant Technology : Formulations and Practical Perspectives. **2014**, *1886*, 1–33. https://doi.org/10.1007/s11104-013-1956-x.
4. Berendsen, R. L.; Pieterse, C. M. J.; Bakker, P. A. H. M. The Rhizosphere Microbiome and Plant Health. *Trends Plant Sci.* **2012**, *17* (8), 478–486. https://doi.org/10.1016/j.tplants.2012.04.001.
5. Massart, S.; Margarita, M. M.; Jijakli, M. H. Biological Control in the Microbiome Era: Challenges and Opportunities. *Biol. Control* **2015**, *89*, 98–108. https://doi.org/10.1016/j.biocontrol.2015.06.003.
6. Marchesi, J. R.; Ravel, J. The Vocabulary of Microbiome Research: A Proposal. *Microbiome* **2015**, *3* (1), 1–3. https://doi.org/10.1186/s40168-015-0094-5.
7. Compant, S.; Samad, A.; Faist, H.; Sessitsch, A. A Review on the Plant Microbiome : Ecology , Functions , and Emerging Trends in Microbial Application. **2019**, *19*, 29–37. https://doi.org/10.1016/j.jare.2019.03.004.
8. Lemanceau, P.; Blouin, M.; Muller, D.; Moënne-Loccoz, Y. Let the Core Microbiota Be Functional. *Trends Plant Sci.* **2017**, *22* (7), 583–595. https://doi.org/10.1016/j.tplants.2017.04.008.
9. Vandenkoornhuyse, P.; Quaiser, A.; Duhamel, M.; Le Van, A.; Dufresne, A. The Importance of the Microbiome of the Plant Holobiont. *New Phytol.* **2015**, *206* (4), 1196–1206. https://doi.org/10.1111/nph.13312.
10. Trivedi, P.; Leach, J. E.; Tringe, S. G.; Sa, T.; Singh, B. K. Plant–Microbiome Interactions: From Community Assembly to Plant Health. *Nat. Rev. Microbiol.* **2020**, *18* (11), 607–621. https://doi.org/10.1038/s41579-020-0412-1.
11. Beckers, B.; Op, M.; Beeck, D.; Weyens, N.; Acker, R. Van; Montagu, M. Van; Boerjan, W. Lignin Engineering in Field-Grown Poplar Trees Affects the Endosphere Bacterial Microbiome. **2016**, *113* (8). https://doi.org/10.1073/pnas.1523264113.

12. Cohen, B. D. Evaluating the Core Microbiota in Complex Communities: A Systematic Investigation. **1996**.
13. Yeoh, Y. K.; Dennis, P. G.; Paungfoo-Lonhienne, C.; Weber, L.; Brackin, R.; Ragan, M. A.; Schmidt, S.; Hugenholtz, P. Evolutionary Conservation of a Core Root Microbiome across Plant Phyla along a Tropical Soil Chronosequence. *Nat. Commun.* **2017**, *8* (1). https://doi.org/10.1038/s41467-017-00262-8.
14. Coleman-Derr, D.; Desgarennes, D.; Fonseca-Garcia, C.; Gross, S.; Clingenpeel, S.; Woyke, T.; North, G.; Visel, A.; Partida-Martinez, L. P.; Tringe, S. G. Plant Compartment and Biogeography Affect Microbiome Composition in Cultivated and Native Agave Species. *New Phytol.* **2016**, *209* (2), 798–811. https://doi.org/10.1111/nph.13697.
15. de Graaff, M. A.; Hornslein, N.; Throop, H. L.; Kardol, P.; van Diepen, L. T. A. Effects of Agricultural Intensification on Soil Biodiversity and Implications for Ecosystem Functioning: A Meta-Analysis, 1st Ed.; Elsevier Inc., 2019; vol 155. https://doi.org/10.1016/bs.agron.2019.01.001.
16. Postma-Blaauw, M. B.; de Goede, R. G. M.; Bloem, J.; Faber, J. H.; Brussaard, L. Agricultural Intensification and De-Intensification Differentially Affect Taxonomic Diversity of Predatory Mites, Earthworms, Enchytraeids, Nematodes and Bacteria. *Appl. Soil Ecol.* **2012**, *57*, 39–49. https://doi.org/10.1016/j.apsoil.2012.02.011.
17. Mokrani Slimane, E. H. Advances in Plant Microbiome and Sustainable Agriculture, 19th Ed.; In *Microorganisms for Sustainability*; Yadav, A. N., Rastegari, A. A., Yadav, N., Kour, D., Eds.; Springer Singapore: Singapore, 2020; vol 19. https://doi.org/10.1007/978-981-15-3208-5.
18. Gupta, R.; Patil, R. Phyllospheric Microbes: Diversity, Functions, Interaction, and Applications in Agriculture, 2021; pp 301–323. https://doi.org/10.1007/978-981-15-6949-4_13.
19. Ghiasian, M. Endophytic Microbiomes: Biodiversity, Current Status, and Potential AgriculturalApplications,2020;pp61–82.https://doi.org/10.1007/978-981-15-3208-5_3.
20. Rossmann, M.; Pérez-Jaramillo, J. E.; Kavamura, V. N.; Chiaramonte, J. B.; Dumack, K.; Fiore-Donno, A. M.; Mendes, L. W.; Ferreira, M. M. C.; Bonkowski, M.; Raaijmakers, J. M.; Mauchline, T. H.; Mendes, R. Multitrophic Interactions in the Rhizosphere Microbiome of Wheat: From Bacteria and Fungi to Protists. *FEMS Microbiol. Ecol.* **2020**, *96* (4). https://doi.org/10.1093/femsec/fiaa032.
21. Pattnaik, S. S.; Busi, S. Rhizospheric Fungi: Diversity and Potential Biotechnological Applications, 2019; pp 63–84. https://doi.org/10.1007/978-3-030-10480-1_2.
22. Geisen, S.; Tveit, A. T.; Clark, I. M.; Richter, A.; Svenning, M. M.; Bonkowski, M.; Urich, T. Metatranscriptomic Census of Active Protists in Soils. *ISME J.* **2015**, *9* (10), 2178–2190. https://doi.org/10.1038/ismej.2015.30.
23. Gao, Z.; Karlsson, I.; Geisen, S.; Kowalchuk, G.; Jousset, A. Protists: Puppet Masters of the Rhizosphere Microbiome. *Trends Plant Sci.* **2019**, *24* (2), 165–176. https://doi.org/10.1016/j.tplants.2018.10.011.
24. Sun, A.; Jiao, X.; Chen, Q.; Trivedi, P.; Li, Z.; Li, F.; Zheng, Y.; Lin, Y.; Hu, H.; He, J. Fertilization Alters Protistan Consumers and Parasites in Crop-associated Microbiomes. *Environ. Microbiol.* **2021**, *23* (4), 2169–2183. https://doi.org/10.1111/1462-2920.15385.
25. Lee, S. M.; Ryu, C. M. Algae as New Kids in the Beneficial Plant Microbiome. *Front. Plant Sci.* **2021**, *12*, 1–18. https://doi.org/10.3389/fpls.2021.599742.
26. Olugbemi, T. A. A. O. M. Soil Algae : A Case Study of Two Vegetable Farmlands in Lagos and Ogun States, Southwest Nigeria. *Ife J. Sci.* **2016**, *17* (3), 765–772.

27. Zhu, H.; Li, S.; Hu, Z.; Liu, G. Molecular Characterization of Eukaryotic Algal Communities in the Tropical Phyllosphere Based on Real-Time Sequencing of the 18S RDNA Gene. *BMC Plant Biol.* **2018**, *18* (1), 365. https://doi.org/10.1186/s12870-018-1588-7.
28. Lundberg, D. S.; Lebeis, S. L.; Paredes, S. H.; Yourstone, S.; Gehring, J.; Malfatti, S.; Tremblay, J.; Engelbrektson, A.; Kunin, V.; Rio, T. G. Del; Edgar, R. C.; Eickhorst, T.; Ley, R. E.; Hugenholtz, P.; Tringe, S. G.; Dangl, J. L. Defining the Core Arabidopsis Thaliana Root Microbiome. *Nature* **2012**, *488* (7409), 86–90. https://doi.org/10.1038/nature11237.
29. Chaparro, J. M.; Badri, D. V.; Vivanco, J. M. Rhizosphere Microbiome Assemblage Is Affected by Plant Development. *ISME J.* **2014**, *8* (4), 790–803. https://doi.org/10.1038/ismej.2013.196.
30. Yachi, S.; Loreau, M. Biodiversity and Ecosystem Productivity in a Fluctuating Environment: The Insurance Hypothesis. *Proc. Natl. Acad. Sci.* **1999**, *96* (4), 1463–1468. https://doi.org/10.1073/pnas.96.4.1463.
31. Louca, S.; Jacques, S. M. S.; Pires, A. P. F.; Leal, J. S.; Srivastava, D. S.; Parfrey, L. W.; Farjalla, V. F.; Doebeli, M. High Taxonomic Variability despite Stable Functional Structure across Microbial Communities. *Nat. Ecol. Evol.* **2017**, *1* (1), 0015. https://doi.org/10.1038/s41559-016-0015.
32. Mehmood, U.; Inam-ul-haq, M.; Saeed, M.; Altaf, A.; Azam, M. F. A Breif Review On Plant Growth Promoting Rhizobacteria (Pgpr): A Key Role In Plant Growth Promotion. *Plant Prot.* **2018**, *02* (02), 77–82.
33. Thilagar, G.; Appa, D. J. B.; Podile, R.; Rao, P. Bacillus Sonorensis , a Novel Plant Growth Promoting Rhizobacterium in Improving Growth , Nutrition and Yield of Chilly ( Capsicum Annuum L .). *Proc. Natl. Acad. Sci. India Sect. B Biol. Sci.* **2016**. https://doi.org/10.1007/s40011-016-0822-z.
34. Ahemad, M. Mechanisms and Applications of Plant Growth Promoting Rhizobacteria : Current Perspective. *J. KING SAUD Univ. Sci.* **2013,** *26* (1). https://doi.org/10.1016/j.jksus.2013.05.001.
35. Shiva, S.; Prasad, K.; Giovanardi, D. Stefani, E. Plant Growth Promoting and Biocontrol Activity of Streptomyces Spp. as Endophytes. *Int. J. Mol. Sci.* **2018,** *19* (4), 952. https://doi.org/10.3390/ijms19040952.
36. Growth, P.; Resistance, I. D. Plant Growth-Promoting Fungi ( PGPF ) Instigate Plant Growth and Induce Disease Resistance in Capsicum Annuum L . upon Infection with Colletotrichum Capsici ( Syd .) Butler & Bisby, 2020; pp 4–6.
37. Castro, R. O.; Campos, J.; José, G.; Bucio, L. Pseudomonas Putida and Pseudomonas Fluorescens Influence Arabidopsis Root System Architecture Through an Auxin Response Mediated by Bioactive Cyclodipeptides. *J. Plant Growth Regul.* **2019**. https://doi.org/10.1007/s00344-019-09979-w.
38. Bhatt, K.; Kumar, D. Zinc Solubilizing Bacteria ( Bacillus Megaterium ) with Multifarious Plant Growth Promoting Activities Alleviates Growth in Capsicum Annuum L . *3 Biotech* **2020**. https://doi.org/10.1007/s13205-019-2033-9.
39. Rincón-molina, C. I.; Martínez-romero, E.; Ruiz-valdiviezo, V. M.; Velázquez, E.; Ruiz-lau, N.; Rogel-hernández, M. A. Plant Growth-Promoting Potential of Bacteria Associated to Pioneer Plants from an Active Volcanic Site of Chiapas (Mexico). *Appl. Soil Ecol.* **2019**, *146* (4), 103390. https://doi.org/10.1016/j.apsoil.2019.103390.

40. Numan, M.; Bashir, S.; Khan, Y.; Mumtaz, R. Plant Growth Promoting Bacteria as an Alternative Strategy for Salt Tolerance in Plants : A Review. *Microbiol. Res.* **2018**, 209.
41. Wang, W.; Wu, Z.; He, Y.; Huang, Y.; Li, X.; Ye, B. Ecotoxicology and Environmental Safety Plant Growth Promotion and Alleviation of Salinity Stress in Capsicum Annuum L . by Bacillus Isolated from Saline Soil in Xinjiang. *Ecotoxicol. Environ. Saf.* **2018,** *164*, 520–529. https://doi.org/10.1016/j.ecoenv.2018.08.070.
42. He, Y.; Wu, Z.; Wang, W.; Ce, B.; Furong, Y.; Xiaochen, Z. Different Responses of Capsicum Annuum L . Root and Shoot to Salt Stress with Pseudomonas Putida Rs-198 Inoculation. *J. Plant Growth Regul.* **2019,** *38*, 799–811. https://doi.org/10.1007/s00344-018-9891-y.
43. Gomes, I. P.; Dayanne, A.; Nietsche, S.; Xavier, A.; Costa, M. R.; Gomes, W. S.; Cristian, M. Auxin Production by Endophytic Bacteria Isolated from Banana Trees. *Braz. Arch. Biol. Technol.* **2017**, *60*, e17160484.
44. Zea, M.; N, D. S. Biofrontiers Effect of Azotobacter Chroococcum (PGPR) on the Growth of Bamboo (Bambusa Bamboo), 2009; vol *1*.
45. Mwajita, M. R.; Murage, H.; Tani, A.; Kahangi, E. M. Evaluation of Rhizosphere , Rhizoplane and Phyllosphere Bacteria and Fungi Isolated from Rice in Kenya for Plant Growth Promoters, 2013; pp 1–9.
46. Moreno-Ramírez, L.; González-Mendoza, D.; Cecena-Duran, C.; Grimaldo-Juarez, O. Molecular Identification of Phosphate-Solubilizing Native Bacteria Isolated from the Rhizosphere of Prosopis Glandulosa in Mexicali Valley. *Genet. Mol. Res.* **2015,** *14* (1), 2793–2798. https://doi.org/10.4238/2015.March.31.9.
47. Muleta, D.; Assefa, F.; Börjesson, E.; Granhall, U. Phosphate-Solubilising Rhizobacteria Associated with Coffea Arabica L. in Natural Coffee Forests of Southwestern Ethiopia. *J. Saudi Soc. Agric. Sci.* **2013,** *12* (1), 73–84. https://doi.org/10.1016/j.jssas.2012.07.002.
48. Behera, B. C.; Singdevsachan, S. K.; Mishra, R. R.; Sethi, B. K. Phosphate Solubilising Bacteria from Mangrove Soils of Mahanadi River Delta, Odisha, India. *World J. Agric. Res.***2016,** *4* (1), 18–23. https://doi.org/10.12691/wjar-4-1-3.
49. Misra, H. S.; Rajpurohit, Y. S.; Khairnar, N. P. Pyrroloquinoline-Quinone and Its Versatile Roles in Biological Processes. *J. Biosci.* **2012,** *37* (2), 313–325. https://doi.org/10.1007/s12038-012-9195-5.
50. Sujatha, E.; Gunaswetha, K.; Bramhachari, P. V. Current Perspectives on Phosphate-Solubilizing Endophytic Fungi: Ecological Significances and Biotechnological Applications, 2020; pp 79–96. https://doi.org/10.1007/978-3-030-38453-1_3.
51. Kusari, S.; Hertweck, C.; Spiteller, M. Chemical Ecology of Endophytic Fungi: Origins of Secondary Metabolites. *Chem. Biol.* **2012,** *19* (7), 792–798. https://doi.org/10.1016/j.chembiol.2012.06.004.
52. Mahmud, K.; Makaju, S.; Ibrahim, R.; Ali, M. Current Progress in Nitrogen Fixing Plants and Microbiome Research, 2020; pp 1–17.
53. Vanderlinde, E. M.; Hynes, M. F.; Yost, C. K. Homoserine Catabolism by Rhizobium Leguminosarum Bv . Viciae 3841 Requires a Plasmid-Borne Gene Cluster That Also Affects Competitiveness for Nodulation. *Environ. Microbiol.* **2014,** *16*, 205–217. https://doi.org/10.1111/1462-2920.12196.
54. Martínez-Hidalgo, P.; Hirsch, A. M. The Nodule Microbiome: N 2 -Fixing Rhizobia Do Not Live Alone. *Phytobiomes J.* **2017**, *1* (2), 70–82. https://doi.org/10.1094/PBIOMES-12-16-0019-RVW.

55. Scavino, A. F.; Pedraza, R. O. The Role of Siderophores in Plant Growth-Promoting Bacteria. In *Bacteria in Agrobiology: Crop Productivity*; Springer Berlin Heidelberg: Berlin, Heidelberg, 2013; pp 265–285. https://doi.org/10.1007/978-3-642-37241-4_11.
56. Kraemer, S. M. Iron Oxide Dissolution and Solubility in the Presence of Siderophores. *Aquat. Sci. Res. Across Boundaries* **2004,** *66* (1), 3–18. https://doi.org/10.1007/s00027-003-0690-5.
57. Kramer, J.; Özkaya, Ö.; Kümmerli, R. Bacterial Siderophores in Community and Host Interactions. *Nat. Rev. Microbiol.* **2020,** *18* (3), 152–163. https://doi.org/10.1038/s41579-019-0284-4.
58. Gu, S.; Wei, Z.; Shao, Z.; Friman, V.-P.; Cao, K.; Yang, T.; Kramer, J.; Wang, X.; Li, M.; Mei, X.; Xu, Y.; Shen, Q.; Kümmerli, R.; Jousset, A. Competition for Iron Drives Phytopathogen Control by Natural Rhizosphere Microbiomes. *Nat. Microbiol.* **2020,** *5* (8), 1002–1010. https://doi.org/10.1038/s41564-020-0719-8.
59. Lewis, R. W.; Islam, A.; Opdahl, L.; Davenport, J. R.; Sullivan, T. S. Comparative Genomics, Siderophore Production, and Iron Scavenging Potential of Root Zone Soil Bacteria Isolated from 'Concord' Grape Vineyards. *Microb. Ecol.* **2019,** *78* (3), 699–713. https://doi.org/10.1007/s00248-019-01324-8.
60. Joseph, B.; Patra, R. R.; Lawrence, R. Characterization of Plant Growth Promoting Rhizobacteria Associated with Chickpea (Cicer Arietinum L.). *Int. J. Plant Prod.* **2007,** *1* (2).
61. Chanclud, E.; Morel, J. Plant Hormones: A Fungal Point of View. *Mol. Plant Pathol.* **2016,** *17* (8), 1289–1297. https://doi.org/10.1111/mpp.12393.
62. Suebrasri, T.; Harada, H.; Jogloy, S.; Ekprasert, J.; Boonlue, S. Auxin-Producing Fungal Endophytes Promote Growth of Sunchoke. *Rhizosphere* **2020,** *16*, 100271. https://doi.org/10.1016/j.rhisph.2020.100271.
63. Hasan, H. A. H. Gibberellin and Auxin-Indole Production by Plant Root-Fungi and Their Biosynthesis under Salinity-Calcium Interaction. *Acta Microbiol. Immunol. Hung.* **2002,** *49* (1), 105–118. https://doi.org/10.1556/amicr.49.2002.1.11.
64. Cohen, A. C.; Travaglia, C. N.; Bottini, R.; Piccoli, P. N. Participation of Abscisic Acid and Gibberellins Produced by Endophytic Azospirillum in the Alleviation of Drought Effects in Maize. *Botany* **2009,** *87* (5), 455–462. https://doi.org/10.1139/B09-023.
65. Ahmad, N.; Hamayun, M.; Khan, S. A.; Khan, A. L.; Lee, I. J.; Shin, D. H. Gibberellin-Producing Endophytic Fungi Isolated from Monochoria Vaginalis. *J. Microbiol. Biotechnol.* **2010,** *20* (12), 1744–1749. https://doi.org/10.4014/jmb.1005.05018.
66. Spaepen, S.; Vanderleyden, J. Auxin and Plant-Microbe Interactions. *Cold Spring Harb. Perspect. Biol.* **2011,** *3* (4), a001438–a001438. https://doi.org/10.1101/cshperspect.a001438.
67. Pohare, M. B.; Wagh, S. G.; Udayasuriyan, V. Bacillus Thuringiensis as Potential Biocontrol Agent for Sustainable Agriculture, 2021; pp 439–468. https://doi.org/10.1007/978-981-15-6949-4_18.
68. Knief, C. Analysis of Plant Microbe Interactions in the Era of next Generation Sequencing Technologies. *Front. Plant Sci.* **2014,** *5*. https://doi.org/10.3389/fpls.2014.00216.
69. Wei, Z.; Gu, Y.; Friman, V. P.; Kowalchuk, G. A.; Xu, Y.; Shen, Q.; Jousset, A. Initial Soil Microbiome Composition and Functioning Predetermine Future Plant Health. *Sci. Adv.* **2019,** *5* (9), 1–12. https://doi.org/10.1126/sciadv.aaw0759.
70. Mendes, R.; Kruijt, M.; De Bruijn, I.; Dekkers, E.; Van Der Voort, M.; Schneider, J. H. M.; Piceno, Y. M.; DeSantis, T. Z.; Andersen, G. L.; Bakker, P. A. H. M.; Raaijmakers, J.

M. Deciphering the Rhizosphere Microbiome for Disease-Suppressive Bacteria. *Science* **2011,** *332* (6033), 1097–1100. https://doi.org/10.1126/science.1203980.
71. Kwak, M.; Kong, H. G.; Choi, K.; Kwon, S.; Song, J. Y.; Lee, J.; Lee, P. A.; Choi, S. Y.; Seo, M.; Lee, H. J.; Jung, E. J.; Park, H.; Roy, N.; Kim, H.; Lee, M. M.; Rubin, E. M.; Lee, S.; Kim, J. F. Rhizosphere Microbiome Structure Alters to Enable Wilt Resistance in Tomato. *Nat. Biotechnol.* **2018**. https://doi.org/10.1038/nbt.4232.
72. Xiong, W.; Song, Y.; Yang, K.; Gu, Y.; Wei, Z.; Kowalchuk, G. A.; Xu, Y.; Jousset, A.; Shen, Q.; Geisen, S. Rhizosphere Protists Are Key Determinants of Plant Health. *Microbiome* **2020,** *8* (1), 27. https://doi.org/10.1186/s40168-020-00799-9.
73. Chaudhary, V.; Prasanna, R.; Nain, L.; Dubey, S. C.; Gupta, V.; Singh, R.; Jaggi, S.; Bhatnagar, A. K. Bioefficacy of Novel Cyanobacteria-Amended Formulations in Suppressing Damping off Disease in Tomato Seedlings. *World J. Microbiol. Biotechnol.* **2012,** *28* (12), 3301–3310. https://doi.org/10.1007/s11274-012-1141-z.
74. Prasanna, R.; Chaudhary, V.; Gupta, V.; Babu, S.; Kumar, A.; Singh, R.; Shivay, Y. S.; Nain, L. Cyanobacteria Mediated Plant Growth Promotion and Bioprotection against Fusarium Wilt in Tomato. *Eur. J. Plant Pathol.* **2013,** *136* (2), 337–353. https://doi.org/10.1007/s10658-013-0167-x.
75. Righini, H.; Roberti, R. Algae and Cyanobacteria as Biocontrol Agents of Fungal Plant Pathogens. In *Plant Microbe Interface*; Springer International Publishing: Cham, 2019; pp 219–238. https://doi.org/10.1007/978-3-030-19831-2_9.
76. El-Ansary, M. S. M.; Hamouda, R. A. Biocontrol of Root-Knot Nematode Infected Banana Plants by Some Marine Algae. *Russ. J. Mar. Biol.* **2014,** *40* (2), 140–146. https://doi.org/10.1134/S1063074014020047.
77. Haas, D.; Keel, C. Regulation Of Antibiotic Production In Root-Colonizing Pseudomonas Spp. And Relevance For Biological Control Of Plant Disease. *Annu. Rev. Phytopathol.* **2003,** *41* (1), 117–153. https://doi.org/10.1146/annurev.phyto.41.052002.095656.
78. Mazurier, S.; Corberand, T.; Lemanceau, P.; Raaijmakers, J. M. Phenazine Antibiotics Produced by Fluorescent Pseudomonads Contribute to Natural Soil Suppressiveness to Fusarium Wilt. *ISME J.* **2009,** *3* (8), 977–991. https://doi.org/10.1038/ismej.2009.33.
79. Vacheron, J.; Moënne-Loccoz, Y.; Dubost, A.; Gonçalves-Martins, M.; Muller, D.; Prigent-Combaret, C. Fluorescent Pseudomonas Strains with Only Few Plant-Beneficial Properties Are Favored in the Maize Rhizosphere. *Front. Plant Sci.* **2016,** *7*. https://doi.org/10.3389/fpls.2016.01212.
80. Konappa, N.; Krishnamurthy, S.; Dhamodaran, N.; Arakere, U. C.; Ramachandrappa, N. S.; Chowdappa, S. Beneficial Effects of Trichoderma on Plant–Pathogen Interactions: Understanding Mechanisms Underlying Genes, **2020**; pp 41–85. https://doi.org/10.1007/978-3-030-54758-5_3.
81. Guigón López, C.; García Ramírez, H. A.; Muñoz Castellanos, L. N. Control of Pepper Powdery Mildew Using Antagonistic Microorganisms: An Integral Proposal. *Prog. Biol. Control Plant Defence Biol. Control* **2020,** 385–420. https://doi.org/10.1007/978-3-030-51034-3_15.
82. Sylla, J.; Alsanius, B. W.; Krüger, E.; Reineke, A.; Strohmeier, S.; Wohanka, W. Leaf Microbiota of Strawberries as Affected by Biological Control Agents. *Phytopathology* **2013,** *103* (10), 1001–1011. https://doi.org/10.1094/PHYTO-01-13-0014-R.
83. Wang, Z.; Li, Y.; Zhuang, L.; Yu, Y.; Liu, J.; Zhang, L.; Gao, Z.; Wu, Y.; Gao, W.; Ding, G.; Wang, Q. A Rhizosphere-Derived Consortium of Bacillus Subtilis and Trichoderma

Harzianum Suppresses Common Scab of Potato and Increases Yield. *Comput. Struct. Biotechnol. J.* **2019,** *17*, 645–653. https://doi.org/10.1016/j.csbj.2019.05.003.
84. Gururani, M. A.; Venkatesh, J.; Upadhyaya, C. P.; Nookaraju, A.; Pandey, S. K.; Park, S. W. Plant Disease Resistance Genes: Current Status and Future Directions. *Physiol. Mol. Plant Pathol.* **2012,** *78*, 51–65. https://doi.org/10.1016/j.pmpp.2012.01.002.
85. Rodriguez, P. A.; Rothballer, M.; Chowdhury, S. P.; Nussbaumer, T.; Gutjahr, C.; Falter-Braun, P. Systems Biology of Plant-Microbiome Interactions. *Mol. Plant* **2019,** *12* (6), 804–821. https://doi.org/10.1016/j.molp.2019.05.006.
86. Salas-Marina, M. A.; Isordia-Jasso, M. I.; Islas-Osuna, M. A.; Delgado-SÃ¡nchez, P.; JimÃ©nez-Bremont, J. F.; RodrÃguez-Kessler, M.; Rosales-Saavedra, M. T.; Herrera-Estrella, A.; Casas-Flores, S. The Epl1 and Sm1 Proteins from Trichoderma Atroviride and Trichoderma Virens Differentially Modulate Systemic Disease Resistance against Different Life Style Pathogens in Solanum Lycopersicum. *Front. Plant Sci.* **2015,** *6*. https://doi.org/10.3389/fpls.2015.00077.
87. Bakker, P. A. H. M.; Pieterse, C. M. J.; van Loon, L. C. Induced Systemic Resistance by Fluorescent Pseudomonas Spp. *Phytopathology* **2007,** *97* (2), 239–243. https://doi.org/10.1094/PHYTO-97-2-0239.
88. Huang, A. C.; Jiang, T.; Liu, Y.-X.; Bai, Y. C.; Reed, J.; Qu, B.; Goossens, A.; Nützmann, H. W.; Bai, Y.; Osbourn, A. A Specialized Metabolic Network Selectively Modulates Arabidopsis Root Microbiota. *Science* **2019,** *364* (6440), eaau6389. https://doi.org/10.1126/science.aau6389.
89. Kosová, K.; Vítámvás, P.; Prášil, I. T.; Renaut, J. Plant Proteome Changes under Abiotic Stress — Contribution of Proteomics Studies to Understanding Plant Stress Response. *J. Proteomics* **2011,** *74* (8), 1301–1322. https://doi.org/10.1016/j.jprot.2011.02.006.
90. Hirayama, T.; Shinozaki, K. Research on Plant Abiotic Stress Responses in the Post-Genome Era: Past, Present and Future. *Plant J.* **2010,** *61* (6), 1041–1052. https://doi.org/10.1111/j.1365-313X.2010.04124.x.
91. Pitman, M. G.; Läuchli, A. Global Impact of Salinity and Agricultural Ecosystems. In *Salinity: Environment - Plants - Molecules*; Kluwer Academic Publishers: Dordrecht; pp 3–20. https://doi.org/10.1007/0-306-48155-3_1.
92. Mukhtar, S.; Malik, K. A.; Mehnaz, S.; Khaliq, Z. Mitigation Strategies for Abiotic Stress Tolerance in Plants Through Stress-Tolerant Plant Growth-Promoting Microbes, 2021; pp 325–351. https://doi.org/10.1007/978-981-15-6949-4_14.
93. Mukhtar, S.; Hirsch, A. M.; Khan, N.; Malik, K. A.; Humm, E. A.; Pellegrini, M.; Shi, B.; Briscoe, L.; Huntemann, M.; Clum, A.; Foster, B.; Foster, B.; Roux, S.; Palaniappan, K.; Varghese, N.; Mukherjee, S.; Reddy, T. B. K.; Daum, C.; Copeland, A.; Ivanova, N. N.; Kyrpides, N. C.; Shapiro, N.; Eloe-Fadrosh, E. A.; Maymon, M.; Mirza, M. S.; Mehnaz, S. Impact of Soil Salinity on the Cowpea Nodule-Microbiome and the Isolation of Halotolerant PGPR Strains to Promote Plant Growth under Salinity Stress. *Phytobiomes J.* **2020,** *4* (4), 364–374. https://doi.org/10.1094/PBIOMES-09-19-0057-R.
94. Yuan, Y.; Brunel, C.; van Kleunen, M.; Li, J.; Jin, Z. Salinity-Induced Changes in the Rhizosphere Microbiome Improve Salt Tolerance of Hibiscus Hamabo. *Plant Soil* **2019,** *443* (1–2), 525–537. https://doi.org/10.1007/s11104-019-04258-9.
95. Marasco, R.; Rolli, E.; Ettoumi, B.; Vigani, G.; Mapelli, F.; Borin, S.; Abou-Hadid, A. F.; El-Behairy, U. A.; Sorlini, C.; Cherif, A.; Zocchi, G.; Daffonchio, D. A Drought Resistance-Promoting Microbiome Is Selected by Root System under Desert Farming. *PLoS One* **2012,** *7* (10), e48479. https://doi.org/10.1371/journal.pone.0048479.

96. Xu, L.; Naylor, D.; Dong, Z.; Simmons, T.; Pierroz, G.; Hixson, K. K.; Kim, Y.-M.; Zink, E. M.; Engbrecht, K. M.; Wang, Y.; Gao, C.; DeGraaf, S.; Madera, M. A.; Sievert, J. A.; Hollingsworth, J.; Birdseye, D.; Scheller, H. V.; Hutmacher, R.; Dahlberg, J.; Jansson, C.; Taylor, J. W.; Lemaux, P. G.; Coleman-Derr, D. Drought Delays Development of the Sorghum Root Microbiome and Enriches for Monoderm Bacteria. *Proc. Natl. Acad. Sci.* **2018,** *115* (18). https://doi.org/10.1073/pnas.1717308115.
97. Naylor, D.; DeGraaf, S.; Purdom, E.; Coleman-Derr, D. Drought and Host Selection Influence Bacterial Community Dynamics in the Grass Root Microbiome. *ISME J.* **2017,** *11* (12), 2691–2704. https://doi.org/10.1038/ismej.2017.118.
98. Santos-Medellín, C.; Liechty, Z.; Edwards, J.; Nguyen, B.; Huang, B.; Weimer, B. C.; Sundaresan, V. Prolonged Drought Imparts Lasting Compositional Changes to the Rice Root Microbiome. *Nat. Plants* **2021,** *7* (8), 1065–1077. https://doi.org/10.1038/s41477-021-00967-1.
99. Song, Y.; Haney, C. H. Drought Dampens Microbiome Development. *Nat. Plants* **2021,** *7* (8), 994–995. https://doi.org/10.1038/s41477-021-00977-z.
100. Cernava, T. How Microbiome Studies Could Further Improve Biological Control. *Biol. Control* **2021,** *160*, 104669. https://doi.org/10.1016/j.biocontrol.2021.104669.
101. Singh, B. K.; Trivedi, P.; Egidi, E.; Macdonald, C. A.; Delgado-Baquerizo, M. Crop Microbiome and Sustainable Agriculture. *Nat. Rev. Microbiol.* **2020,** *18* (11), 601–602. https://doi.org/10.1038/s41579-020-00446-y.
102. Favero, V. O.; Carvalho, R. H.; Motta, V. M.; Leite, A. B. C.; Coelho, M. R. R.; Xavier, G. R.; Rumjanek, N. G.; Urquiaga, S. Bradyrhizobium as the Only Rhizobial Inhabitant of Mung Bean (Vigna Radiata) Nodules in Tropical Soils: A Strategy Based on Microbiome for Improving Biological Nitrogen Fixation Using Bio-Products. *Front. Plant Sci.* **2021,** *11*. https://doi.org/10.3389/fpls.2020.602645.
103. Babalola, O. O.; Fasusi, O. A.; Amoo, A. E.; Ayangbenro, A. S. Genomic Analysis of a Pseudomonas Strain with Multiple Plant Growth Promoting Properties. *Rhizosphere* **2021,** *18*, 100342. https://doi.org/10.1016/j.rhisph.2021.100342.
104. French, E.; Kaplan, I.; Iyer-pascuzzi, A.; Nakatsu, C. H.; Enders, L. Emerging Strategies for Precision Microbiome Management in Diverse Agroecosystems. *Nat. Plants* **2021,** *7* (March). https://doi.org/10.1038/s41477-020-00830-9.
105. Snelders, N. C.; Rovenich, H.; Petti, G. C.; Rocafort, M.; van den Berg, G. C. M.; Vorholt, J. A.; Mesters, J. R.; Seidl, M. F.; Nijland, R.; Thomma, B. P. H. J. Microbiome Manipulation by a Soil-Borne Fungal Plant Pathogen Using Effector Proteins. *Nat. Plants* **2020,** *6* (11), 1365–1374. https://doi.org/10.1038/s41477-020-00799-5.
106. Passera, A.; Vacchini, V.; Cocetta, G.; Shahzad, G.-I.-R.; Arpanahi, A. A.; Casati, P.; Ferrante, A.; Piazza, L. Towards Nutrition-Sensitive Agriculture: An Evaluation of Biocontrol Effects, Nutritional Value, and Ecological Impact of Bacterial Inoculants. *Sci. Total Environ.* **2020,** *724*, 138127. https://doi.org/10.1016/j.scitotenv.2020.138127.
107. Karabulut, F. Emerging Trends for Harnessing Plant Metabolome and Microbiome for Sustainable Food Production. *Micro Environ.* **2021,** *1* (01), 33–53. https://doi.org/10.54458/mev.v1i01.6672.
108. Seybold, H.; Demetrowitsch, T. J.; Hassani, M. A.; Szymczak, S.; Reim, E.; Haueisen, J.; Lübbers, L.; Rühlemann, M.; Franke, A.; Schwarz, K.; Stukenbrock, E. H. A Fungal Pathogen Induces Systemic Susceptibility and Systemic Shifts in Wheat Metabolome and Microbiome Composition. *Nat. Commun.* **2020,** *11* (1), 1910. https://doi.org/10.1038/s41467-020-15633-x.

109. Badri, D. V.; Zolla, G.; Bakker, M. G.; Manter, D. K.; Vivanco, J. M. Potential Impact of Soil Microbiomes on the Leaf Metabolome and on Herbivore Feeding Behavior. *New Phytol.* **2013,** *198* (1), 264–273. https://doi.org/10.1111/nph.12124.
110. Li, H.; Shi, H.; Xu, P.; Yu, D. Metabolomics and Microbiome Reveal Potential Root Microbiota Affecting the Alkaloidal Metabolome in Aconitum Vilmorinianum Kom. *BMC Microbiol.* **2022,** *22* (1), 70. https://doi.org/10.1186/s12866-022-02486-1.
111. Arif, I.; Batool, M.; Schenk, P. M. Plant Microbiome Engineering: Expected Benefits for Improved Crop Growth and Resilience. *Trends Biotechnol.* **2020,** *38* (12), 1385–1396. https://doi.org/10.1016/j.tibtech.2020.04.015.
112. Orozco-Mosqueda, M. del C.; Rocha-Granados, M. del C.; Glick, B. R.; Santoyo, G. Microbiome Engineering to Improve Biocontrol and Plant Growth-Promoting Mechanisms. *Microbiol. Res.* **2018,** *208*, 25–31. https://doi.org/10.1016/j.micres.2018.01.005.
113. Pane, C.; Sorrentino, R.; Scotti, R.; Molisso, M.; Di Matteo, A.; Celano, G.; Zaccardelli, M. Alpha and Beta-Diversity of Microbial Communities Associated to Plant Disease Suppressive Functions of On-Farm Green Composts. *Agriculture* **2020,** *10* (4), 113. https://doi.org/10.3390/agriculture10040113.
114. Bellini, A.; Ferrocino, I.; Cucu, M. A.; Pugliese, M.; Garibaldi, A.; Gullino, M. L. A Compost Treatment Acts as a Suppressive Agent in Phytophthora Capsici – Cucurbita Pepo Pathosystem by Modifying the Rhizosphere Microbiota. *Front. Plant Sci.* **2020,** *11*. https://doi.org/10.3389/fpls.2020.00885.
115. Deng, X.; Zhang, N.; Shen, Z.; Zhu, C.; Li, R.; Salles, J. F.; Shen, Q. Rhizosphere Bacteria Assembly Derived from Fumigation and Organic Amendment Triggers the Direct and Indirect Suppression of Tomato Bacterial Wilt Disease. *Appl. Soil Ecol.* **2020,** *147*, 103364. https://doi.org/10.1016/j.apsoil.2019.103364.
116. Chihaoui, S.-A.; Trabelsi, D.; Jdey, A.; Mhadhbi, H.; Mhamdi, R. Inoculation of Phaseolus Vulgaris with the Nodule-Endophyte Agrobacterium Sp. 10C2 Affects Richness and Structure of Rhizosphere Bacterial Communities and Enhances Nodulation and Growth. *Arch. Microbiol.* **2015,** *197* (6), 805–813. https://doi.org/10.1007/s00203-015-1118-z.
117. Sheth, R. U.; Cabral, V.; Chen, S. P.; Wang, H. H. Manipulating Bacterial Communities by in Situ Microbiome Engineering. *Trends Genet.* **2016,** *32* (4), 189–200. https://doi.org/10.1016/j.tig.2016.01.005.
118. Ke, J.; Wang, B.; Yoshikuni, Y. Microbiome Engineering: Synthetic Biology of Plant-Associated Microbiomes in Sustainable Agriculture. *Trends Biotechnol.* **2021,** *39* (3), 244–261. https://doi.org/10.1016/j.tibtech.2020.07.008.

## CHAPTER 6

# Biostimulants from Algae Applied in Agriculture

KATTY SHCARLEN HERNÁNDEZ-GAYTÁN[1],
CELESTINO GARCÍA-GÓMEZ[1], JULIA MARIANA MÁRQUEZ-REYES[1],
JUAN ANTONIO VIDALES-CONTRERAS[1], and
SANTIAGO IVÁN SUÁREZ-VÁZQUEZ[2]

[1]*Universidad Autónoma de Nuevo León, Facultad de Agronomía, General Escobedo, Nuevo León, México*

[2]*Universidad Autónoma de Nuevo León, Facultad de Ingeniería Civil, San Nicolás de los Garza, Nuevo León, México*

## ABSTRACT

Since the 19th century, the use of micro and macroalgae as fertilizers in the agricultural area has been implemented. Previously, they were applied directly to the soil and their benefits on the physiology of the plant were observed in a short time. In recent years, micro and macroalgae have been used as plant biostimulants and soil improvers. A plant biostimulant is a substance that improves the efficiency of nutrition and increases tolerance to abiotic stress, yield, and crop production. The chemical composition of micro and macroalgae is directly related to growth conditions, such as nutrient availability, light, and pollution. On the other hand, the use of algae extracts has the advantages of a higher germination rate, development of the root system, greater leaf area, number of leaves, among others. Algae compounds are uniquely defined as natural biostimulants, due to their chemical composition, which include proteins, carbohydrates, phytohormones, and secondary metabolites such as terpenes and polyphenols, which enhance plant growth. Elaborate blends of seaweed extracts add to or enhance their effects. The

Sustainable Agricultural Production Systems: Traditional and Modern Perspectives.
Romeo Rojas, Guillermo Cristian Guadalupe Martínez-Ávila, Julia Mariana Márquez-Reyes, and Ma del Carmen Ojeda-Zacarías (Eds.)

use of biostimulants is a promising option within sustainable agriculture that seeks to reduce the application of synthetic fertilizers. In this chapter we will address topics on the applications of biostimulants and their effects in agriculture.

## 6.1 INTRODUCTION

Algae are classified primarily by their life cycle, pigmentation, and cell structure. There are about 800,000 species of algae, and 600,000 have been studied. Due to the large number of species there is a very wide range of possible uses. Algae are divided into four categories based on their morphological characteristics and size. The classification includes unicellular, multicellular, colonial, and filamentous algae. Based on this classification they are mainly divided into two groups: macro and microalgae. Macroalgae are multicellular, macroscopic organisms with a length of up to 65 m. On the other hand, microalgae reach a size from 1 to 900 μm; they are unicellular and microscopic photosynthetic organisms. They can be produced in fresh water as wastewater, which reduces their cost of production. The most commercial microalgae species are *Chlorella spp* and *Arthrospira spp.*[1]

Algae have the ability to absorb sunlight to carry out photosynthesis, that is, they transform light energy into chemical energy, capturing $CO_2$ and releasing oxygen ($O_2$). This is why algae are considered the true lungs of our planet, even above the Amazon rainforest.[2] The biomass produced from microalgae is used in the production of food, fuel, and crops. The color of macroalgae is very varied; however, they all contain chlorophyll, a pigment that gives plants their green color and is found inside the chloroplasts. The coloration or pigmentation of an algae is due to the combination of photosynthetic pigments that are present in the cells, which is why a classification of algae is due to their coloration; the color of the obtained pigment is shown in Figure 6.1.[3] Photosynthetic pigments can be classified by their color, for example:

1. Red macroalgae belong to the Rhodophyta phylum, contain chlorophyll a and some carotenoids as photosynthetic pigments;
2. Green macroalgae are included in the Chlorophyta phylum and have a pigmentation identical to that of vascular plants, that is, carotenoids and chlorophylls a and b;
3. Brown macroalgae belong to the Ochrophyta phylum and are grouped in the Phaeophyceae class, contain carotenoid pigments and chlorophylls a and c, predominantly Fucoxanthin, which gives its brownish color;

4. Finally, the blue-green macroalgae are included in the Cyanobacteria phylum; their pigments are due to carotenoids, chlorophyll a, phycocyanin, and phycoerythrin.[3]

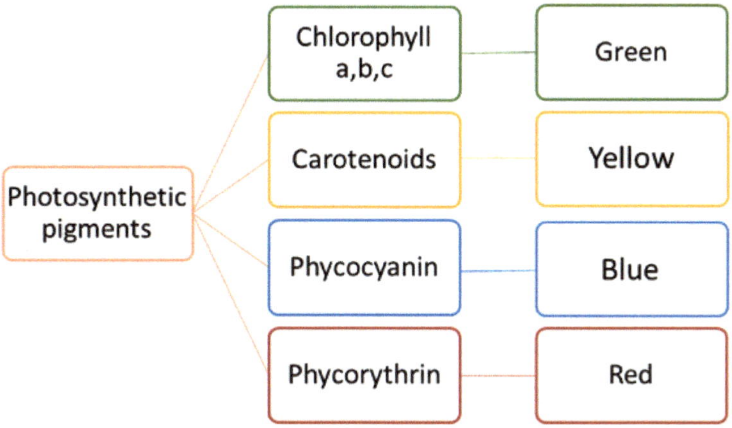


**FIGURE 6.1** Pigments obtained from photosynthetic microorganisms such as microalgae, macroalgae, and cyanobacteria.

Macro and microalgae have a high commercial value. Japan and China are the main consumers in the world. Macroalgae are used for different purposes such as food, feed, medicine, pharmaceuticals, and in agriculture.[2]

Since the 1950s, algae have been used in agriculture to obtain higher yields in crops, and even prevent and treat certain diseases. On the other hand, the growing demand for food due to the increase in the world population leads to the use of agrochemicals such as fertilizers, seeking to increase production; however, the misuse of agrochemicals results in the loss of the fertile layer of the soils, decreased biodiversity, and has even eliminated the natural predators of pests.[4] The need to protect the environment has led us to seek efficient alternatives with zero toxicity toward it.

## 6.2 EXPERIMENTAL METHODS AND MATERIALS

Within the productivity of crops, the soil is fundamental in the face of drought and heavy rains; however, intensive work within agriculture has caused concern, due to the degradation of soil water, in addition to the release of nutrients. The compaction, erosion, and loss of organic matter in the soil are the changes that affect the physical and chemical properties of the soil, the state in which the nutrients are found, and the stability of the aggregates. There are key indicators of soil quality, which are water-stable aggregates: (1) protect organic matter from decomposition, (2) increase water-holding capacity, (3) pore spaces for water infiltration and root growth, and (4) resistance to erosion, finally, to avoid the formation of crust on the surface, in addition to leaching, which leads to groundwater contamination. The depletion of soil organic material (OM) is due to the intensity with which the soil is plowed, using high concentrations of minerals. The OM is responsible for storing nutrients and maintaining the soil structure, which generates the loss of nutrients to the bodies of water. This phenomenon has been aggravated due to the erosion of the layer where the nutrients are found, in countries of the United Kingdom there are continuous farmlands with an average rate of 9.5 ton per ha, this leads to the loss of the finest particles such as the OM. Due to this, a dependence on organic fertilizers is generated, such as vegetable waste, animal manure, biochar, among others. The use of animal manure has caused ecological contamination related to veterinary grade antibiotics and metals such as copper (Cu) and zinc (Zn). Using the classic chemical fertilizers has financial advantages; however, they present a significant toxicological risk that affects human health and the integrity of the biota. There are alternative sources of fertilizers that can provide micro and macronutrients to plants, in addition to benefiting the soil, one of them is algae. Algae occurs in water bodies and is stimulated by mineral nutrients.[6]

Fertilization from algae biomass dates back to the earliest civilizations. This demonstrates its importance and how beneficial it can be in the agricultural area; today, it is known that algae biomass modifies soil pH by acting as a nitrogen fixer and OM source.[7]

Algae have been used to amend the soil, increasing nitrogen content by up to 6 times, compared to mineral nitrogen.[8]

As anthropogenic pollution increases, algae formation increases, and climate change is expected to aggravate the problem. One option is to cultivate and harvest the biomass of microalgae, because it can be used as an organic fertilizer, providing nutrients and carbon to the soil and even improving crop growth and nutrition. Large-scale algae production has been increasing in recent years to produce biofuels, animal feed, and as a nutrient remover in wastewater treatment.[6]

Although the wastewater contains high percentages of nitrogen (N) and phosphorus (P), it cannot be used as irrigation, because it contains contaminants such as antibiotics, analgesics, and antihypertensives, which generates a risk to crops and the environment.[9]

This paved the way for the development of algae-based fertilizers. Some species of microalgae such as *Spirulina* and *Chlorella* are widely used for the removal of nutrients (N and P) in wastewater treatment, which makes them a great option as soil conditioners, since it has been reported that Spirulina improves the soil macronutrient availability. *Spirulina* is a cyanobacterium that contains amino acids, carotenoids, xanthophyll and is a natural source of vitamin B12.[6]

Other species used as biofertilizers are the species *Ulva reticulata, Gelidium crinale, Turbina ornata, Brassica napus, and Ulva Lactuca.* It has been reported that some marine algae are composed of bioactives and secondary metabolites that have been shown to be effective against nematodes and phytopathogens.[7]

### *6.2.1 BIOCARBON AS AN ALTERNATIVE SOIL AMENDMENT*

On the other hand, biomass production can be transformed into biochar. Biochar is a carbon-rich solid that is obtained from pyrolysis or carbonization. Biochar improves crop productivity and improves soil quality, making it profitable. Biochar has been proposed as a soil conditioner to improve moisture retention, microbial activity, nutrient availability, and soil aggregation. Biochar from algae has the ability to prevent nutrient loss through a slow release of mainly nitrogen and phosphorous. In addition to providing substances such as amino acids, vitamins, and carotenoids that can improve the quality of the fruits, biochar can improve physicochemical properties of the soil, such as residual carbon, improves pH and electrical conductivity, and increases nitrogen. The interaction of the soil microbial population and toxic metals generates mutagenesis, which are changes in its functioning and its genome; however, it is known that adding carbon as an energy source maintains or improves the microbial abundance of the soil, and even in soils contaminated with metals. It can also be used for power generation and carbon storage, which has an impact on mitigating climate change in the control of greenhouse gases (Amin et al., 2016).

Biochar has recently been considered a remedial agent for soils contaminated by metals. Bacteria play an important role in biogeochemical processes in the soil, such as the decomposition of organic matter. The application of biochar in the soil generates an abundance of members of the *Bacteroidetes phylum*. Obtaining a complete physicochemical characterization of biochar is usually complicated; however, certain values are known for agricultural use, for example, the carbon content is greater than 50%, 1% nitrogen and 15% phosphorus. Biochar from *Chlorella vulgaris* contains 9.76% nitrogen, 61.32% carbon, has a pH of 8.1, and a moisture content of 3.81%, which is beneficial for the soil.[11,12]

Obviously, biochar can be widely used to improve the quality and productivity of agricultural soil, since it affects the properties and functions of the soil.

### 6.2.2 EFFECTS ON VEGETABLES

The food demand leads agriculture to look for alternatives without degrading the environment, looking for natural fertilizers instead of synthetic ones. Biostimulants, vitamins, polysaccharides, phytohormones, etc., are substances from natural materials that promote plant growth.[13] Biostimulants have the ability to provide a renewable option to improve crop yields, because they are friendly to the environment and allow greater absorption of phosphorus, nitrogen, and potassium, which are applied in large volumes,[14] instead biostimulants are applied in small quantities.

Microalgae are a source of biofertilizers and biostimulants for crop production, because they contain lipids, phytohormones, polysaccharides, and proteins. [13]

Using biomass from algae as fertilizer has been implemented by Spain, Great Britain, France, and East Asia (China and Japan). Recently, algae derivatives have aroused interest as a biostimulant.[15]

Said interest in biostimulants from algae has increased in sustainable agriculture, increasing growth and yield in crops. The cyanobacteria *Spirulina plantesis* and the microalgae *Senedesmus quadricauda* are examples of growth promoters in roots and leaves.[16]

Algal biomass is composed of significant amounts of plant hormones, betaines, polyamides, auxins, and cytokinins.[17]

On the other hand, fertilization applied directly to the soil may not meet the requirements of the crop; therefore, foliar fertilization is an alternative to achieve high yields. In recent years, the foliar application of organic substances has increased agricultural sustainability. There is a positive effect when fertilizing foliarly with microalgae in crops such as pepper, tomato, and garlic, obtaining a faster response when applying foliarly compared to soil application.[1]

Plants have defense mechanisms against pathogens, mechanisms that are induced by applying elicitor molecules either directly or systemically. Algae are rich in bioactive compounds that improve plant health, protecting a wide variety of pathogens.[18]

Applying biostimulants in vegetables increases the chlorophyll content, which is more attractive to consumers.[19] Vegetables such as potatoes, tomatoes, eggplants, and peppers are from the *Capsicum* genus, part of the *Solanaceae* family, with a high commercial value, native to a wide area of America, including Mexico.[20]

Tomato (*Solanum lycopersicum L.*) is a crop rich in β-carotene, vitamins, lycopene, minerals, and anticancer compounds; it is considered a vegetable of high economic value and great nutritional importance.[13] Consuming tomato reduces the risk of cancer, thanks to its high content of antioxidants, lycopene, and carotenoids. Its flavor is related to growth conditions and cultivation techniques.[21]

Due to the increase in food demand, it is necessary to address fertilization practices without degrading the environment. On the other hand, using the microalgae Acutodesmus dimorphus in tomato increases the height of the plant, branches per plant, and number of flowers; applying the same microalgae to the seeds decreases the germination time.[1]

Fertilizing foliarly using Spirulina spp. improves the size of the tomato, increases the weight of the roots, in addition to the number of nodes.[22] Some of the algae reported as biostimulants are listed in Table 6.1.

**TABLE 6.1** Algae Used as Biostimulants in Vegetable Crops.

| Algae | Crops | Reference |
|---|---|---|
| *Laminaría digitata* | Banana | [32] |
| *Lessonia vadose* | Wheat | [23] |
| *Padina gymnospora* | Bean | [24] |
| *Gigartina acicularis* | Tabacco | [25] |
| *U. lactuca* | Tomatoe | [24] |
| *Ulva spp.* | Bean | [26] |

Another important crop within agriculture is wheat, since it is a dietary component worldwide. It is one of the most important cereals for its nutritional value, for its source of protein, vitamins, minerals, and carbohydrates. In addition, they contain phenolic compounds, anthocyanidins, flavonoids, quinones, and aminophenolic compounds important for health. Eating cereals prevents type 2 diabetes, cardiovascular and cerebrovascular diseases, and some types of cancer. In addition, it has antiviral, antiaging, and antiallergic bioactivity.[27]

Due to this there is a growing demand and yields are limited. Therefore, it is sought to increase production using algae biostimulants.[29] Biostimulants, in addition to the positive effects against drought, promote the growth of shoots and roots, generate resistance to pests and diseases, and improve water use and nutrient absorption.[30]

## 6.3 OBTAINING EXTRACTS FROM ALGAE

### *6.3.1 BENEFICIAL EXTRACTS FOR AGRICULTURE*

Much has been studied about the economic viability of algae in the health sector, including the area of cosmetics, while environmental protection has been neglected The manufacture of biostimulants from algae is a very attractive option.[31] Algae are an important source of bioactives in agricultural applications.[32] Mainly they have been applied to the soil as a fertilizer; however, advantageous effects have recently been observed when applying crude extracts in a foliar way on crop growth.[33]

Methods for algae biomass production to extract metabolites range from ocean agriculture to photobioreactors, or a number of system options such as open ponds. The extracts provide multiple direct and indirect benefits to both humans and plants, since they contain bioactives, as well as macro- and micronutrients.[31] Extracts from algae are widely used in the agricultural area as biostimulants in various crops. The extracts are an alternative to reduce and even replace chemical fertilizers. Algae are a source of extracts due to their chemical composition, which, as already mentioned, includes proteins, carbohydrates, phytohormones, polysaccharides, lipids, and secondary metabolites such as polyphenols and terpenes, improving plant growth.[34] In addition, they have the ability to form Zn (ll) ions,[35] essential to form chelates. Chelates (boron atoms) are considered an important plant nutrient, which can be formed from oligosaccharides (ulvans), from algae.[36]

The most used algae to produce biostimulants are brown, due to their abundance, large size, and chemical composition. The first patented alkaline algal extract was from *Macrocystis pyrifera* and over the years different species of algae from various producing countries have been used.[37] The brown alga *Ascophyllum nodosum* is the most used to produce biostimulants, along with others from the genera Laminaria, Turbinaria, Fucus, Sargassum, and Ecklonia. There are biostimulants produced with green and red algae, for example, *Kappaphycus alvarezii*, a red algae, and *Ulva Lactuca*, a green algae.[38] The mechanisms of action of the extracts can have positive or negative effects, depending directly on the concentration and components applied to the seeds and plants, since each plant has a different sensitivity threshold for bioactive molecules.[39]

### 6.3.2 METHODS FOR OBTAINING SEAWEED EXTRACTS

Various extraction methods are used to formulate biostimulants. There are biostimulants such as polysaccharides, alginates, carrageenan, laminaran, and fucoidans, as well as acetone and ethanolic extracts that are obtained from algae and increase plant growth.[40]

Generally, the biomass obtained during the exponential growth phase is recovered by microfiltration and centrifugation, and the most common preservation process is pasteurization or dehydration, then it is ground and finally stored for later use.[41]

The active components of the algae are found within the cell walls and/or linked to structures. To achieve their release, the biomass must undergo various processes, such as cryoprocessing, where the previously frozen biomass passes through a mill and is cooled with liquid nitrogen. To obtain very fine particles, high-pressure treatments can achieve cell rupture, releasing its content. Another method is the enzymatic one, using cellulase, protease, xylanase, glucanase, glucosidase, and pectinase. (Gil et al., 2013).

Various extraction techniques can be used to extract compounds from algae, for example, solid–liquid or liquid–liquid extraction, extractions using $CO_2$, pasteurized liquid extraction, ultrasonication or microwave-assisted extractions. The extraction method with super critical fluids (pressure ~ 500 bar, temperature ~ 50°C, and ~ 100 kg of $CO_2$ per kg of biomass) attracts a lot of interest since the extract obtained is free of solvents.[1] A representation of the extraction processes to obtain biomolecules is shown in Figure 6.2.

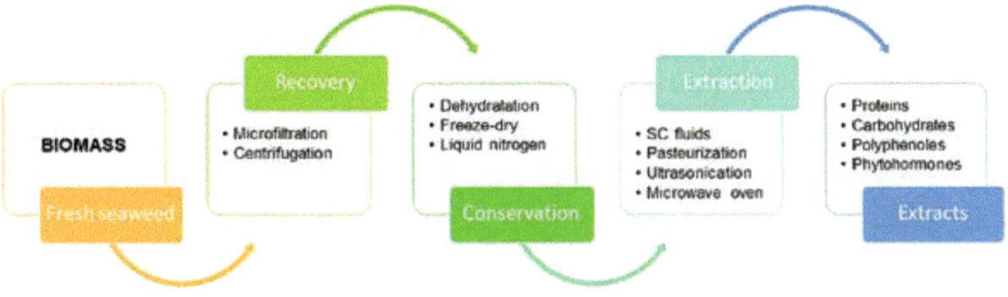


**FIGURE 6.2** Representation of the extraction process of biomolecules from algae and cyanobacteria.

The extracts vary in odor, composition, viscosity, and color (from colorless to black). The extraction methods must be focused on the metabolite that is desired to be obtained, it is not reasonable to expect that all the metabolites of the alga can be extracted using the same method. For example, the extraction of proteins is achieved by aqueous, acid, and alkaline methods, and to recover them they are centrifuged, precipitated, or microfiltered. In search of increasing the efficiency to extract proteins, polysaccharidases (alginates, galactans, starch, cellulose, fucoidans, xylans, and laminarin) are applied to degrade enzymes and improve cell wall decomposition. It is possible to obtain a 98% protein yield by high-pressure cell disruption.[31] On the other hand, aqueous extraction to obtain polysaccharides and organic solvents such as methanol, ethanol, acetone, and ethyl acetate to obtain low molecular weight compounds is of the utmost importance to formulate mixtures and obtain extracts that stimulate plant growth. Polysaccharides have important functions within the plant; they act as a hormone by hydrolyzing oligosaccharides, controlling plant development (Trouvelot et al., 2014).

In recent years, algal polysaccharides have been evaluated in crops such as grapes, corn, beans, and barley, obtaining increased dry weight in plants, improved tolerance in environmental issues, increased germination percentage and plant growth (Kannan et al., 2014).

Algae-derived polysaccharides have a protective effect on plants, inducing defense events against microbial pathogens, such as activating secondary metabolic pathways. As an alternative tool to control diseases in plants, elicitors have been considered and, since algal extracts contain certain polysaccharides, they can be used to control diseases in crops. *Ulva armuricana* (Northern France) is composed of sulfated polysaccharides, which protect alfalfa (*Medicago truncatula*) against *C. trifolii*, by expressing genes that are related to pathogenesis, such as PR-10 (Cluzet et al. 2004).

In microalgae, the extraction of polysaccharides began to produce biofuels; however, other metabolites such as chlorophyll, carotenoids, and carbohydrates are currently extracted.[31]

As already mentioned, algae are an important raw material for the formulation of biostimulants, since they can provide various nutrients and protective functions. There are eight categories of biostimulants: (1) Inorganic salts, (2) Humic substances, (3) Algae extracts, (4) Inorganic compounds, (5) Chitosans and other biopolymers, (6) Fungi, (7) Hydrolyzed proteins, and (8) Bacteria.[34]

Algae can be used as humic substances when added to the soil, they are organic matter, and they accumulate inorganic salts. They contain proteins and amino acids that produce secondary metabolites such as phlorotannins, which have recently been tested as biostimulants.[40]

In addition, after obtaining and using the extracts as liquid fertilizer, the industrialization and study of the effects of extracts from algae as biostimulants on plant growth began. This has generated various industries that are dedicated to obtaining and marketing algal biostimulants.[29]

## 6.4 ABIOTIC STRESS PRESENT IN VEGETABLE CROPS

### *6.4.1 CONDITIONS GENERATING ABIOTIC STRESS*

Seaweed derivatives or extracts are typically applied to the soil as a slow-release fertilizer and to improve the physical and biological properties of the soil. Interest in biostimulants has recently been aroused with a main objective, the response to abiotic stress and water deficit in crops.[31]

Plants are often under unfavorable conditions, which is called environmental stress. The factors that lead the plant to stress are divided into two types: biotic and abiotic (physical, chemical, and physical-chemical). Those of the abiotic type cover environmental factors such as temperature, radiation, water, chemical substances, and others. Abiotic stress includes factors such as salinity and extreme temperatures, causing great losses in agricultural production worldwide.[42]

### *6.4.2 SOIL SALINITY*

Soil salinity can be caused by natural causes such as coastal waters, salty rain, ocean salts, and parental rocks; however, poor agricultural practices have also increased the concentration of salts in the rhizosphere. It is estimated that salinization in arable soils will cause 30% soil loss in the next 25 years and up to 50% in the next 40 years.[43] The salinity of the soil affects the development of the plants, which generates low yields. In a saline soil, the availability of atmospheric $CO_2$ is restricted, because the stomata are closed and the consumption of nicotinamide adenine dinucleotide phosphate (NADPH) is reduced by the Calvin cycle.[44]

There are three negative effects of salinity on the physiology of plants: the first is the inhibition of protein synthesis caused by the toxic effects of sodium and chloride ions, alteration of the structure of enzymes affects respiration and photosynthesis. Second, it causes a nutrient imbalance because the uptake and transport of nutrients to the plant shoot is decreased. And finally, the difficulty of water absorption by the roots, caused by the osmotic potentials of the soil, due to the accumulated salts, leads to a physiological drought of the plants.[45]

### 6.4.3 WATER STRESS

Water and thermal stress depress root development and nodulation processes. Exposing plants to water stress has a negative effect on plant productivity and growth. Drought conditions and high temperatures promote changes at the cellular level on proteins due to the increase in free radicals (ROS) and changes in the plant defense system. Inactivating ROS impacts crop yield.[46]

Among the products used to counteract the effects of stress are biostimulants.[47]

### 6.4.4 BIOSTIMULANTS AS AN ALTERNATIVE AGAINST ABIOTIC STRESS

Using biostimulants has the advantage of moderating the synthesis of endogenous phytoregulators, which are responsible for regulating crop development. Maintaining the balance of endogenous levels under stress conditions favors the setting of flowers. Biostimulant extracts are a source of antioxidants such as folsistein and cofactors such as ribofavin, thiamin, pyridoxine, niacin, and vitamins B1, B2, B3, and B12.[48] Biostimulants are a key tool for crop management in adverse environments. The potential of the extracts as biostimulants for salinity is little explored, one of the most important types of abiotic stress in terms of damage in the agricultural area. The effectiveness of the extracts against salinity depends directly on the moment in which it is applied. Applications can be in 3 different stages: before the crop is damaged, during stress conditions, and after stress, that is, when the plants show signs. It is necessary to use algae extracts that contain molecules that can activate the endogenous plant biosynthesis of protective metabolites and physiological adapters related to stress. It is important to understand that the result of the application of the extracts is related to the concentration and time of application and also that not all crops respond in the same way.[49]

On the other hand, a paradigm recently emerged that plants are not independent of their environment; in other words, there is a partnership between plants and microorganisms (bacteria and fungi). Plants host diverse microbiota and these associations either inside or outside their tissues allow them to adapt and respond to abiotic stress.[50] Obviously, if these associations are optimized, protection against stress in plants is strengthened.

In recent years, algae extracts have been studied to control stress in plants caused by cold. Using extracts rich in zinc and manganese in maize cultivation improves tolerance to low temperatures, due to improved ROS responses and in terms of micronutrients, these play a cofactor role in anti-oxidant enzymes.[51]

The key to this tolerance is genes. The lipophilic ones are rich in butyric acid, palmitic acid, linoleic acid, oleic acid, and sterol fucosterol, extracts that increase the concentration of total soluble sugars, which influences tolerance to low temperatures (freezing).[52] In the same way, cytokinins from algal extracts have been used to improve tolerance to heat and drought stress in ornamental, horticultural, and tree crops. In spinach (*Spinacia oleracea L.*) and lettuce (*Lactuca sativa L.*) plants under drought stress, the ferrous ion chelation capacity is reduced and cotyledon growth is improved, respectively. Foliar application of isopropanol extracts to grape plants (*Vitis vinifera L*) under drought stress conditions increases stomatal conductance.[53] The limited availability of water and climate change mean that the promise of biostimulants to increase tolerance to abiotic stress has great potential from an agronomic point of view.

## 6.5 CONCLUSION

The effects of algae extracts and biostimulants on plants and soil have increased the interest of industries for their production. Currently, there is a diversity of extracts and innovative biostimulants from different micro and macro algae. Its formulation must be directly related to the chemical composition of the algae, because the percentage of proteins, lipids, and carbohydrates varies by species. In this way, the application is planned based on the requirements of the crop or the needs to improve the structure or fertility of the soil. The application of biostimulants or extracts ranges from vegetables to fodder, improving yields and fruit quality, even reducing germination time. On the other hand, the application of algae to the soil can be direct or after some treatment. For example, the biocarbon that comes from the pyrolysis of algae improves soil structure, electrical conductivity, and nutrient availability. It is clear that the application of algae in agriculture has great potential, and there is much more to investigate, thanks to the wide variety of algal species.

# KEYWORDS


- **biostimulant**
- **agriculture**
- **algae**


# REFERENCES


1. Ronga, D.; Biazzi, E.; Parati, K.; Carminati, D.; Carminati, E.; Tava, A. Microalgal Biostimulants and Biofertilisers in Crop Productions. *Agron J.* **2019,** *9* (4), 146–163.
2. Gaspar, R.; Fonseca, R.; Pereira, L. Illustrated Guide to the Macroalgae of Buarcos Bay, Figueira da Foz, Portugal, 1st ed.; MARE UC, DCV, FCT: Coimbra, Portugal, 2020; p 128.
3. Pereira, L.; Correia, F. In *Algas Marinhas da Costa Portuguesa—Ecologia, Biodiversidade e Utilizações*, 1st ed.; Nota de Rodapé Editores: Paris, France, 2015; p 341.
4. Battacharyya, D.; Babgohari, M. Z.; Rathor, P.; Prithiviraj, B. Seaweed Extracts as Biostimulants in Horticulture. *Sci. Hortic.* **2015,** *30* (196), 39–48.
5. Graves, A. R.; Morris, J.; Deeks, L. K.; Rickson, R. J.; Kibblewhite, M. G.; Harris, J. A.; Farewell, T. S.; Truckle, I. The Total Costs of Soil Degradation in England and Wales. *Ecol. Econ.* **2015,** *119*, 399–413.
6. Alobwede, E.; Leake, J. R.; Pondhal, J. Circular Economy Fertilization: Testing Micro and Macro Algal Species as Soil Improvers and Nutrient Sources for Crop Production in Greenhouse and Field Conditions. *Geoderma* **2019,** *334*, 113–123.
7. Nekvapil, F.; Ganea, I. V.; Ciorît˘a, A.; Hirian, R.; Tomši´c, S.; Martonos, I; Pinzaru, S. C. A New Biofertilizer Formulation With Enriched Nutrients Content from Wasted Algal Biomass Extracts Incorporated in Biogenic Powders. *Sustainability* **2021,** *13*, 8777.
8. Mulbry, W.; Kondrad, S.; Pizarro, C. Biofertilizers from Algal Treatment of Dairy and Swine Manure Effluents. *J. Veg. Sci.* **2007,** *12*, 107–125.
9. Avagyan, A. B. In *Algae to Energy and Sustainable Development, Technologies, Resources, Economics and System Analyses. New Design of Global Environmental Policy and Live Conserve Industry*; Amazon, ISBN, 2018; vol 13, p 978.
10. Ahmad, M.; Lee, S. S.; Dou, X.; Mohan, D.; Sung, J. K.; Yang, J. E.; Ok, Y. S. Effects of Pyrolysis Temperature on Soybean Stover- and Peanut Shell-Derived Biochar Properties and TCE Adsorption in Water. *Bioresour. Technol.* **2012,** *118*, 536–544.
11. Santos, F. M.; Pires, C. M. José. Nutrient Recovery from Wastewater by Microalgae and its Potential Application as Bio-char. *Bioresour. Technol.* **2018,** *267*, 725–731.
12. Ahmad, M.; Ok, Y. S.; Kim, B. M.; Ahn, J. N.; Lee, Y. H.; Zhong, M.; Moon, D. H.; Al-Wabel, M. I. Impacto of Soybean Stover-and Pine Needle-Derived Biochar Son Pb and As Mobility, Microbial Comminity, and Carbon Stability in a Contaminated Agricultural Soil. *J. Environ. Manag.* **2016,** *166*, 131–139.

13. Supraja, K. V.; Bunoshree, B.; Balasubramanion, P. Efficacy of Ectracts as Biostimulants Through Seed Treatment and Foliar Spray for Tomato Cultivation. *Ind. Crops Prod.* **2020,** *151*, 12453.
14. Kapoore, R. V.; Wood, E.; Llewellyn, C. A. Algae Biostimulants: A Critical Look at Microalgal Biostimulants for Sustainable Agricultural Practices. *Biotechnol. Adv.* **2021,** *49*, 107754.
15. Nedumaran, T.; Arulbalachandran, D. Seaweeds: A Promising Source for Sustainable Development. In *Environmental Sustainability: Role of Green Technologies*; Thangavel, P., Sridevi, G., Eds.; Springer: New Delhi, India, 2015; pp 65–88.
16. Gemin, L. G.; Mógor, Á. F.; Amatussi, J. D. O.; Mógor, G. Microalgae Associated to Humic Acid as a Navel Biostimulant Improving Onion Growth and Yield. *Sci. Hortic.* **2019,** *256*, 108560.
17. Stirk, W. A.; Ördög, V.; Novák, O.; Rolčík, J.; Strnad, M.; Bálint, P.; Staden, J. Auxin and Cytokinin Relationships in 24 Microalgal Strains. *J. Phycol.* **2013,** *49*, 459–467.
18. Gunupuru, L. R.; Patel, J. S.; Sumarah, M. W.; Renaud, J. B.; Mantin, E. G.; Prithiviraj, B. A Plant Biostimulant Made from the Marine Brown Algae Ascophyllum Nodosum and Chitosan Reduce Fusarium Head Blight and Mycotoxin Contamination in Wheat. *PLoS ONE* **2019,** *14* (9), e0220562.
19. Ferrante, A.; Incrocci, L.; Maggini, R.; Serra, G.; Tognoni, F. Colour Changes of Fresh-Cut Leafy Vegetables During Storage. *J. Food Agri. Environ.* **2004,** *2* (3&4), 40–44.
20. Ashour, M.; Hassan, S. M.; Elshobary, M. E.; Ammar, G. A. G.; Gaber, A.; Alsanie, W. F.; Mansour, A. T.; El-Shenody, R. Impact of Commercial Seaweed Liquid Extract (TAM®) Biostimulant and Its Bioactive Molecules on Growth and Antioxidant Activities of Hot Pepper (Capsicum annuum). *Plants* **2021,** *10*, 1045.
21. Higo, M.; Azuma, M.; Kamigoshihara, Y.; Kanda, A.; Tatewaki, Y.; Isobe, K. Impacto of Phosphorus Fertilization on Tomato Growth and Arboscular Mycorrhizal Fungal. *Microorhanisms* **2020,** *81*, 178.
22. Elarroussia, H.; Elmernissia, N.; Benhimaa, R.; El Kadmiria, I. M.; Bendaou, N.; Smouni, A.; Wahbya, I. Microalgae Polysaccharides a Promising Plant Growth Biostimulant. *J. Algal Biomass Utln.* **2016,** *7*, 55–63.
23. Chandía, N. P., Matsuhiro, B.; Mejías, E.; Moenne, A. Alginic Acids in Lessonia Vadosa: Partial Hydrolysis and Elicitor Properties of the Polymannuronic Acid Fraction. *J. Appl. Phycol.* **2004,** *16*, 127 133.
24. Hernández-Herrera, R. M., Santacruz-Ruvalcaba, F. Zañudo-Hernández, J. Hernández-Carmona, G. Activity of Seaweed Extracts and Polysaccharide-Enriched Extracts form Ulva Lactuca and Padina Gymnospora as Growth Promoters of Tomato and Mung Bean Plants. *J. Appl. Phycol.* **2016,** *28*, 2549–2560.
25. Mercier, L.; Lafitte, C.; Borderies, G.; Briand, X.; Esquerré-Tugayé, M. T. The Algal Polysaccharide Carrageenans can Act as an Elicitor of Plant Defence. *New Phytologist.* **2001,** *149*, 43–51.
26. Borsato, L. C.; Di Piero, R. M.; Stadnik, M. J. Mecanismos de defensa eliciados por ulvana contra Uromyces appendiculatus em tres cultivares de feijoeiro. *Trop. Plant Pathol.* **2010,** *35* (5), 318–322.
27. Abd El-Baky, H. H.; Hussein, M. M.; El-Baroty, G. S. Algal Extracts Improve Antioxidant Defense Abilities and Salt Tolerance of Wheat Plant Irrigated with Sea Water. *Afr. J. Biochem. Res.* **2008,** *2*, 151–164.

28. Liyana-Pathirana, C. M.; Shahidi, F. Antioxidant and Free Radical Scavenging Activities of Whole Wheat and Milling Fractions. *Food Chem.* **2007,** *101*, 1151–1157.
29. Sharma, S. H. S.; Fleming, C.; Selby, C.; Rao, J. R.; Martin, T. Plant Biostimulants: A Review on the Processing of Macroalgae and Use of Extracts for Crop Management to Reduce Abiotic and Biotic Stresses. *J. Appl. Phycol.* **2014,** *26*, 465–490.
30. Stadnik, M. J.; de Freitas, M. B. Algal Polysaccharides as Source of Plant Resistance Inducers. *Trop. Plant Pathol.* **2014,** *39* (2), 111–118.
31. Carrillo, P.; Ciarmello, L. F.; Woodrow, P.; Corrado, G.; Chiaiese, P.; Roophael, Y. Enhacing Sistainability by Improving Plant Salt Tolerance Through Macro and Micro-Algal Biostimulants. *Biology* **2020,** *9*, 253.
32. Aziz, A.; Poinssot, B.; Daire, X.; Adrian, M.; Bézier, A.; Lambert, B.; Joubert, J. M.; Pugin, A. Laminarin Elicits Defense Responses in Grapevine and Induces Protection Against Botrytis Cinerea and Plasmopara Viticola. *Mol. Plant Microbe Interact.* **2003,** *16*, 1118–1128.
33. Chandía, N. P.; Matsuhiro, B. Characterization of a Fucoidan from Lessonia Vadosa (Phaeophyta) and its Anticoagulant and Elicitor Properties. *Int. J. Biol. Macromol.* **2008,** *42*, 235–240.
34. du Jardin, P. Review: Plant Biostimulants: Definition, Concept, Main Categories and Regulation. *Sci. Hort.* **2015,** *196*, 3–14.
35. Michalak, I.; Chojnacka, K.; Saeid, A. Plant Growth Biostimulants, Dietary Feed Supplements and Cosmetics Formulated with Supercritical CO2 Algal Extracts. *Molecules* **2017,** *22* (1), 1–17.
36. Dembitsky, V. M.; Smoum, R.; Al-Quntar, A.; Ali, H. A.; Pergament, I.; Srebnik, M. Natural Occurrence of Boron-Containing Compounds in Plants, Algae and Microorganisms. *Plant. Sci.* **2002,** *163*, 931–942.
37. Khan, W.; Rayirath, U. P.; Subramanian, S.; Jithesh, M. S.; Rayorath, P.; Hodges, D. M.; Critchley, A. T.; Craigie, J. S.; Norrie, J.; Prithiviraj, B. Seaweed Extracts as Biostimulants of Plant Growth and Development. *J. Plant. Growth Regul.* **2009,** *28*, 386–399.
38. Castellanos-Barriga, L. G.; Santacruz-Ruvalcaba, F.; Hernández-Carmona, G.; Ramírez-Briones, E.; Hernández-Herrera, R. M. Effect of Seaweed Extracts from *Ulva lactuca* on Seedling Growth of Mung Bean (*Vigna radiata*). *J. Appl. Phycol.* **2017,** *29*, 2479–2488.
39. Ertani, A.; Francioso, O.; Tinti, A.; Schiavon, M.; Pizzeghello, D.; Nardi, S. Evaluation of Seaweed Extracts from *Laminaria* and *Ascophyllum nodosum* spp. as Biostimulants in *Zea mays* L. Using a Combination of Chemical, Biochemical and Morphological Approaches. *Front. Plant. Sci.* **2018,** *9*, 428.
40. Rengasamy, K. R. R.; Kulkarni, M. G.; Papenfus, H. B.;van Staden, J. Quantification of Plant Growth Biostimulants, Phloroglucinol and Eckol, in Four Commercial Seaweed Liquid Fertilizers and Some by-Products. *Algal Res.* **2016,** *20*, 57–60.
41. Shaaban, M. M.; El-Saady, A. K. M.; El-Sayed, A. E. K. B. Green Microalgae Water Extract and Micronutrients Foliar Application as Promoters to Nutrient Balance and Growth of Wheat Plants. *J. Am. Sci.* **2010,** *6*, 631–636.
42. Mittler, R.; Blumwald, E. Genetic Engineering for Modern Agriculture: Challenges and Perspectives. *Annu. Rev. Plant. Biol.* **2010,** *61*, 443–462.
43. Porcel, R.; Aroca, R.; Rui´z-Lozano, J. M. Salinity Stress Alleviation Using Arbuscular Mycorrhizal Fungi. A Review. *Agron. Sustain. Dev.* **2012,** *32*, 181–200.

44. Miller, G.; Suzuki, N.; Ciftci-Yilmaz, S.; Mittler, R. Reactive Oxygen Species Homeostasis and Signalling During Drought and Salinity Stress. *Plant Cell Environ.* **2010,** *33*, 453–467.
45. Evelin, H.; Kapoor, R.; Giri, B. Arbuscular Mycorrhizal Fungi in Alleviation of Salt Stress: A Review. *Ann. Bot.* **2009,** *104*, 1263–1280.
46. Boote, K.; Allen, Jr. L. H.; Vara Prasad, P. V.; Baker, J. T.; Gesch, R. W.; Snyder, A. M.; Pan, D.; Thomas, M. G. Elevated Temperature and CO2 Impacts on Pollination, Reproductive Growth, and Yield of Several Globally Important Crops. *J. Agric. Meteorol.* **2005,** *60*, 469–474.
47. Van Oosten, M. J.; Pepe, O.; De Pascale, S.; Silletti, S.; Maggio, A. The Role of Biostimulants and Bioeffectors as Alleviators of Abiotic Stress in Crop Plants. *Chem. Biol. Technol. Agric.* **2017,** *4* (5), 1–12.
48. Liu, F.; Jensen, C. R.; Andersen, M. N. Drought Stress Effect on Carbohydrate Concentration in Soybean Leaves and Pods During Early Reproductive Development: Its Implication in Altering Pod Set. *Field Crops Res.* **2004,** *86*, 1–13.
49. Carillo, P.; Cirillo, C.; De Micco, V.; Arena, C.; De Pascale, S.; Rouphael, Y. Morpho-anatomical, Physiological and Biochemical Adaptive Responses to Saline Water of Bougainvillea SpectabilisWilld. Trained to Diferent Canopy Shapes. *Agric. Water Manag.* **2019,** *212*, 12–22.
50. Vandenkoornhuyse, P.; Quaiser, A.; Duhamel, M.; Le Van, A.; Dufresne, A. The Importance of the Microbiome of the Plant Holobiont. *New Phytol.* **2015,** *206* (4), 1196–206.
51. Bradáčová, K.; Weber, N. F.; Morad-Talab, N.; Asim, M.; Imran, M.; Weinmann, M.; et al. Micronutrients (Zn/Mn), Seaweed Extracts, and Plant Growth-Promoting Bacteria as Cold-Stress Protectants in Maize. *Chem. Biol. Technol. Agric.* **2016,** (1), 19.
52. Nair, P.; Kandasamy, S.; Zhang, J.; Ji, X.; Kirby, C.; Benkel, B.; et al. Transcriptional and Metabolomic Analysis of Ascophyllum Nodosum Mediated Freezing Tolerance in Arabidopsis Thaliana. *BMC Genom.* **2012,** *13* (1), 643.
53. Xu, C.; Leskovar, D. I. Effects of A. Nodosum Seaweed Extracts on Spinach Growth, Physiology and Nutrition Value Under Drought Stress. *Sci. Hortic.* **2015,** *12* (183), 39–47.

# CHAPTER 7

# Biological Response of *Anethum graveolens* Exposed to Cadmium

JULIA MARIANA MÁRQUEZ-REYES[1], CELESTINO GARCÍA-GÓMEZ[2], and VANIA NOEMÍ HERNÁNDEZ CÁRCAMO[1]

[1]*Universidad Autónoma de Nuevo León, Facultad de Agronomía, Laboratorio de Biotecnología Microbiana, General Mariano Escobedo, N.L., México*

[2]*Universidad Autónoma de Nuevo León, Facultad de Agronomía, Laboratorio de Remediación Ambiental y Análisis de Planta, General Mariano Escobedo, N.L., México*

## ABSTRACT

Heavy metal contamination in ecosystems is becoming more frequent due to its use in industrial activities and poor waste management. In ecosystems, we can find much heterogeneity of heavy metals in various concentrations and their interaction with other specific compounds, and the different environmental conditions define whether they become an essential or toxic element for plants. Plants exposed to distinct concentrations of heavy metals have developed mechanisms to minimize their toxicity and be able to coexist daily. The processes through which a plant acts on a particular contaminant depend on the degree of contamination, the physicochemical characteristics of the contaminant and the polluting site, and the phenotypic and genotypic properties of each plant species. Phytoremediation uses the phenotypic and genotypic characteristics of each plant species mentioned above to remove organic and inorganic contaminants from water, soil, and air. The objective of this chapter is to identify current problems. As well as current trends

Sustainable Agricultural Production Systems: Traditional and Modern Perspectives.
Romeo Rojas, Guillermo Cristian Guadalupe Martínez-Ávila, Julia Mariana Márquez-Reyes, and Ma del Carmen Ojeda-Zacarías (Eds.)

in heavy metal phytoremediation and evaluate responses of dill (*Anethum graveolens)* in a closed and open system in the presence of cadmium.

## 7.1 INTRODUCTION

Removing of heavy metals and other pollutants from wastewater with the help of aquatic plants is an effective and low-cost treatment. In this work, the physiological behavior of *Anethum graveolens,* also known as dill, a herbaceous plant of the genus *Anethum*, belonging to the family Umbelliferae, was studied. The evaluated aquatic systems, two open-flow and closed-flow with synthetic wastewater, were established to assess the behavior of *A. graveolens* exposed to cadmium.

### *7.1.1 BIOREMEDIATION*

Bioremediation when microorganisms, fungi, plants, or enzymes accumulate, immobilize, or transform contaminants. Phytoremediation refers to several technologies that rely on plants to clean up or restore contaminated environments.[53] In this technique, plants act as traps or biological filters that decompose contaminants and stabilize these substances in soil and water. The plants fixing them in their roots and stems or metabolizing them as microorganisms finally convert them into less dangerous and more stable compounds, such as carbon dioxide, water, and mineral salts.[37] There is information on many plant species capable of eliminating different pollutants, such as the case of *Pteris vittata, Phytolacca Americana, Haumaniastrum robertii, Eleocharis acicularis, Achillea millefolium, Medicago sativa, Brassica juncea, Brassica nigra* among many others.[22,24,28,36,44,51] There are different processes within phytoremediation, such as extraction, immobilization in roots or part area of the plant species, and transformation of compounds. The root system is the central entry tissue of contaminants into the plant. By diffusion processes, heavy metals can enter the roots because they have negative charges coming from carboxyl groups of the pectic acid; this differential of behavior with the positive heavy metals promotes a dynamic cation exchange toward the interior of the cell has a hydrophilic cell wall. Subsequently, heavy metals can be transported by apoplastic and symplastic routes.[8]

Plants have different mechanisms to counteract the adverse effects of exposure to heavy metals. The main molecules involved in this action are antioxidants, phenolic compounds, and proteins. Antioxidants are compounds that can delay or inhibit the oxidation of lipids or other molecules by inhibiting or propagating oxidation chain reactions.[13] Phenolic compounds are secondary metabolites produced by plants as a defense against ultraviolet radiation and pathogen attack. They are mainly classified into phenolic acids, flavonoids, stilbenes, and lignans according to the number of phenolic rings they contain and the chemical groups attached to them.[16] Flavonoids are the most abundant phenolic compounds in the human diet. They are classified into flavones, flavonols, isoflavones, flavanones, anthocyanins, and flavanols.[13,17] Enzymes are proteins with an antioxidant capacity that are not consumed when reacting with free radicals and are dependent on specific cofactors, generally metals such as copper, iron, magnesium, zinc, or selenium. The most important antioxidant enzymes are catalase (CAT), superoxide dismutase (SOD), and glutathione peroxidase (GPX). They represent the first barrier against free radical production.[49]

Excessing toxic heavy metals forms free radicals or reactive oxygen species (ROS), characterized by presenting an unpaired electron and being highly reactive with superoxide ion or hydrogen peroxide, which causes oxidative stress. This stress is understood as the state of cells in which the intracellular oxide-reduction homeostasis is altered due to ROS production or the deficiency of antioxidant mechanisms, causing severe metabolic dysfunctions and damage to biological macromolecules. However, plants possess a defense mechanism to oxidative stress formed by enzymatic and non-enzymatic compounds. These include antioxidant enzymes such as superoxide dismutase (SOD), peroxidase (POD), catalase (CAT), guaiacol peroxidase (GPX), ascorbate peroxidase (APX), glutathione reductase (GR), monodehydroascorbate reductase, superoxide dismutase (MDHAR), and low molecular mass antioxidants. They are essential for aerobic cells, as they are the ones that maintain ROS concentrations at stable levels.[26]

Plants generate root exudates which are low molecular weight and high molecular weight compounds.[9] The former is released in favor of the concentration gradient, and the latter is released by active transport mechanisms or cell lysis. The low molecular weight exudates include organic acids, sugars, phenols, amino acids, gases, and high molecular weight exudates, on the other hand, are products of lysates, secretions, and mucilages. Such compounds favor the availability or complexation of heavy metals.[10,32]

A charged metal ion with different oxidation states forms stable complexes with organic ligands that can react to varying points in the cell.[4] The plasma membrane has a fine molecular arrangement capable of selective uptake of substances, such as heavy metals, using pumps and transporters. Excess heavy metals lead to plant toxicity by creating three possible effects at the membrane level that inhibit plant growth.[14] First, decreased membrane integrity and impairing ions' flow affect the cytoplasm and organelles, reducing photosynthesis and respiration.[11] Second, it modifies the binding to SH group of enzymes or substitutes an essential ion in the coenzyme or prosthetic group, causing alterations in enzymatic activity. Third, in addition to the inhibition of growth, mainly in the plant's roots, it causes chlorosis or necrosis in its leaves and later senescence and abscission.[4] *Anethum graveolens L.* (dill) (Apiaceae) is one of the most popular culinary herbs in the Mediterranean area, Europe, and Central Southern Asia. The seeds of *A. graveolens* have been extensively used in Ethnopharmacological/traditional medicine for digestive disorders as carminative, as lactation stimulating, antispasmodic, antihyperlipidemic, and antihypercholesterolemic.[34] They are also known as dill, herbaceous plant of the genus *Anethum,* belonging to the Umbelliferae family. It is an annual herb up to 50 cm high. It has a long, axonomorphic taproot. The stem is erect, fistulous, with abundant white pith and white and green striations. Glaucous leaves, tri pinnatisect , two or three at each node, divided into filiform laciniae, deltoid, or rhombic in outline, with a well-developed sheath of up to 30 × 5 mm and petiolate. The corolla is formed by five petals, entire, suborbicular, and the apex curved inward. The fruit is dark brown, smooth, and compressed on the back. It is composed of two mericarps with five ribs so that the two lateral ones are dilated in the form of yellowish wings. The seeds are flat on one side, oval, and have a bitter taste. Flowering is usually from June onward, but fruiting is delayed under lower temperatures and abundant rainfall conditions.

### 7.1.2 WATER CONTAMINATION

The water is the universal solvent, and in nature, is never found alone; other compounds are always present.[21] A pollutant is any form of matter and energy that exceeds the initial concentrations of an ecosystem. Therefore, contamination is the presence of contaminants that cause a negative effect on the ecosystem in which they are present. The World Health Organization (WHO) considers that water is polluted when its composition or natural state is modified to lose the conditions suitable for its intended uses. Polluted water shows physical (temperature, color, density, suspensions, radioactivity, etc.), chemical (dissolved substances, composition, etc.), or biological alterations so that it cannot fulfill its ecological functions.[52] Water pollution can have a natural origin or an anthropic origin; the various activities of humans cause the latter. The sources of pollution of natural origin are much dispersed and do not cause high concentrations of pollutants, except in specific places related to specific mineral deposits or natural events.[52] The primary sources of water pollution that modify its quality are urban and industrial waste, agricultural and mining drainage, erosion, spills of toxic substances, poorly confined and treated waste, and the dragging of pollutants by rain or wind.[21]

Wastewater is the water of varied composition from discharges of municipal, industrial, commercial, service, agricultural, livestock, and domestic uses, including subdivisions and any other use and the mixture of these. Wastewater comprises 99.9% water in its drinking water state and 0.1% solids, either dissolved or suspended.[29] Therefore, the average wastewater has the following components: odor, color, suspended and dissolved solids, temperature, carbohydrates, fats and oils, proteins, surfactants, alkalinity, chlorides, nitrogen, phosphorus, sulfur, hydrogen sulfide, methane, and microorganisms.[29]

### *7.1.3 CADMIUM*

Cadmium and its derivatives, especially cadmium hydroxide, are also used in electroplating and electroplating processes in the manufacture of alkaline accumulators and steam lamps; to moderate the fission rate of radioactive substances in nuclear reactors, among other processes. Cadmium sulfide is used to give yellow or orange colorations. Selenium and cadmium sulfides are used to achieve pink, red, and brown paints; these pigments benefit applications requiring high temperatures or pressures. They are commonly used in manufacturing plastics, ceramics, colors, and coats. Cadmium pigments are also used in traffic signs and glass bottles; cadmium carboxylates are used as polyvinyl chloride (PVC) stabilizers.[1] Cd has no known biological function, but plants actively accumulate it. In its most common cationic form ($Cd^{2+}$), this element presents great chemical analogies with other essential elements. This similarity explains why its transport and storage within the plant are intimately linked to other cations' specific transport proteins such as $Ca^{2+}$, $Zn^{2+}$, $Fe^{2+}$, $Mg^{2+}$, $Mn^{2+}$, $Pb^{2+}$, $Cu^{2+}$, and $Ni^{2+}$.[12] Furthermore, excessive accumulation of Cd alters the uptake and distribution of other essential plant metals through a competitive mechanism.[39] Thus, it is responsible for nutritional deficiencies and imbalances and reduces plant growth by altering its metabolic processes.[43] Aquatic plants are natural buffers for heavy metals and other nutrients.[30] The removal of heavy metals and other pollutants from wastewater with the help of aquatic plants is reported as an effective and low-cost treatment.[1] Cadmium at the molecular level, known as a calcium blocker in the plasma membrane, causes the displacement of calcium and zinc in specific proteins, in addition to causing oxidative stress generating severe damage to DNA and lipids.[32] Emergent plants such as *Typha domingensis* and *Phragmites australis* are used in wetlands because they can accumulate metal in their tissues due to adsorption through cell walls.[47]

## 7.2 MATERIAL AND METHODS

### 7.2.1 ANETHUM GRAVEOLENS

The seeds of *A. graveolens* were acquired in a commercial house specializing in selling plant material. The germination phase was carried out in 500 mL containers where the seeds were submerged in water for 14 days at temperature of 25°C. This methodology was implemented based on the data found by Villanueva,[50] to favor more excellent germination and growth of *A. graveolens* seedlings. After elapsed that time, the seedlings were transplanted in germination trays containing soil at 25°C a 24 h photoperiod and 1500 ($\mu$mol/m$^2$s$^1$) light intensity and maintained under the same conditions until their use in hydroponic systems.

### 7.2.2 HYDROPONICS

#### 7.2.2.1 CLOSED SYSTEM

Five *A. graveolens* plants per treatment were placed in plastic containers with their respective three replicates. To evaluate the effect of cadmium on the plants, different concentrations (0.6 mg/L, 1.5 mg/L, 3 mg/L, and 4.5 mg/L) were tested in a mixture with synthetic wastewater (8 g/L $NH_4Cl$; 2.5 g/L $NaNO_3$; 1.5g/L $K_2HPO_4$; 7.5 g/L $MgSO_4$•$7H_2O$; 30 g/L $NaHCO_3$; 5g/L $Na_2$•EDTA; 10 g/L $FeCl_3$•$6H_2O$; 4 g/L $MnCl_2$•$4H_2O$; 5 g/L $CuSO_4$•$5H_2O$; 5g/L $ZnSO_4$•$7H_2O$; 0.2 g/L CoC12•6H2O; 0.4 g/L Na2Mo$O_4$•$2H_2O$ y 0.24 g/L glucose). The experiment lasted for 18 days at 25°C, with a 24 h photoperiod of 1500 $\mu$mol/m$^{2}$·s light intensity. Periodic solution changes are maintained every 8 days, and the plant material is collected on the initial and final days for analysis. Chemical Oxygen Demand COD and metals were quantified from the solution, while for *A. graveolens,* metals, pigments, proteins, and catalase quantified.

#### 7.2.2.2 OPEN SYSTEM

A plastic container (55 cm × 43 cm × 16 cm) was used, which established as a sub-surface flow artificial wetland. An outlet hole was installed 5 cm from the base and contained a 5 cm bed of volcanic rock substrate. It was kept recirculating synthetic wastewater at a flow rate of 0.138 L/h and a fixed Cd concentration of 5 mg/L. The wetland was seeded with 150 *A. graveolens* plants and maintained for 18 days at 25°C, with a 24 h photoperiod at 1500 μmol/$m^2$ s light intensity. COD and metals were determined in solution at the end of the experiment, while for A. *graveolens*, metals, pigments, protein, and catalase.

### 7.2.3 PROTEIN AND ENZYME EXTRACTS

A phosphate buffer (0.1 mM) used with 200 mg of plant material for protein and enzyme extraction. It was macerated and then centrifuged at 10,000 rpm for 15 min at 4°C. The extracts were kept refrigerated until further use.

### 7.2.4 QUANTIFICATION OF PROTEIN, CATALASE (CAT), AND PIGMENTS

Protein determined by the Bradford method at 595 nm.[6] For catalase determination, 180 μL of 30 mM peroxide was used with 100 μL of the extract and quantified at 240 nm. For chlorophyll, 0.2 g of the plant material taken for pigment extraction with 80% acetone; then, the samples were filtered and quantified at 663, 647, and 470 nm. Chlorophyll and carotenoid concentrations (mg/L) calculated using the mathematical expressions Lichtenthaler[25] proposed.

### 7.2.5 ABTS QUANTIFICATION

The 0.5 g of plant material was weighed, macerated with methanol, and allowed to stand for 24 h in complete darkness. Tanticharakunsiri et al.[48] were used 25 μL of the sample with 1 mL of ABTS and then protected from light for 45 min, after which time it quantified at 734 nm.

### 7.2.6 DETERMINATION OF HEAVY METALS AND COD

Previously dried the samples at 105°C, 0.2 g of plant material was weighed, placed in crucibles, and incinerated at 500°C for 4 h. Subsequently, 2 mL of HCl were added and allowed to evaporate, and then 5 mL of 1 N HCl have added again. The determination carried out by atomic absorption spectroscopy. Finally, COD was determined using the APHA[3] method.

## 7.3 RESULTS AND DISCUSSION

### *7.3.1 CHEMICAL OXYGEN DEMAND*

Both systems showed similar behavior decrease in concentrations. This behavior may be due to the accumulation of organic matter within the systems, due to damage to the plant material, an increase of uncontrolled microbial biomass in the experiments, and a decrease in the volume of the solution due to evaporation and evapotranspiration, generating a higher concentration in the solution, as reported in other works. In the open system, there was the presence of fungal biomass growth on the rock surface, and the secretion of organic and inorganic compounds increased the COD concentration, which reflected in a decrease in the system's pH and a low removal rate 11%. Quipuzco[38] indicates that COD concentration increases organic matter accumulation due to the detachment of roots or biofilms that degrade in the medium and increase the concentration. Reduction in the concentration of organic and inorganic load has been observed in the first days, followed by an increase (Fig. 7.1) for the closed. Similar behaviors were in other wetlands reaching removal percentages higher than 50%.[41]

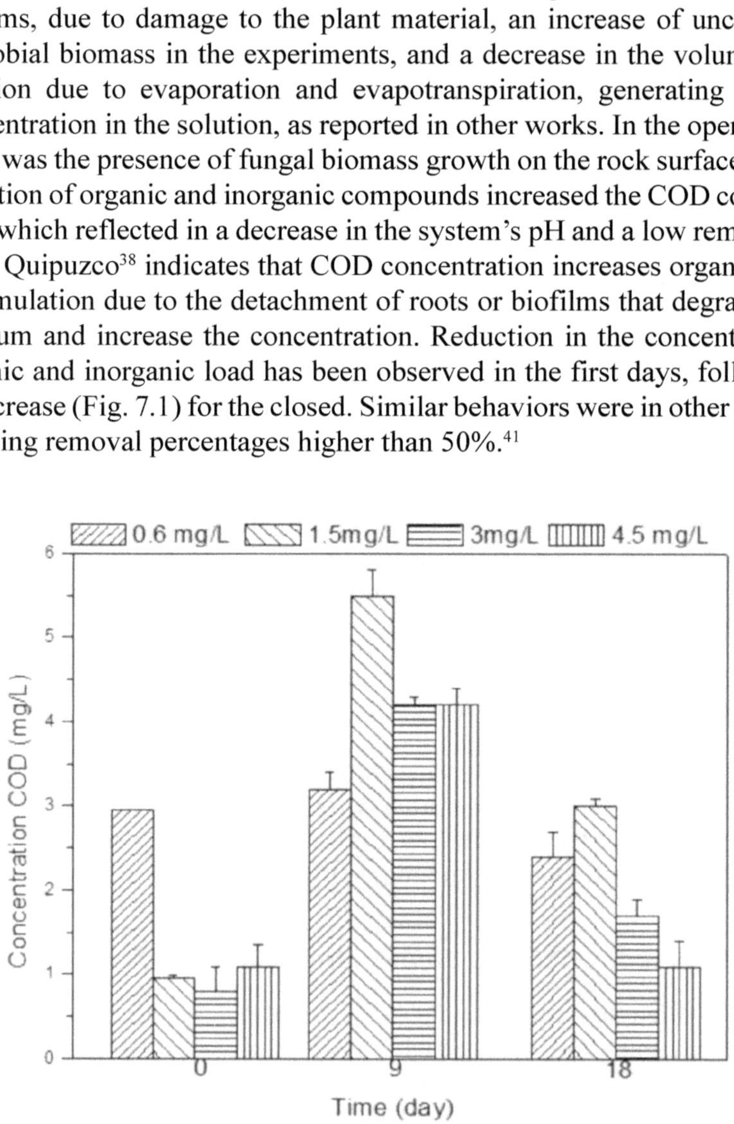


**FIGURE 7.1** Concentration of chemical oxygen demand in the closed system with the presence of *A. graveolens*.

### 7.3.2 CADMIUM

The Cd concentrations in the solution increased concerning the initial concentration for the closed system. The stabilization of the metal in the roots, but under stress conditions caused by the contaminant. The plant stops secreting organic acids that participate in these reactions, and the metal begins to mobilize again. In *A. graveolens,* the higher the cadmium concentration, the higher the concentration of the dissolved metal, the higher and the cadmium concentration was quantified (Fig. 7.2).

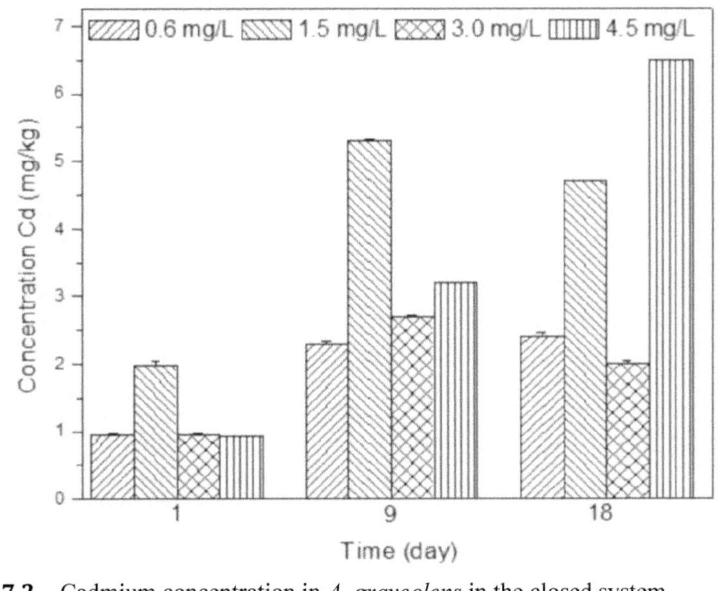


**FIGURE 7.2** Cadmium concentration in *A. graveolens* in the closed system.

For the open system, the cadmium concentration in the solution remained in a range of 0.12 mg/L–0.26 mg/L. The control plants of *A. graveolens* did not present cadmium at the beginning of the experiment, but 0.31 mg Cd/Kg quantified at the end of the experiment. It is important to note that cadmium accumulation in the plant is more significant in an open system than a closed system. This phenomenon may be mainly due to other interactions completely controlled within the open system, such as the development of microorganisms that promote the decrease in the soluble metal concentration and adsorption on active sites, among other phenomena. The behavior observed in plants is generated by cadmium toxicity as it alters the homeostasis of Fe, Zn, and Cu, increasing their concentration, producing membrane damage and oxidative damage to proteins.[40,42,43,45]

### 7.3.3 PIGMENTS AND ABTS

The pigment concentrations in the closed system increased concerning the control (Table 7.1). Among the treatments, 1.5 mg Cd/L has the highest chlorophyll A, B, total and carotenes, followed by 1.5 mg Cd/L, 0.6 mg Cd/L, and 4.5 mg Cd/L; these last two have similar concentrations, under similar conditions this behavior reported by Flores.[20] In all treatments, decreasing trends in the concentration of all pigments. The concentration of the quantified pigments shows significant differences concerning all the established treatments ($p < 0.05$).

**TABLE 7.1** Pigment Concentration in the Closed System.

| Cd (mg/L) | Cll A (mg/g) | Cll B (mg/g) | Cll total (mg/g) | Carotenoids (mg/g) |
|---|---|---|---|---|
| 0 | $0.86 \pm 0.21^{c}$ | $0.34 \pm 0.07^{c}$ | $2.41 \pm 0.56^{a}$ | $0.48 \pm 0.11^{c}$ |
| 0.6 | $0.50 \pm 0.18^{a}$ | $0.22 \pm 0.07^{a,b}$ | $1.45 \pm 0.50^{a,b}$ | $0.28 \pm 0.10^{a}$ |
| 1.5 | $0.75 \pm 0.14^{b,c}$ | $0.29 \pm 0.05^{c}$ | $2.08 \pm 0.39^{c,d}$ | $0.41 \pm 0.06^{c,d}$ |
| 3 | $0.68 \pm 0.10^{a,b}$ | $0.27 \pm 0.03^{a,b,c}$ | $1.91 \pm 0.26^{b,c}$ | $0.38 \pm 0.03^{b,c}$ |
| 4.5 | $0.50 \pm 0.16^{a}$ | $0.20 \pm 0.08^{a}$ | $1.42 \pm 0.47^{a}$ | $0.30 \pm 0.07^{a,b}$ |

*Note*: Mean ± standard deviation. The letters (a, b, c, d) represent the differences between treatments.

When comparing the chlorophylls and carotenoids of the treatments with the blank, observed that ClB had the lowest concentration when the Cd concentration increased; these results are similar to studies carried out with cadmium in *B. juncea L.*, where the content of chlorophyll B was decrease, but simultaneously increased its levels of anthocyanins as a defense substance.[2] Furthermore, other works with prolonged Cd treatments detected strong chlorosis,[35] which was also characteristic of the *A. graveolens*. The decrease of photosynthetic pigments can be induced by the presence of the metal, causing iron deficiency and inhibition of enzymes involved in biosistensis that generates a reduction of RubisCO production and/or modification of their activities.[15,19,23,27] Majdoub et al. (2017), reports lower concentrations in *A. graveolens* exposed to Zn.

For the open system, the concentrations of CllB and total Cl of the final uptake increased concerning the initial uptake, CllA and carotenoids decreased

(Table 7.2). When comparing both systems, the chlorophylls that showed differences were ClIA and ClIB, since in the closed system, ClIB decreased to the blank and in the open system, it increased to the initial uptake.

**TABLE 7.2** Pigment Concentration in the Open System.

| Time (day) | Cll A (mg/g) | Cll B (mg/g) | Cll total (mg/g) | Carotenoids (mg/g) |
|---|---|---|---|---|
| **0** | 0.47 ± 0.09 | 0.17 ± 0.03 | 1.29 ± 0.24 | 0.29 ± 0.05 |
| **18** | 0.44 ± 0.05 | 0.23 ± 0.07 | 1.34 ± 0.23 | 0.27 ± 0.04 |

*Note*: Mean ± standard deviation.

Some plants under stress modify their antioxidant activity as a defense mechanism. This behavior was reflected in *A. graveolens* when exposed to different cadmium concentrations. For concentrations similar to 1.5 mg Cd/L, the antioxidant activity remained stable (142.4 µmol TE/g); when the metal concentration increased, the activity decreased to 116.4 µmol TE/g (Fig. 7.3).

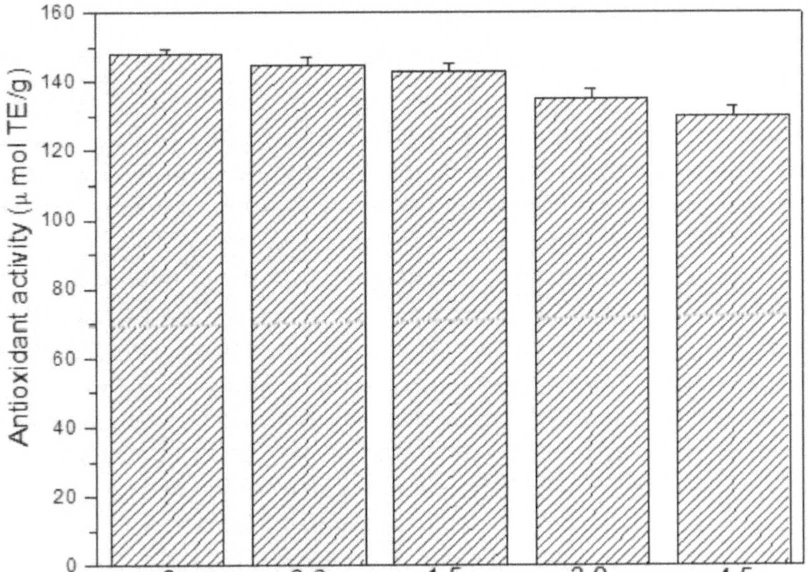


**FIGURE 7.3** Antioxidant activity of *A. graveolens* in the presence of different cadmium concentrations.

While in the open system, no statistical differences ($p < 0.05$) found in the antioxidant concentration over time (225 μmol TE/g). The differences in both systems could be due to the concentration of cadmium, which could regulate the production of oxidative stress substances. In the closed system at low metal concentrations, the antioxidant activity remains similar to plants not exposed to cadmium. *A. graveolens* leaf has been shown to have high antioxidant activity and percentage inhibition in lipid peroxidation of the linoleic acid system.[46] However, there is little information on other antioxidant compounds that *A. graveolens* may contain, which could help counteract the negative effects of metal exposure. The antioxidant compounds inhibit the growth of *Staphylococus aureus,* coliforms, and fungi.[7]

### 7.3.4 CATALASE

CAT activity increased as the concentration of the metal in the solution increased, and this is because the enzyme is one of the mechanisms through which organisms tolerate the presence of contaminants; it is the activation of defense systems that suppress or eliminate reactive oxygen species (ROS). CAT can convert $H_2O_2$ into $H_2O$ and $O_2$, reducing stress conditions in the plant that produces cellular homeostasis.[31] The presence of cadmium in the medium increased CAT activity. In the open system, CAT increased from 0.38 ± 0.004a UA/mg protein to 0.88 ± 0.12b UA/mg protein at the end. While in the closed system, the activity increased with increasing concentration of the metal in the solution (Fig. 7.4). In both cases, CAT increased as a metabolic response to counteract the negative effect of oxidation in *A. graveolens.* The activities of the antioxidant enzymes superoxide dismutase (SOD), glutathione reductase (GR), ascorbate peroxidase (APX), peroxidases (POD), and (CAT), responsible for cellular defense against ROS, experience reductions or increases in their activity depending on the concentration of Cd included in the growth medium, the duration of treatment, the type of tissue and the plant species involved.[5,45] Several studies on the mechanisms involved in the production of radicals in response to Cd have demonstrated the indirect production of ROS in different cellular compartments (plasma membrane, mitochondria, and peroxisomes) with NADPH oxidase being one of the primary sources of generation of these radicals.[18,33,40]

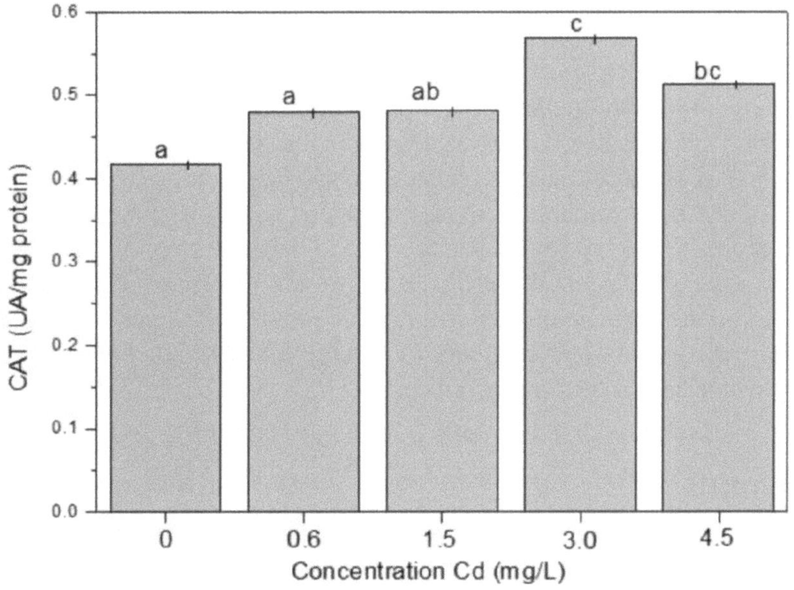


**FIGURE 7.4** Catalase enzyme activity quantified in *A. graveolens* exposed to cadmium.

## 7.4 FUTURE TRENDS

Using treated effluent as irrigation water is a viable option for the supply of liquid for the cultivation of different plant species. However, it is necessary to monitor water quality to evaluate the various phenomena that may occur due to the presence of a pollutant. It has been shown that plants are an essential tool to be used for ecosystem restoration as they have the capacity for stabilization, bioaccumulation, extraction, and volatilization. There is still a lack of information on the metabolic pathways that contaminants follow within plant species, as well as which genes are expressed in stress situations, which opens an area of knowledge to continue developing and applying it to have better proposals for bioremediation strategies. It is convenient to continue evaluating the response of plants with enzymatic or non-enzymatic variables to complement the new information generated. It is not be forgotten that the incredible plant biodiversity on the planet leaves us a long way to go to discover how plants behave in different environmental conditions. It is emphasized that the help of genetic engineering is increasing and will help clarify many mechanisms of action of plants.

## 7.5 CONCLUSION

The *A. graveolens* plant can accumulate cadmium, but the metal concentration in the solution defines the responses of antioxidant compounds, pigments, and enzymes. The flow of water in the system decreases the toxicity of the metal because there are other mechanisms to minimize the negative effect of the contaminant. In a closed system at higher concentrations, the plant decreases its antioxidant activity and increases its catalase activity. The toxicity of the metal and the concentration present in the ecosystem will define the defense mechanisms of the plants. However, it is necessary to study the effects and interactions further because not all plant species respond similarly to different stimuli.

## KEYWORDS


- **congo red**
- **modified bentonite**
- **RSM**
- **heavy metals**
- **Anethum graveolens**


## REFERENCES


1. Alzate, A. E. Influencia del pH y el tipo de macrofita en la remoción de materia orgánica y cadmio de un drenaje sintetico de mineria de carbono por medio de humedales contruidos de flujo subsuperficial horizontal a escala piloto. Master Dissertation, Universidad de Antioquia, España, 2015.
2. Ázcon, B. J.; Talón, M. *Fundamentos de fisiología vegetal*; 1er ed.; Universitat de Barcelona: España, 2008.
3. APHA. In *Standard Methods for the Examination of Water and Wastewater*, 21st Ed.; American Public Health Association/American Water Works Association/Water Environment Federation: Washington, DC, 2005; p 525.
4. Barceló, J.; Poschenrieder, C. Respuesta de las plantas a la contaminación por metales pesados. *Suelo y plantas.* **1992,** *2*, 345–361.
5. Benavides, M. P.; Gallego, S. M.; Tomaro, M. Cadmium Toxicity in Plants. *Braz. J. Plant Physiol.* **2005,** *17*, 21–34.

6. Bradford, M. M. A Rapid and Sensitive Method for the Quantitation of Microgram Quantities of Protein Utilizing the Principle of Protein-Dye Binding. *Anal. Biochem.* **1976,** *72* (1-2), 248–254.
7. Castro, D.; Pantoja, A.; Gomajoa, H. Evaluación in vitro de la capacidad antimicrobiana del aceite esencial de eneldo anethum graveolens como inhibidor del crecimiento de *Staphylococcus aureus*, coliformes y hongos presentes en la carne de trucha. *Revista de la Facultad de Medicina Veterinaria y de Zootecnia* **2017,** *64* (2), 44–51.
8. Clemens, S.; Palmgren, M. G.; Krämer, P. A. Long Way Ahead: Understanding and Engineering Plant Metal Accumulation. *Trends Plant Sci.* **2002,** *7*, 309–315.
9. Clemens S. Molecular Mechanisms of Plant Metal Tolerance and Homeostasis. *Planta* **2001,** *212*, 475–486.
10. Clemens S. Toxic Metal Accumulation, Responses to Exposure and Mechanisms of Tolerance in Plants. *Biochimie.* **2006,** *88*, 1707–1719.
11. Colangelo, E. P.; Guerinot, M, L. Put the Metal to the Petal: Metal Uptake and Transport Throughout Plants. *Curr. Opin. Plant Biol.* **2006,** *9*, 322–330.
12. Conn, S.; Gilliham, M. Comparative Physiology of Elemental Distributions in Plants. *Ann. Bot.* **2010,** *105*, 1081–1102.
13. Cui, X.; Lin, Q.; Liang, Y. Plant-Derived Antioxidants Protect the Nervous System From Aging by Inhibiting Oxidative Stress. *Front. Aging Neurosci.* **2020,** *12*, 209.
14. Demidchik, V.; Bowen, H. C.; Maathuis, F. J. M.; Shabala, S. N.; Tester, M. A.; White, P. J.; Davies, J. M. Arabidopsis Thaliana Root Non-selective Cation Channels Mediate Calcium Uptake and are Involved in Growth. *Plant J.* **2002,** *32*, 799–808.
15. Di Baccio; D.; Tognetti, R.; Minnocci; A.; Sebastiani, L. Responses of the Populus x euramericana clone I-214 to Excess Zinc: Carbon Assimilation, Structural Modifications, Metal Distribution and Cellular Localization. *Environ. Exp. Bot.* **2009,** *67* (1), 153–163.
16. Dos Santos Souza, C.; Grangeiro, M. S.; Lima Pereira, E. P.; Dos Santos, C. C.; da Silva, A. B.; Sampaio, G. P.; Ribeiro, D. D.; David, J. M.; Amaral, V. D.; Butt, A. M.; Costa, S. L. Agathisflavone, A Flavonoid Derived from Poincianella pyramidalis (Tul.), Enhances Neuronal Population and Protects Against Glutamate Excitotoxicity. *Neurotoxicology* **2018,** *65*, 85–97.
17. Duthie, G.; Crozier, A. Plant-Derived Phenolic Antioxidants. *Curr. Opin. Clin. Nutr. Metab. Care.* **2020,** *3*, 447–451.
18. Garnier, L.; Simon-Plas, F.; Thuleau, P.; Agnel, J. P.; Blein, J. P.; Ranjeva, R.; Montillet, J. L. Cadmium Affects Tobacco Cells by a Series of Three Waves of Reactive Oxygen Species that Contribute to Cytotoxicity. *Plant Cell Environ.* **2006,** *29*, 1956–1969.
19. Fernàndez-Martínez, J.; Zacchini, M.; Fernández-Marín, B.; García-Plazaola, J. I.; Fleck, I. Gas-exchange, Photo- and Antioxidant Protection, and Metal Accumulation in I-214 and Eridano Populus sp. clones Subjected to Elevated Zinc Concentrations. *Environ. Exp. Bot.* **2014,** *107*, 144–153.
20. Flores, C. M. Caracterización de los mecanismos de respuesta antioxidante de Medicago sativa a metales pesados. PhD Dissertation, Universidad Autónoma de Madrid, España, 2013.
21. Jiménez, C. B. E. In *La contaminación ambiental en México: Causas, efectos y tecnología apropiada México*, 1er Ed.; Limusa: México, 2001; p 926.
22. Kalve, S.; Sarangi, B. K.; Pandey, R. A.; Chakrabarti, T. Arsenic and Chromium Hyperaccumulation by an Ecotype of Pteris Vittata–Prospective for Phytoextraction from Contaminated Water and Soil. *Curr. Sci. India.* **2011,** *100*, 888–894.

23. Islam, F.; Yasmeen, T.; Riaz, M.; Arif, M. S.; Ali, S.; Raza, S. H. Proteus Mirabilis Alleviates Zinc Toxicity by Preventing Oxidative Stress in Maize (Zea mays) Plants. *Ecotoxicol. Environ. Saf.* **2014,** *110*, 143–152.
24. Koptsik, G. Problems and Prospects Concerning the Phytoremediation of Heavy Metal Polluted Soils: A Review. *Eurasian Soil Sci.* **2014,** *47*, 923–939.
25. Lichtenthaler, H. K. Chlorophylls and Carotenoids: Pigments of Photosynthetic Biomembranes. *Plant Cell Membranes* **1987,** 350–382.
26. Mansoor, S.; Wani, O. A.; Lone, K. F.; Manhas, S.; Kour, N.; Alam, P.; Ahmad, A.; Ahmad, P. Reactive Oxygen Species in Plants: From Source to Sink. *Antioxidants* **2022,** *11* (2), 225.
27. Marichal, A.; Dallal, S.; Ouerghemm, S.; Houcine H.; Casabianca, H.; Hosni, K. Responses of Nigella sativa L. to Zinc Excess: Focus on Germination, Growth, Yield and Yield Components, Lipid and Terpene Metabolism, and Total Phenolics and Antioxidant Activities. *J. Agric. Food Chem.* **2016,** *64* (8), 1664–1675.
28. Marques, A. P.; Rangel, A. O.; Castro, P. M. Remediation of Heavy Metal Contaminated Soils: Phytoremediation as a Potentially Promising Clean-Up Technology. *Crit. Rev. Env. Sci. Technol.* **2009,** *39*, 622–654.
29. Metcalf and Eddy. In *Wastewater Engineering, Treatment and Reuse*, 4th Ed.; McGraw-Hill Inc., 2004; pp 253–587.
30. Mukhopadhyay, S.; Maiti, S. Phytoremediation of Metal Mine Waste. *Appl. Ecol. Environ. Res.* **2010,** *8* (3), 207–222.
31. Murillo, E.; Lombo, O.; Tique, M.; Méndez, J. Potencial Antioxidante de la *Bauhinia KalbreyeriHarms. (FABACEAE). Inf. Tecnol.* **2007,** *8* (6), 65–74.
32. Navarro-Aviñó, J.; Aguilar, A. I.; López, J. Aspectos bioquímicos y genéticos de la tolerancia y acumulación de metales pesados en plantas. *Ecosistemas* **2007,** *16* (2), 10–25.
33. Olmos, E.; Martinez-Solano, J. R.; Piqueras, A.; Hellin, E. Early Steps in the Oxidative Burst Induced by Cadmium in Cultured Tobacco Cells (BY-2 line). *J. Exp. Bot.* **2003,** *54*, 291–301.
34. Oshaghi, E. A.; Khodadadi, I.; Mirzaei, F.; Khazaei, M.; Tavilani, H.; Goodarzi, M. T. Methanolic Extract of Dill Leaves Inhibit AGEs Formation and Shows Potential Hepatoprotective Effects in CCl4 Induced Liver Toxicity in Rats. *J. Pharm.* **2017,** 1–9.
35. Paradiso, A.; Berardino, R.; de Pinto, M.; di Toppi, L.; Storelli, M., Tommasi, F.; De Gara, L. Increase in Ascorbate–Glutathione Metabolism as Local and Precocious Systemic Responses Induced by Cadmium in Durum Wheat Plants. *Plant Cell* **2008,** 362–374.
36. Peng, K.; Luo, C.; You, W.; Lian, C.; Li, X.; Shen, Z. Manganese Uptake and Interactions with Cadmium in the Hyperaccumulator Phytolacca Americana L. *J. Hazard Mater.* **2008,** *154*, 674–681.
37. Peña Varón, M.; Van Ginneken, M.; Madera P., C. A. Humedal de flujo subsuperficial una alternativa natural para el tratamiento de aguas residuales domésticas en zonas tropicales. *Ingeniería y competitividad* **2003,** *5*, 1 25–35.
38. Quipuzco, E. Evaluación de comportamiento de dos pantanos artificiales instalados en serie con Phragmites australis para el tratamiento de agias residuales domésticas. *Rev. Inst. Investig. Esc. Mnas Metal Cienc. Geogr.* **2002,** *5*, 52–57.
39. Ramos, I.; Esteban, E.; Lucena, J. J.; Gárate, A. Cadmium Uptake and Subcellular Distribution in Plants of Lactuca sp. Cd-Mn Interaction. *Plant Sci.* **2002,** *162*, 761–767.

40. Romero-Puertas, M. C.; Rodríguez-Serrano, M.; Corpas, F. J.; Gómez, M.; del Río L. A.; Sandalio, L. M. Cadmium Induced Subcellular Accumulation of $O^{2-}$ and H2O2 in Pea Leaves. *Plant Cell Environ.* **2004,** *27*, 1122–1134.
41. Romero, M.; Colìn, A; Sánchez, E. Tratamiento de aguas residuales por un sistema piloto de humedales artificiales: evaluación de la remoción de la carga orgánica. *Rev. Int. Contam. Ambient.* **2009,** *25* (3), 157–167.
42. Romero-Puertas, M. C.; Palma, J. M., Gómez, M.; del Río, L. A.; Sandalio, L. M. Cadmium Causes the Oxidative Modification of Proteins in Pea Plants. *Plant Cell Environ.* **2002,** *25*, 677–686.
43. Sagardo, C. R. Estudio de la homeostasis de Zn y Cd en plantas superiores. Zaragoza. PhD Dissertation, Consejo Superior de Investigaciones Científicas, España, 2011.
44. Sakakibara, M.; Ohmori, Y.; Ha, N. T. H.; Sano, S.; Sera, K. Phytoremediation of Heavy Metal-Contaminated Water and Sediment by Eleocharis acicularis. *Clean Soil Air Water* **2011,** *39*, 735–741.
45. Sandalio, L. M.; Dalurzo, H. C.; Gomez, M.; Romero-Puertas, M. C., del Río, L. A. Cadmium-Induced Changes in the Growth and Oxidative Metabolism of Pea Plants. *J. Exp. Bot.* **2001,** *52*, 2115–2126.
46. Sebenem, S. I.; Sgiroglu, A. Potencial antioxidate de diferentes extractos de hojas de eneldo (Anethum Graveolens L-). *Int. J. Food Prop.* **2011,** *14* (4), 894–902.
47. Sheoran, A.; Sheoran, V. Heavy Metal Removal Mechanism of Acid Mine Drainage in Wetlands: A Critical Review. *Miner. Eng.* **2006,** *19*, 105–116.
48. Tanticharakunsiri, W.; Mangmool, S.; Wongsariya, K.; Ochaikul, D. Characteristics and Upregulation of Antioxidant Enzymes of Kitchen Mint and Oolong Tea Kombucha Beverages. *J. Food Biochem.* **2021,** *45*, e13574.
49. Turan, M.; Nikerel, E.; Kaya, K.; Kitir, N.; Gunes, A.; Mokhtari, N. E. P.; Tüfenkçi, S.; Çimrin, K. M.; Karaman, M. R. Enzyme Dynamic in Plant Nutrition Uptake and Plant Nutrition, Cap. 10. 1st Ed.; In *Enzyme Inhibitors and Activators*; Intechopen, 2017. https://www.intechopen.com/books/5540
50. Villanueva, M. C. Eliminación de latencia en semillas de eneldo bouque (*Anethum graveolens*) y arugula astro (eruca vesicaria), utilizando tratamientos físicos químicos y mecánicos en laboratorio e invernadero. PhD Dissertation, Universidad Autónoma Agraria, Antonio Narro, México, 2008.
51. Wang, J.; Feng, X.; Anderson, C. W.; Xing, Y.; Shang, L. Remediation of Mercury Contaminated Sites–A Review. *J. Hazard Mater.* **2012,** *221*, 1–18.
52. WHO, World Health Organization, In *Guidelines for Drinking-Water Quality*, 3rd Ed.; 2006; pp 6–125.
53. Yan, A.; Wang, Y.; Tan, S. N.; Mohd Yusof, M. L.; Ghosh, S.; Chen, Z. Phytoremediation: A Promising Approach for Revegetation of Heavy Metal-Polluted Land. *Front. Plant Sci.* **2020,** *11*, 359.

# CHAPTER 8

# Cartography Models to Identify Areas Vulnerable to the Environmental Interactions Intervening in the Degradation of Pesticides: A Case Study in Huasteca Potosina, México

MELVA GUADALUPE HERRERA-GODINA[1] and
JESÚS ARTURO MUÑIZ-JAUREGUI[2]

[1]*Salud Pública, Universidad de Guadalajara, Guadalajara, Jalisco, México*

[2]*Escuela Nacional de Estudios Superiores, Universidad Nacional Autónoma de México, Juriquilla, Querétaro, México*

## ABSTRACT

The agricultural production of the Huasteca Potosina has repeatedly been affected by the presence of the locust plague, resulting in a constant use of pesticides such as Malathion, Parathion, Cypermethrin, and Fipronil. When facing risks of imminent losses, there may be an excessive use of pesticides, or various pesticides may be applied on the same surfaces, without knowing the effects this may have on the environment and vice versa. The time needed for their degradation and the products resulting from pesticide metabolization are influenced by environmental factors. The use of geographical information systems to analyze the interaction of pesticides with the environmental features during their break down process allows to obtain map models through which we can detect areas that present elements that hinder the metabolization of pesticides, or, on the contrary, facilitate

Sustainable Agricultural Production Systems: Traditional and Modern Perspectives.
Romeo Rojas, Guillermo Cristian Guadalupe Martínez-Ávila, Julia Mariana Márquez-Reyes, and Ma del Carmen Ojeda-Zacarías (Eds.)

their break down into a more toxic and persistent component. The results of this study were, on the one hand, a series of models to identify both areas that are more vulnerable to a faster degradation and those that present convergence of elements that induce a slower degradation of Fipronil; on the other, the creation of other models for the detection of areas that are prone to the formation of more toxic derivatives from the decomposition of Fipronil.

## 8.1 INTRODUCTION

The Huasteca region covers the northern part of the state of Veracruz and smaller portions of the western edges of the states of San Luis Potosí, Hidalgo, and Puebla. This area has been subjected to heavy deforestation and changes in soil use, especially during the decades between 1980 and 2000, a period of intensified destruction of forests and jungles due to projects such as Pujal-Coy .[1] These programs favored extensive livestock production and irrigated farming.[1] According to Reyes et al.,[12] since the start of the Pujal-Coy project in 1973 and until 1985, the area presented a deforestation rate of 11%, causing a 75% loss of the forest cover. The area of the Huasteca Potosina is located in the coastal plain of the Gulf of Mexico, bordered to the north by the Tamesí River, to the south by the foothills of the Trans-mexican Volcanic Belt or Neovolcanic Cordillera, up to the Punta de Villa Rica, and to the west by the Sierra Madre Oriental.

### *8.1.1 HYDROGRAPHY OF THE HUASTECA POTOSINA*

There are three hydrological regions (HR), present in S, the Pánuco Hydrological Region (RH26), a region that covers the entire Huasteca Potosina, to the northwest the El Salado Hydrological Region (RH37), and to the southeast a small portion of the Lerma-Santiago Hydrological Region (RH12). In the Huasteca Potosina area, the Sierra Madre Oriental marks two perfectly distinct zones, one at the southeast with warm and semi-warm, humid and sub-humid climates, a region characterized by abundant rainfall that increases the flow of the Santa María, Moctezuma, and Tampaón rivers.

[1]The Pujal-Coy project corresponds to the irrigation plan for the lower basin of the Pánuco River, which was presumed to be the largest risk district in Latin America, with an area of 720,000 ha, of which 288,000 would be part of the Pujal-Coy district, distributed along the banks of the Tampaón, Moctezuma, and Tamesí rivers. The aim of the project was the transition from extensive livestock production to irrigated agriculture, involving the redistribution of land, the construction of hydraulic infrastructure, the increase in the region's productivity, and incentives to attract farmers from other parts of Mexico. The initial phase of the project was carried out in the municipalities of Ébano and Tamuín, in San Luis Potosí, and in the municipality of Pánuco in Veracruz, resulting in the transformation of 72,000 ha of territory.[12]

On the other hand, a dry and semi-dry climate predominates in the western zone, where water currents are intermittent.

### 8.1.2 CLIMATE IN THE HUASTECA POTOSINA

According to the Síntesis de Información Geográfica and the Carta Estatal de Climas del INEGI,[7] which uses Köppen's classification, modified by Enriqueta García, the predominant climates in the Huasteca Potosina are Group A warm climates and Group C temperate climates.

Warm climates subgroup A is characterized by an average annual temperature of over 22°C and an average annual temperature of the coldest month of over 18°C. The extreme northeast of the Huasteca Potosina presents an $Aw_0$ climate, which has the lowest humidity of all the warm sub-humid types with summer rainfall in Group A, and less than 60 mm of precipitation in the driest month. Characteristically, 5–10% of the total annual precipitation in $Aw_0$ climates occurs in the winter.

Humidity increases toward the west, in the municipality of Ciudad Valle we find an $Aw_1$ climate, and further west the climate changes to $Aw_2$, this being the most humid of the warm sub-humid climates. A little further west, the temperature decreases and Temperate C climates predominate, especially those of the semi-warm climates subgroup (A)C, which are characterized by an average annual temperature of over 18°C and an average temperature in the coldest month that ranges between 8 and 18°C. Of this group, the predominant climate in the Huasteca Potosina is the semi-warm humid climate with abundant rainfall in summer, (A)C(m)(w), with less than 42 mm of precipitation in the driest month and less than 5% of the annual precipitation in the winter.

The most abundant rainfall in the entire state of San Luis Potosi occurs in the Huasteca Potosina, ranging from 3000 to 3500 mm. The northeast is the driest portion, where rainfall ranges between 800 and 1000 mm. The Huasteca Potosina also has the highest temperatures, with an annual regional average of 24–26°C.

### 8.1.3 EDAPHOLOGY OF THE HUASTECA POTOSINA

The Huasteca Potosina presents no dominant soil unit, since there are at least six of them in its territory, including Regosol, Leptosol, Vertisol, Rendzina, Phaeozem, and Luvisol soils. The region includes soils that are rich in organic matter, supporting jungle vegetation communities, oak and pine forests, grasslands, and submontane scrub.

## 8.2 THE LOCUST PLAGUE

### *8.2.1 THE LOCUST PLAGUE AND ITS GEOGRAPHICAL DISTRIBUTION*

The environmental characteristics of the Huasteca Potosina favor the presence of locust and grasshopper pests. Permanent locust breeding areas are characterized by altitudes below or equal to 100 masl; these regions must have two well-defined climatic seasons, the dry season and the rainy season, and the temperature must oscillate between 28 and 32°C throughout the year.[9] Locusts (*S. piceifrons*) belong to the order of Orthoptera and have egg, nymph, and adult or imago stages. Eggs are oviposited in masses known as *otecas*, which are deposited in the soil. During its biological cycle, the female grasshopper buries its eggs at a depth of 1.5–5 cm in the soil, in meadows and pastures, vacant lots, roadsides, canals, and drains. Eggs begin to hatch at the beginning of the rainy season, although hatching may be delayed in dry years.[8] At their young stage, locusts are known as nymphs and go through five or six stages before they reach sexual maturity, when their color changes to a yellowish brown.

The geographical distribution of the locust pest throughout the Huasteca Potosina spreads across the cultivated areas of the 20 municipalities that make up the region. However, the records show that since 1960 the pests have been more frequent in the municipalities of Ébano, Tamuín, and San Vicente Tancuayalab, specifically, in 1963, 1964, 1965, 1989, 1990, 1991, 1996, 1997, and 1998, both the locusts and the grasshoppers attacked. Since 2000, these pests have been attacking various crops; for example, in the municipality of Tamuín they affected pastures and sugarcane crops, and in 2002, also in Tamuín, they attacked citrus crops. In the first decade of the 2000s, in the municipality of Ébano, the pests affected fields of corn, cotton, soybean, sorghum, and citrus, and also pastures. In San Vicente, the crops that were most affected by the locusts and the grasshoppers were the grasslands.[2]

### *8.2.2 LOCUST PEST CONTROL PRACTICES*

Locust control measures are generally taken and implemented during the months of August to September, when the locust is at its adult stage. During this period, the locust forms highly voracious and fast-moving swarms.

---

[2]Data drawn from the database entitled "Chemical Control 2" of the project "Proposal of a methodology to identify areas vulnerable to drought and pests that affect agricultural activities in the Huasteca Potosina, supported by satellite images, geographic information systems, multivariate analysis, and a climate simulation model," directed by Dr. María Guadalupe Galindo, Academic of the Universidad Autónoma de San Luis Potosí. Project in coordination with SAGARPA and CONACYT.

The National Institute of Forestry, Agricultural and Livestock Research (INIFAP) recommends that, to effectively manage the pest, actions should be taken at three stages according to the developmental stages of the locust. The aim of this type of containment is to avoid the formation of swarms in the adult phase, that is why the first stage goes from June to August, when the locust is a first generation nymph; the second stage is to attack the second generation nymphs, during the months of September to November, and finally, the third stage occurs when the locust is in its imago phase, from December to June.[9] Crop control is a method proposed for the management of the locust pest without the use of chemical products. This method consists in harrowing the lands up to at least 30 cm deep, in order to expose the eggs to the weather and leave them at the mercy of their natural enemies. To prevent the formation of pests, the phytosanitary process must be carried out in the gregarious areas.

Another alternative to contain the locust is the use of mycoinsecticides, a control option considered safer. Countries such as Great Britain and Australia were the driving forces in the development of biological insecticides to combat the locust, developing Green Muscle and Green Guard, whose active ingredient is a hyphomycete fungus of the genus *Metarhizum*.

When applying biological insecticides in the Huasteca Potosina, the best results have been obtained with those whose active ingredient is the *Metarhizium anisopliae* fungus, of the *acridum* variety, in concentrations of $1.2 \times 10^{12}$ conidia per hectare; its effectiveness has been of 100% in 10 days after its application.

With regard to chemical control practices in the Huasteca Potosina, insecticides of various chemical groups have been used, for example, organophosphate insecticides such as Malathion and Parathion. Pyrethroid chemicals have also been used, and, in recent years, Fipronil has been applied more frequently.

### 8.2.2.1 FIPRONIL, COMPOSITION AND INTERACTIONS WITH THE ENVIRONMENT

INIFAP[9] evaluated the effectiveness of several insecticides, their safety and toxicity interval, their toxicological management of resistance, and the costs of use and management. With the data obtained, they concluded that Malathion and Fipronil are the most convenient insecticides, due to their effectiveness, residuality, costs, and low environmental impact due to their selectivity. In this process, it was determined that the most appropriate dose for application would be 5 g of active ingredient per hectare (g a.i/ha) for Fipronil and 1000 g a.i/ha for Malathion.

Fipronil has become widely known as an insecticide that acts selectively by inhibiting the passage of ions through the GABA channel, which is a receptor of the nervous system that controls the activity of neurons by regulating the passage of negative chloride ions. Gunasekara et al.,[6] mentions that Firponil shows selective toxicity toward insects (where the $LC_{50}$ can be less than 24.8 nM or ≈11.7 mg/L) by having a strong affinity to bind more readily to the GABA channel in insects than in mammals.

#### *8.2.2.1.1 Chemical Origin of Fipronil*

Fipronil belongs to the group of insecticides called pyrazoles, specifically to the phenylpyrazoles class. Pyrazoles comprise both single aromatic ring organic compounds of the heterocyclic series characterized by a 5-element ring structure composed of three carbon atoms and two adjacent position nitrogen atoms, and the substituents of the parent compound.

Pyrazoles are produced synthetically by the reaction of alpha aldehydes, beta-unsaturated with hydrazine, and subsequent hydrogenation. The phenylpyrazole group of insecticides consists of Acetropole, Ethiprole, Pyrafluprole, and Fipronil.

Fipronil is an insecticide that was discovered and developed by Rhône Poulenc Agro in 1987[3,11]; it was introduced in the market by 1993 and registered as a pesticide in the United States by 1996; Fipronil is used to control mites and insects.[16]

#### *8.2.2.1.2 Interactions of Fipronil with the Environment*

One of the chemical characteristics of Fipronil is its potential to adhere to soil particles, since its organic carbon coefficient (Koc) corresponds to 803 mL/g of organic carbon and is within the range of 100–1000 mL/g of organic carbon. This range corresponds to the class of moderately adherent to the soil, so it can be deduced that due to this affinity to adhere to the organic carbon particles, there will be no filtration of Fipronil toward phreatic mantles, unless the areas where it has been applied present fractures that facilitate its filtration.

Fipronil, according to Henry's Law constant, has a low capacity to volatilize, since its constant corresponds to $3.7 \times 10^{-5}$ atm $m^3$/mol, whose value is within the low volatility range that goes from $3.7 \times 10^{-7}$ to $1 \times 10^{-5}$ atm $m^3$/

mol, so the probabilities of finding Fipronil airborne particles at the time of application or later are low.

The Kow constant represents an important characteristic of Fipronil to identify whether it is fat-soluble or water-soluble; this constant corresponds to the octanol-water partition (Kow), and the logarithm of this variable determines whether the compound has the capacity to bioaccumulate. Fipronil has a logarithm of the Kow constant of 4.01, a significant value for the compound to bioaccumulate.[3] When interacting with diverse characteristics of the space where it is found, Fipronil can undergo modifications in its chemical composition, by means of degradation or biodegradation processes, and thus give way to other chemical products. Fipronil amide is generated by hydrolysis in water and biodegradation in soil. Fipronil is transformed to Fipronil sulfide by biodegradation in soil, reduction in soil and photolysis in foliage; it is broken up to Fipronil sulfone by oxidation in soil, and Fipronil desulfinyl is produced by photolysis in soil and foliage.

When interacting with the environment, the compounds resulting from the degradation processes of Fipronil present different conditions of adherence capacity to soil, liposolubility or hydrosolubility, and volatilization capacity (see Table 8.1).

**TABLE 8.1** Physicochemical Properties of Substances Derived from Fipronil.

| Compound | Koc (in mL/g of organic carbon) | Henry's lawa (in Pa) | Log Kowb |
|---|---|---|---|
| Fipronil sulfon | 4209 | 7.6 × 10 – 7 | 3.8 |
| Fipronil sulfide | 2719 | 2.3 × 10 – 6 | 3.7 |
| Fipronil desulfinyl | 1290 | 0.4 × 10 – 5 | 3.4 |
| Fipronil amide | 166 | | |

[a]Calculated at 25°C

[b]Estimated at 20°C

*Source*: Own elaboration with FAO data, 2001.

The products derived from the degradation of Fipronil show physical and chemical characteristics that are very similar—if not the same—as those of its parent compound, since they also have the capacity to bioaccumulate, exhibit low volatility, and moderate adherence to soil particles, with the exception of Fipronil amide, a compound that has a much lower Koc constant, and is, therefore, expected to have greater mobility in soil, increasing the possibilities of ending up in phreatic mantles.

#### *8.2.2.1.3 Environmental Variables Interfering With the Time of Degradation and the Fate of Fipronil*

The behavior, fate, and some of the effects caused by Fipronil are the result of a complex interaction between the chemical properties of the insecticide itself and the processes governing the transport, degradation, capture, and response of organisms.[5] They are also influenced by environmental conditions such as temperature, air movement, soil moisture, and the chemical properties of Fipronil.

#### *8.2.2.1.4 Environmental Variables Interfering With the Degradation Time and Products of Fipronil in Soil*

Several studies have been conducted to analyze the elements that intervene in the persistence and degradation of Fipronil, focusing on factors such as organic matter, moisture, amount of clay, pH, soil use, organic carbon content, and type of organic carbon. These elements are closely related, for example, a certain crop will thrive on that soil depending on the pH and textural class. The moisture storage capacity of the soil will also depend on the textural class.

When evaluating the potential of organic matter (OM) and clay content in soil for the adsorption of pesticides, Sheng et al.,[14] mentions that, by considering OM in soil as the dominant adsorbent phase for organic pollutants and insecticides in the soil–water system, the potential contribution of soil minerals in adsorption would be ignored. This effect occurred in the adsorption of Atrazine by different minerals of the montmorillonite clay group at different loadings; the clay with the highest loading was the least adsorbent, so that a higher density loading would result in a loss of adsorption fields. These results indicate that for certain insecticides, expandable clays have the potential to be equally or more absorbent when compared to OM for insecticide retention in soil. Liu and Song[17] add that not only the amount of organic matter present in the soil must be considered to determine the adsorption of pesticides, but also the characteristics of OM itself, as they found that the ancient natural organic matter (NOM) of the surface shows a higher affinity to adsorb organic pollutants than young soils. Adsorption increases with increasing aromaticity and decreasing polarity. Thus, the adsorption coefficient of organic contaminants cannot be accurately predicted without considering the characteristics of the natural OM.

In studies conducted by Oliver et al.,[10] to inquire into the role of pH and other soil characteristics in the adsorption of insecticides, he found that there were no significant differences in the adsorption of Thiacloprid, Diuron, and Imidacloprid on tropical soils in the Philippines and temperate soils in Australia. Adsorption was not even significantly related to soil characteristics, in particular organic carbon (OC) content, clay content, and pH. When the data were classified by land use, adsorption of the three chemicals was highly correlated with OC for soils in Philippine rice paddies and in temperate soils. This study shows that soil origin (temperate versus tropical) has no significant effect on adsorption behavior, however, soil use did significantly affect the adsorption behavior of the three insecticides.

Similarly, it has been shown that soil use plays an important role in the time it will take for Fipronil to break down, as well as in the products that will derive from this degradation. In this sense, the values reported by Connelly[3] for the dissipation time of Fipronil in soils without vegetation ranged from 33 to 75 days, while in soils with vegetation it was 12–15 days. This duration is similar to that reported by the U.S. Environmental Protection Agency (1996), which states that the half-life for Fipronil is 1.1–1.5 months in soils without vegetation and 0.4–0.5 months for soils with vegetation.

The type of crop, in conjunction with moisture conditions, not only interferes with the degradation time of the insecticide, but also with the formation of degradation by-products, as happened in the cultivation of rice, where Desthiofipronil was formed in water and Fipronil sulfide in soil, possibly as a result of oxygen exclusion in soil due to flooding, while Fipronil sulfone and Fipronil amide were less abundant.[4]

Studies by Ying and Kookana,[18] also showed that moisture content has a strong effect on the rate of degradation and the type of degradation product. High moisture content (above 50%) favored the formation of Fipronil sulfide, while low moisture content (below 50%) and well-aerated conditions favored the formation of Fipronil sulfone by oxidation.

Studies conducted by Gunasekara et al.[6] determined that the half-life of Fipronil is affected by the type of soil in which it is applied, for example in loam soil the half-life was 34 days, while in sandy loam soils, the half-life ranged from 30 to 194 days.

Interest in colloid behavior has increased with recent studies showing how organic and inorganic contaminants are transported by colloid particles.[13] Colloids not only retain plant nutrient ions, but also water molecules, biomolecules, viruses, toxic metals, insecticides, and hosts of other minerals and organic substances.[2]

## 8.3 METHODS AND MATERIALS

The study area is comprised of three of the 20 municipalities that make up the Huasteca Potosina. Together, the municipalities of Ébano, Tamuín, and San Vicente Tancuayalab cover an area of 3058.37 km$^2$, a region that is continually and highly affected by the locust plague, demanding the use of chemical products to control it. When in contact with the environment, these insecticides can modify their chemical composition, and therefore, alter their route of exposure.

In the development of cartographic models to identify the interactions of insecticides with the environment, not only were considered the environmental variables that accelerate the metabolization process, but also those that prolong the half-life of Fipronil in the environment, as well as their influence on the generation of compounds resulting from metabolization. Temperature, soil texture, pH, colloid content, soil use, and humidity were included as variables that influence the time of degradation, which, depending on their observed value, accelerate or slow down the metabolization process.

Data on pH, textural class, and colloids were obtained from samples taken from the study area and analyzed in the laboratory. Soil profiles were not taken for sample collection because, according to the literature reviewed, there are no records of Fipronil exceeding a depth of 5–6 inches, therefore, the samples were taken from the surface of the soil.

The process for pH measurement consisted in crushing part of the sample material in a mortar and sieving it through 10 mesh sieve, then 10 g were weighed and mixed with 20 mL of distilled water. The mixtures were left to stand for 30 min, during which the samples were shaken twice. After 30 min, they were subjected to the potentiometer.

To identify the textural class of the samples, 50 g of each sample was pulverized and mixed with 100 mL of sodium hexametaphosphate and 40 mL of hydrogen peroxide. They were left to stand for 10 min and then shaken in a blender for 3 min. The mixture was poured into a Bouyoucos cylinder, and water was added to 1100 mL. The next step was to shake the mixture for 1 min and let it stand for 40 s to take the first temperature measurement and the soil–water density ratio with the Bouyoucos densimeter. After 2 h, the measurements were taken again to calculate the sand, clay, and silt content.

The calculation began by making a temperature correction; if in the first measurement the temperature exceeded 20°C, the exceeding degrees were multiplied by 0.36 and added to the first reading of the Bouyoucos densimeter. Conversely, if the first reading was below 20°C, these missing degrees were multiplied by 0.36 and the result was subtracted from the initial measurement. The resulting amount was multiplied by 2 and subtracted from

100% to obtain the sand content in percent. The procedure prior to the second reading was repeated to obtain the clay content in percent. To calculate the silt content of the sample, the percentage of sand and clay was added, the sum was subtracted from 100 percent, and the result was the silt content.

Example of the calculation performed for sampling point 476 in corn fields belonging to the El Laurel ejido in the municipality of San Vicente:

First reading: Temperature = 23°C; Bouyoucos densimeter = 43°B

Correction for temperature:

(23°C – 20°C) = (3°C) × (0.36) = 1.08
43°B + 1.08 = 44.08
44.08 × 2 = 88.16
First reading = (100 – 88.16) = 11.84% sand
Second reading: Temperature = 23°C; Bouyoucos densimeter = 24°B
Correction for temperature:
(23°C – 20°C) = (3°C) × (0.36) = 1.08
24°B + 1.08 = 25.08
25.08 × 2 = 50.16
Second reading = 50.16% clay
Calculation of silt content
Silt = 100 – (% sand + % clay) = 100 – (11.84 + 50.16) = 100 – 62 = 38% silt.

At the end of the texture analysis, and due to the high clay content, the samples were left to rest for 15 days to determine the existence of colloids. After 15 days, in more than 80% of the samples there was still material in suspension in the Bouyoucos test tubes. Therefore, material from the samples was prepared to be exposed to hydrochloric acid, since the greater the calcareous material, the greater the reaction to hydrochloric acid.

The normalized difference vegetation index (NDVI) was used to determine soil moisture. According to Tieszen et al.,[15] the NDVI is a calculation based on several spectral bands of photosynthetic output in a satellite image pixel. Some of the variables that can be derived from this index are leaf water content, net vegetation productivity, leaf chlorophyll content, phenological dynamics, and potential evaporation, among others. For this study, the formula proposed by Chuvieco (1996) was used to obtain the NDVI.

$$NDVI = 0 + C4 - C3/C4 + C3 \times 256$$

Where: C4 is band 4 (close infrared) and C3 is band 3 (visible red) of our original image, while value 256 is the display levels, that is, the grey levels at which the result will be displayed. We choose 256 to match the display levels of the original 8-bit satellite image (0–255).

### 8.3.1 GEOGRAPHIC INFORMATION SYSTEMS

GIS were used as a tool for the development of a cartographic model that highlights the interactions between Fipronil and the environment where it is applied, identifying the effect that environmental characteristics exert on the time and manner in which Fipronil breaks down.

The spatial distribution of the variables was performed with GIS support through interpolation of the sampled data. Interpolation tools create a continuous surface of discrete sample values, such as elevation or chemical concentration. There are many interpolation tools, and each has parameters that influence the resulting surfaces. These tools are based on the analysis of surface properties by neighboring locations. In spatial terms, attributes that are closer to a point tend to be more similar in relation to more distant points, a principle that governs spatial analysis, so considering a point far away from the point to be predicted can be detrimental, as it may be located in an area that is drastically different from the latter.

For the interpolation of the variables included in the model, the inverse distance weighted (IDW) method was used, whose interpolated values are an average of the weight of the values of a set of neighboring points, where the influence of the nearby points is greater than that of the distant points.

For the development of the models, GIS Boolean algebra processes were used, which allow us to perform mathematical operations on raster files. The Boolean operators & and not were used, in combination with relational operators such as >=, <=, ==, <, and <. As several operators of different orders were used, several strategies were applied to guarantee the order of execution and the weight of each variable. In order not to affect or alter the order of execution, parentheses were used to separate each Boolean operator used in the expression, in addition to the fact that all the layers of information entered were in raster format, since the aim was also to ensure that all the variables included had the same weight in the expression.

## 8.4 RESULTS

The first cartographic model was designed to highlight the areas where, thanks to the interaction with environmental factors, Fipronil degradation will occur more rapidly. This model considered temperature above 30°C, soil texture with high clay content, alkaline pH, and non-urban land use (see Fig. 8.1).

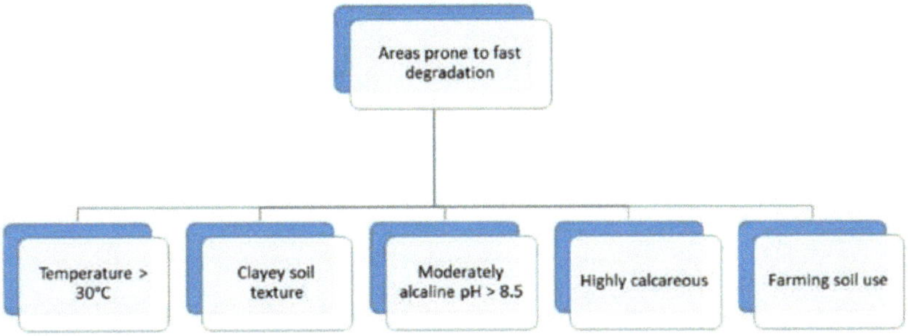


**FIGURE 8.1** Accelerated fipronil degradation model due to environmental interactions.

Model for areas prone to rapid degradation by environmental interactions (MRDEI).

MRDEI = (T ≥ 30) and (TS==Ac) and (pH > 8.5) and (CO==AC) and (US not U)

where
T = Temperature
TS = Soil texture
Ac = Clay soil texture
pH = Soil alkalinity degree
CO = Colloids
AC = Highly calcareous
US = Soil use
U = Urban land use

The model to identify areas where the degradation time of Fipronil is increased by environmental interactions considered temperatures below 30°C, sandy soils with low colloid content, pH tending to neutral or acid, and urban land use (see Fig. 8.2).

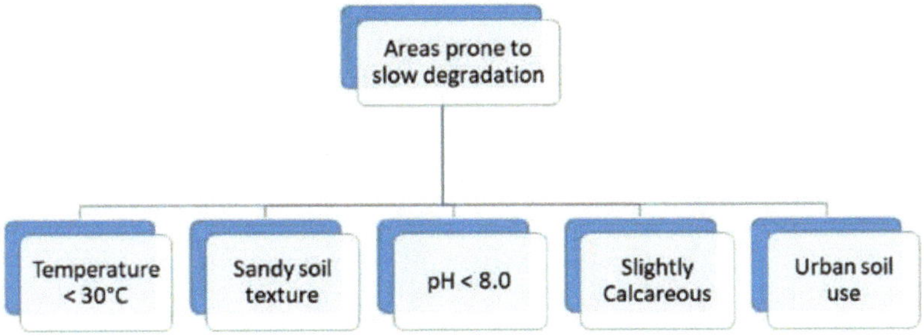


**FIGURE 8.2** Slow fipronil degradation model due to environmental interactions.

Model for areas prone to slow degradation by environmental interactions (MSDEI).
MSDEI = (T < 30) and (TS==A) and (pH ≤ 8.0) and (CO==LC) and (US==U)

where
T = Temperature
TS = Soil texture
A = Sandy soil texture
pH = Soil alkalinity degree
CO = Colloids
LC = Slightly Calcareous
SU = Soil Use
U = Urban land use

After seeing the effect that soil moisture had on the degradation of Fipronil, NDVI was used to design models to identify the generation of compounds derived from the metabolization of the insecticide (see Figs. 8.3 and 8.4).

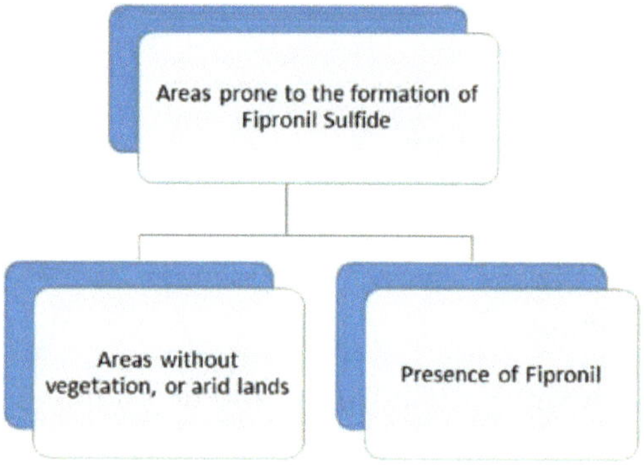


**FIGURE 8.3** Fipronil sulfide detection model.

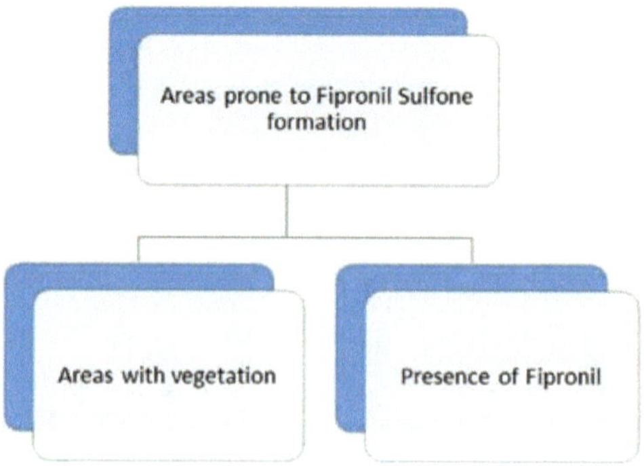


**FIGURE 8.4** Fipronil sulfone detection model.

Model for the detection of Fipronil sulfide formation (MDFFS)
MDFFS = (NDVI $\leq$ 0) and (F $\geq$ 1)
Model for the detection of Fipronil sulfone formation (MDFFSa)
MDFFSa = (NDVI $\geq$ 0.1) and (F $\geq$ 1)
where
F = Fipronil
NDVI = Normalized Difference Vegetation Index

## 8.5 CONCLUSION

The interactions of insecticides with the physical and chemical characteristics of the environment where they are applied need to be considered when designing and developing strategies for pest control in agricultural areas that resort to chemical products. Although the effects that the insecticide could have on the environment and on the health of the population are known from the safety datasheet of each compound, we must consider that these behaviors are not immovable, since they can be modified by the environmental conditions where the products are applied.

Cartographic models and GIS are very useful in the study and analysis of the variables necessary to understand the behavior of insecticides in the environment, since they provide the most appropriate tool for the analysis of a wide variety of elements and, above all, the possibility of studying them spatially.

Through these models, we can identify areas where environmental conditions converge that may favor a slower degradation of Fipronil, which translates into greater persistence in the environment, as well as its counterpart, that is, areas where environmental characteristics favor a rapid metabolization. It is important to emphasize that the areas where Fipronil degradation occurs more rapidly require special attention, since the presence of Fipronil-derived products is more likely to occur, and these may in turn be much more toxic than Fipronil due to their superior ability to adhere to the GABA channel.

Likewise, it is not only relevant to know the degradation time of an insecticide, but also the interactions that occur during the metabolization process, since the type of products that will derive from this degradation will depend on the former, and these can be equally or more toxic and equally or more persistent than the parent compound. For example, the lethal dose (LD) of Fipronil for the species Oncorhynchus mykiss (Rainbow trout) is $LD_{50(acute)}$

248 μg i.a./L, $LD_{50(acute)}$ 39 μg i.a./L of Fipronil sulfone, and $LD_{50(acute)}$ 29.5 μg i.a./L of Fipronil sulfide.

In addition to the chemical interactions that occur between the insecticide and the environment, we must consider that there are other non-environmental elements that intervene in the degradation processes of the compounds, such as interactions with other insecticides. For example, when Fipronil interacts with Cypermethrin, the latter inhibits its degradation, making it more persistent. Similarly, the frequency and concentration at which the insecticide is applied will interfere in the metabolization process.

Knowledge of the interactions between insecticides and the environment provides elements to determine the potential zones of occurrence of dangerous phenomena such as the persistence of chemical compounds in the environment. Moreover, it allows to identify the areas that are vulnerable to the concentration of highly toxic by-products, due to the convergence of factors that affect the metabolization processes. In addition, it will serve as a tool to identify the route of exposure to these compounds, which will benefit the planning of strategies for pest control using chemical compounds and will provide elements to consider in the prevention of health risks.

## KEYWORDS

- **fipronil**
- **environment**
- **GIS**
- **spatial analysis**
- **hydrography**

## REFERENCES

1. Ávila, A. A Dónde va La Huasteca? Patterson, J. Z., Ed.; *Estudios Agrarios* **1996,** *2*, 9–30.
2. Brady, N. C.; Weil, R. R. In *The Nature and Properties of Soils*; 13th Ed.; Pearson: New Jersey, USA, 2002.
3. Connelly, P. Environmental Fate of Fipronil; California Environmental Protection Agency, 2001.

4. Doran, G.; Eberbach, P.; Helliwell, S. The Sorption and Degradation of the Rice Pesticides Fipronil and Thiobencarb on Two Australian Rice Soils. *Aust. J. Soil Res.* **2006,** *44* (6), 599–610. DOI: 10.1071/SR05173.
5. Freed, V. H.; Chiou, C. T.; Haquet, R. Chemodynamics: Transport and Behavior of Chemicals in the Environment A Problem in Environmental Health. *Environ. Health Perspect.* **1977,** *20*, 55–70.
6. Gunasekara, A. S.; Troung, T. Environmental Fate of Fipronil; California Environmental Protection Agency, 2007; pp 1–28.
7. INEGI. Síntesis de Información Geográfica Del Estado de San Luis Potosi, 2002.
8. INIFAP; SAGARPA. *El Chapulin Melanoplus Sp. y Su Manejo En La Planicie Huasteca*; Folleto ténico num. 11, 2005a.
9. INIFAP; SAGARPA. *La Langosta Schistocerca Piceifrons Piceifrons y Su Manejo de Planicie Huasteca*; Folleto técnico num. 12, 2005b.
10. Oliver, D. P.; Kookana, R. S.; Quintana, B.; Csiro Land, Glen Osmond, and South Australia.Sorption of Pesticides in Tropical and Temperate Soils from Australia and the Philippines. *J. Agric. Food Chem.* **2005,** *53*, 6420–6425. DOI: 10.1021/jf050293l.
11. Pan-UK.Fipronil; *Pesticide News* 48, 2000.
12. Hernández, H. R.; Robledo, M. A.; Rivera, J. R. A.; Vázquez, I T. Cambios En La Cubierta Vegetal y Uso Del Suelo En El Área Del Proyecto Pujal-Coy, San Luis Potosí, México, 1973-2000. *Invest. Geog.* **2006,** (59), 26–42.
13. Goldberg, S.; Lebron, I.;Suarez, D. L. Soil Colloidal Behavior; In *Handbook of Soil Science Properties and Processes*; Pan, M. H., Li, Y., Malcolm, E. S., Ed.; Taylor & Francis: Florida, USA, 2011; pp 195–240.
14. Sheng, G.; Johnston, C. T.; Teppen, B. J.; Boyd, S. A.Potential Contributions of Smectite Clays and Organic Matter to Pesticide Retention in Soils. *J. Agric. Food Chem.* **2001**. DOI: 10.1021/jf001485d.
15. Tieszen, L. L.; Reed, B. C.; Bliss, N. B.; Wylie, B. K.;DeJong, D. D. NDVI, C3 AND C4 Production, and Distributions in Great Plains Grassland Land Cover Classes. *Ecol. Appl.* **1997,** *7* (1), 59–58. DOI: 10.2307/2269407.
16. Wood, A. Compendium of Pesticide Common Names [Online] 2006. https://pesticidecompendium.bcpc.org/ (accessed Mar 18, 2022).
17. Liu, X. B.; Song, C. Y.; Xing, B. S. Characteristics of the Natural Organic Matter Sorption Affects of Organic Contaminants. *J. Environ. Sci. (China)* **2002,** *14* (3), 351–356.
18. Ying, G. G.; Kookana, R. Laboratory and Field Studies on the Degradation of Fipronil in a Soil. *Soil Res.* **2002,** *40* (7), 1095–1102. DOI: 10.1071/SR02018.

# CHAPTER 9

# Nuclear and Plastid Engineering in Agricultural Crops

BRENDA JULIÁN-CHÁVEZ[1], STEPHANIE SOLANO-ORNELAS[1], QUINTÍN RASCÓN-CRUZ[1], TANIA SAMANTA SIQUEIROS-CENDÓN[1], BLANCA FLOR IGLESIAS-FIGUEROA[1], SIGIFREDO ARÉVALO-GALLEGOS[1], SUGEY RAMONA SINAGAWA-GARCÍA[2], JORGE ARIEL TORRES-CASTILLO[3], CARMEN DANIELA GONZÁLEZ-BARRIGA[4], HÉCTOR SILOS ESPINO[5], and EDWARD ALEXANDER ESPINOZA-SÁNCHEZ[1]

*[1]Laboratorio de Biotecnología I, Facultad de Ciencias Químicas, Universidad Autónoma de Chihuahua, Circuito Universitario s/n Nuevo Campus Universitario, Chihuahua, México*

*[2]Facultad de Agronomía, Laboratorio de Biotecnología, Universidad Autónoma de Nuevo León, Francisco Villa S/N Col. Ex hacienda El Canadá, General Escobedo, Nuevo León, México*

*[3]Universidad Autónoma de Tamaulipas, Instituto de Ecología Aplicada. Col. Libertad, Ciudad Victoria, Tamaulipas, México*

*[4]Laboratorio de Cultivo de Tejidos, División de Ingeniería y Ciencias, Tecnológico de Monterrey, Av. Heroico Colegio Militar, Nombre de Dios, Chihuahua, México*

*[5]Laboratorio de Biotecnología Aplicada, Instituto Tecnológico El Llano, Aguascalientes, Km 18 carretera Aguascalientes-San Luis Potosí, México*

## ABSTRACT

Human nutrition has been a constant challenge. This challenge is exacerbated by a projected increase of more than one billion people in the next three

Sustainable Agricultural Production Systems: Traditional and Modern Perspectives.
Romeo Rojas, Guillermo Cristian Guadalupe Martínez-Ávila, Julia Mariana Márquez-Reyes, and Ma del Carmen Ojeda-Zacarías (Eds.)

decades, imposing an unprecedented demand to increase food production by 110%. In recent years, strategies have been suggested to increase production; however, it has been observed that in ~40% of our major growing areas, yield is no longer improving. Therefore, biotechnological crops capable of resisting pests, adverse climatic conditions, or increasing the quality and quantity of food have emerged as an alternative. Although biotechnological crops have been tested previously for the expression of proteins of pharmaceutical and energy interest, their use to secure the food supply seems promising. However, despite the advantages offered by modified crops, this technology is still associated with environmental problems. In recent years, the elimination of side effects in plant modification has been explored by inserting genes into the chloroplastid genome. Although this has increased the production of proteins of interest, simultaneously eliminating position effects and environmental contamination problems, this technology has yet to solve several challenges for its widespread adoption in the market. As various technologies are deployed to boost food production, it is clear that efforts must intensify as demand begins to outstrip available food supply.

## 9.1 INTRODUCTION

Plants have been ideal targets for biotechnology to produce large amounts of protein to meet the needs of a growing population (estimated to be 9.8 billion people by 2050 and 11.2 billion in 2100), and although there has been some resistance to the production of genetically modified plants, they constitute an opportunity to obtain plants that exhibit desired characteristics.[1] Different industries have seen in plants the opportunity to produce enriched foods, plants with low cell wall contents components, vaccines, nutraceutical proteins, antibodies, and plants resistant to biotic and abiotic factors.

Obtaining modified plants has been routinely carried out through nuclear transformation, but collateral effects have appeared, such as gene silencing, position effects, and low levels of protein expression. Uncontrolled gene dispersion is also a latent biosafety problem.[2] This is where chloroplast transformation approaches have their utility because they reduce these unwanted effects, allowing the possibility of expressing multiple genes in a single transformation event.[3,4] However, despite plastid transformation being a promising tool of plant biotechnology and being accomplished in many laboratories worldwide on several species, these have been mainly dicot plants, whereas monocots are still considered recalcitrant to plastid transformation. Furthermore, as in nuclear transformation, many collateral effects have been documented in transplastomic lines.[5–7] Although it has been proposed that these depend mainly on the type of gene expressed,[8] it is a fact that further studies are necessary to convert chloroplast transformation into the de facto tool in plant improvement.

The opportunities offered by plant genetic engineering for the expression of proteins of interest have made it an area with accelerated growth, and although it still presents some challenges, there are already crops on the market that exhibit interesting properties to ensure and increase food production. In this chapter, we explore the advances in genetic engineering to produce plants with agronomic traits and the challenges that must be solved to develop genetically modified plants.

## 9.2 NUCLEAR TRANSFORMATION AGAINST BIOTIC AND ABIOTIC STRESSES

Agriculture continues to be the main source of food for the human population, and there is a growing demand worldwide. To satisfy the requirements, plants have been modified to allow large-scale production of different proteins of interest, and since 1995, many countries have approved the cultivation of modified plants. To date, an increase in the area destined for their production has been observed, resulting in substantial economic gains.[9,10] Obtaining modified plants has been possible thanks to the development techniques like microprojectile bombardment, direct protoplast transformation, or *Agrobacterium*-mediated transformation, which is lately the most used in the plant's nuclear transformation. Also other alternative methods have been developed, such as microinjection, cell electroporation, silicon carbide- and liposome-mediated transformation. However, due to lower transformation efficiency, these methods are not widely used.[11,12]

Currently, worldwide research aims to produce high-value "pharming" where plants are used as factories to produce industrial enzymes, antibodies, synthetic drugs, vaccines, nutraceutical proteins, vitamins, or enriched foods.[13] The initial trend in the development of modified crops was the insertion of genes in the nuclear genome to confer resistance to biotic and abiotic stresses, as reflected in Vaeck et al.,[14] McBride et al.,[15] or Daniell et al.[16] This trend has not abated, which can be supported by the approval for commercialization/planting and importation of multiple crops that display herbicide tolerance, fertility restoration, altered lignin content, antibiotic tolerance, pest tolerance, delayed fruit softening, drought stress tolerance, foliar late blight resistance, enhanced photosynthesis/yield, enhanced provitamin A content, male sterility, viral disease resistance, etc.[17]

### 9.2.1 MODIFIED PLANTS TO CONFER PEST RESISTANCE

Plants are an important part of human survival as they provide about 80% of the food that is consumed,[18] but these are exposed to different attackers, and although they can defend themselves from their predators, they often succumb to them. There are ~70,000 species of known pests in the world, and despite that, only 10% of these are considered serious pests.[19] More than 40% of all possible food production is lost to plant-pathogen, weed, and insect pests before harvest, with a subsequent 20% of postharvest losses by other pests,[20] so the use of pesticides has been a recurrent strategy.

Currently, the use of pesticides plays a key role in modern agriculture; their use in cropland increased from 1.55 kg $Ha^{-1}$ in 1990 to 2.69 kg $Ha^{-1}$ in 2019[21]; nevertheless, the widespread use causes annually ~385 million cases of unintentional acute pesticide poisoning worldwide, which indicates that ~44% of farmers are poisoned with pesticides each year, of which 11,000 cases end in death.[22] The problem is not limited to farmers; 50% of pesticides are classified as organophosphates,[23] which can be translocated from the roots to the leaves of plants that are consumed by the general population,[24,25] which could increase cases of intoxication by pesticides.

Considering the health problems associated with pesticides and the resistance cases recorded for many pests to different agrochemicals (>17,000 cases of resistance among 612 species globally by 2020)[26] and thinking of a nonchemical alternative to effective pest controls, multiple genes have been incorporated into plants of agricultural interest for their protection.

In 1901, Ishiwatari[27] reported a silkworm disease that turned out to be caused by *Bacillus thuringiensis*, which attracted interest in it for use in insect control. The result was so promising that in 1938, the first Bt product, named "Sporeine," was commercialized in France.[28] With the impact of *B. thuringiensis* and their insecticidal Cry proteins, in 1987, the first insecticidal gene (*Cry1Ab* gene) was introduced in tobacco.[14] Since then, the Cry proteins have remained the leading biological agent for pest control. More than 700 *cry* genes have been reported, many with pesticide activity (http://www.lifesci.sussex.ac.uk/home/Neil_Crickmore/Bt/) that can be used to control different orders of insects such as Diptera, Coleoptera, Lepidoptera, Orthoptera, Hemiptera, Phthiraptera, and Hymenoptera, as well as nematodes, mites, and protozoa.[29,30]

The potential of Cry proteins has allowed them to be incorporated into different crops with agronomic interest, mainly through nuclear transformation. For example, in 1997, Singsit et al.[31] reported the expression of *Cry1Ac* gene in peanut exhibiting resistance against *Elasmopalpus lignosellus*; in 2000, Shu et al.[32] reported rice with the *cry1Ab* gene exhibiting resistance against the larvae of lepidopteran *Scirpophaga incertulas* (rice yellow stem borer), *Chilo suppressalis* (rice stem borer), *Mycalesis gotama* (Chinese bushbrown), *Cnaphalocrocis medinalis* (rice leafroller), *Parnara guttata* (common straight swift), *Naranga anescens* (rice green caterpillar), *Herpetogramma licarsisalis* (tropical grass webworm), and *Sesamia inferens* (pink stem borer). In 2000, the *cry1Ac* and *cry1C* genes were reported to control *Plutella xylostella* in broccoli and collards,[33,34] and in 2007, these same genes were reported to control *P. xylostella* on Indian mustard.[35]

The impact of Bt crops has been such that since their commercial adoption in 1996, 88% of cotton and 82% of corn grown in the United States is genetically modified to confer resistance to pests.[36] Nevertheless, the continuous and systematic use of *B. thuringiensis* products eventually leads insects to develop resistance;[37–40] a growing number of cases of resistance to Bt toxins in modified crops has been reported, from 3 in 2005 to 16 in 2016, involving crops with Cry34/35Ab, Cry1Ab, Cry1Ac, Cry2Ab, Cry1A.105, Cry3Bb, Cry1Fa, mCry3A, and eCry3.1Ab toxins with an average of 5.2 years for the appearance of resistance.[41,42] These resistances could be increased by cross-resistance, which has been verified through observation of resistance by *Diabrotica virgifera virgifera* to the eCry3.1Ab protein in the field before the recombinant plants that produce that toxin were commercialized.[43,44] Although some strategies have been tested to prevent insect resistance, such as high or ultrahigh doses,[37] the use of multiple toxins,[33,34,45] protein engineering,[46,47] and the compartmentalization of proteins into chloroplast using transit peptides to increase their accumulation.[48–50] In the last instance, the use of refuges for delaying resistance to Bt crops,[51] the Cry protein resistance remains a latent risk.

To remedy the situation of the behavior of insects with respect to Cry proteins, the search for new toxins with high toxicity and broad-spectrum pest control has become a necessity. It has been reported that proteins with pesticide activity from *Bacillus toyonensis,* homologous to *B. thuringiensis* and *Lysinibacillus sphaericus* proteins, exhibited high toxicity against the beetle *Anthonomus grandis* (Coleoptera: Curculionidae) and the caterpillar *Cydia pomonella* (Lepidoptera: Tortricidae), relatively low toxicity against the mosquitoe *Aedes aegypti* (Diptera: Culicidae) and the beetle *Alphitobius diaperinus* (Coleoptera: Tenebrionidae).[52] The expression of lectin to pest control has also been reported,[53–55] and although it has yet to be thoroughly explored, it opens a way to develop plants prepared with a new arsenal against pests.

### 9.2.2 *USE OF RNAI TO PEST RESISTANCE*

Until recently, conferring resistance to plants against insect pests has been addressed by the production of specific proteins. All of the commercialized insect resistance crops are based on insecticidal proteins derived from *B. thuringiensis*,[56] but some problems have been associated with this; therefore, in the search for alternatives, the gene silencing technology in plants (by using interference RNA "RNAi") has recently received attention because it eliminates protein dependency, and from this, there are some advances in the use of hpRNA (long hairpin RNA) and dsRNA (double-stranded RNAs) as innovative strategies to pest control.

The first demonstrations of the potential of RNAi were reported by Baum et al.[57] in maize plants using dsRNA that prevented the expression of a vacuolar ATPase in the beetle *D. virgifera virgifera,* causing larval stunting and increase in mortality, and by Mao et al.[58] that silenced the cytochrome *P450* gene (CYP6AE14), which is highly expressed in the midgut from the lepidopteran *Helicoverpa armigera*, using specific dsRNA decreasing the CYP6AE14 transcripts and retarding the larval growth. Since then, the potential of gene silencing has been envisioned, and new studies have been reported.

In 2014, RNAi targeting the *MpAChE2, V-ATPaseE*, and *TBCD* genes that encode acetylcholinesterase 2, V-type proton ATPase subunit E-like, and tubulin folding cofactor D, respectively, of *Myzus persicae* decreased the transcripts and increased the resistance of the plant to the insect.[59] *M. persicae* also was studied by Bhatia and Bhattacharya[60] and Rauf et al.[61] to silence a cuticular protein (*MyCP* gene) and cathepsin L that encode a cysteine protease, resulting in delayed nymph production, lower protein content in the bodies, decreased fecundity, and mortality. Agrawal et al.,[62] Chen et al.,[63] and Saini et al.[64] reported the silencing of the chitinase gene, *HARP1* gene, and the *HaAce1* gene that encodes the predominant isoform of acetylcholinesterase of *H. armigera* resulting in reduced growth, deformity, cessation of molting, and death of the larvae. The whitefly *Bemisia tabaci* (Hemiptera: Aleyrodidae) also has been studied with RNAi using as a target the glutathione-S-transferase (GST), *BtGSTs5,* and *Bt56* genes obtaining delaying development and decreasing the capacity of ingesting of the whitefly by blocking the capacity of eliciting the salicylic acid- (SA-) signaling pathway.[65,66]

Other examples of the usefulness of RNAi include studies on the Colorado potato beetle (*Leptinotarsa decemlineata*), *Sitobion avenae, Diaphorina citri, P. xylostella,* and *C. suppressalis*. Colorado potato beetle was studied using as a target of RNAi the ecdysone receptor (*EcR*) gene, exhibited reduced *EcR* transcripts, decreased larval weight, and mortality.[67]

The silencing of *SaZFP* gene of *S. avenae* increased the nymphs' mortality.[68] The silencing of trehalose-6-phosphate synthase (*TPS* gene) of *D. citri* decreased the TPS expression in the host, increased the mortality and malformation in nymphs.[69] The silencing of the arginine kinase (*PxAK*) and integrin $\beta_1$ subunit (*Pxβ*) genes of *P. xylostella* showed longer developmental time as well as lower pupal weight and pupation rate,[70] whereas the use of RNAi against *C. suppressalis* in rice caused growth and pupation retardation.[71,72]

These results make silencing by RNAi a very promising technology, and in 2017, the first transgenic maize variety, Smartstax PRO®, expressing a dsRNA of Sucrose non-fermenting 7 (*Snf7* gene) from *D. virgifera virgifera* was approved.[17,73,74] Its commercial launch is planned for 2022, but although RNA technology has been advancing, some aspects should be considered, such as the suitable target gene(s) and the fact that the host's digestive system degrades the RNAi, which would require increasing the amounts of RNAi in the artificial diet to achieve an insecticidal effect.[75] This is a challenge in itself when expression in nucleus is used because the accumulation of RNAi is affected by the cell's RNAi machinery and trapping in acidic bodies within the cell, decreasing the amount of nonprocessed hpRNA or dsRNA available to pest ingestion.[76,77] In addition, it should be considered the expansion of the RNAi delivery methods, besides microinjection and ingestion methods, although the use of topical RNAi has already been reported for this, which appears to be a profitable alternative.[78]

### 9.2.3 PLANTS WITH ABIOTIC STRESS TOLERANCE

Reactive oxygen species (ROS) are forms of oxygen partially reduced produced usually during cellular respiration, photosynthetic processes, as well as under abiotic or biotic stress that have been associated with the alteration of the correct function of the cells. Therefore, decreasing ROS in plants has gained special interest to prevent damage caused by different types of stress, but the decrease of ROS and the improvement of resistance to drought, salinity, and chilly is complicated due to the multiple genes involved. However, several studies have been successful, e.g., the codA gene encoding choline oxidase from *Arthrobacter globiformis* was satisfactorily expressed in tomato,[79] alfalfa,[80] and poplar,[81] conferring tolerance to salt stress, which also has been conferred by the expression of flavonol synthetase gene from *Apocynum venetum* and the aldehyde dehydrogenase gene (*BrALDH7B2*) from *Brassica rapa* in tobacco,[82,83] indicating that the protection of plants to abiotic stress is possible with the expression of different genes.

Water deficit is another stress to which plants are subject, and recently, it was reported that the expression of phosphoenolpyruvate carboxykinase (*ZmPCK2* gene from *Zea mays*) in rice improved the water-deficit tolerance.[84] The use of protective genes is a plausible strategy, and it has been proposed to focus efforts on targeting proteins of interest that enhance the abscisic acid (ABA) signaling pathway, associated with the development, growth, and response to various types of abiotic stress in plants.[85,86] Recently, a receptor of ABA, "PYL9," was expressed in apple (*Malus domestica*), protecting against drought stress,[87] indicating that this way to protect plants is a good bet.

Conferring protection to plants is complicated because many types of stress activate similar metabolic pathways, so a single strategy may not generate enough protection. Therefore, many studies have been reported where they confer reduced oxidative stress,[88] tolerance to waterlogging,[89] drought and salt stress tolerance,[90,91] metal tolerance,[92–94] ROS decrease,[95] and cold stress,[96] but possibly, to produce a plant resistant to multiple types of stress the way forward will be the insertion of multiple genes. However, this should be carefully explored due to the collateral effects of multiple expression of foreign genes on the plant phenotype. Despite conferring protection to plants against biotic and abiotic stresses being quite a challenge, efforts are necessary to carry it out promptly since, in a constantly growing population, agricultural crops are the main source of food.

Plant nuclear transformation is a biotechnology milestone and has maintained the lead in plant genetic engineering by producing plants with new characteristics without a risk to human health. However, the random insertion of genetic material in the nucleus has shown different expression profiles in each transformation event. In addition, gene silencing is still a latent risk,[97,98] and although, to date, significant results have been achieved that can support the plant's nuclear transformation as the use of insulator sequences to increase transgenic expression and avoid position effects,[99] it is an area that has not been yet fully explored. Even when these problems are solved, many reports have indicated that crop-to-crop gene flow is a possibility,[2] and considering that modified crops probably have come to stay, approaches that decrease the environmental side effects must be considered a pressing need.[100,101]

## 9.3 CHLOROPLAST ENGINEERING, THE BET ON PROTEIN OVEREXPRESSION

Chloroplasts are organelles that participate in the main metabolic processes that occur in plant cells, such as metabolism of phytohormones (cytokinins, gibberellins, and abscisic acid), nitrogen and amino acids metabolism, biosynthesis of fatty acids, phenolic compounds, pyrimidines, starch, purines, isoprenoids, as well as synthesis of vitamins and pigments.[102] In recent years, the interest in chloroplasts has increased, so much so that in the RefSeq database from the National Center for Biotechnology Information (NCBI), to date, there are reported 6,768 chloroplast genomes from different species. Of these, more than 1,500 complete genomes were reported in 2021, and more than 200 have been reported so far in 2022. The interest of biotechnology in these organelles lies in the possibility of expressing genes within them being more controllable, efficient, and safer, eliminating ecological problems currently associated with nuclear-modified plants.

Many studies have focused on plant development with resistance to pests and protection against abiotic stress using plastid transformation technology (Table 9.1). Although nuclear transformation has already been carried out in crops of agronomic interest, it seeks to take advantage of the potential benefits offered by the compartmentalization of chloroplasts from the protection of the proteins produced provided by the double membrane of this organelle to the overexpression of proteins of interest.

**TABLE 9.1** Plants Species Transformed by Plastid Engineering.

| Plants | Family | Species | Name | Ref. |
|---|---|---|---|---|
| | | **Algae** | | |
| ***Green algae*** | Haematococcaceae | *Haematococcus pluvialis* | – | 153 |
| | Dunaliellaceae | *Dunaliella tertiolecta* | – | 154 |
| | Chlamydomonadaceae | *Chlamydomonas reinhardtii* | – | 155 |
| | Euglenaceae | *Euglena gracilis* | – | 156 |
| ***Red algae*** | Porphyridiophyceae | *Porphyridium* sp. UTEX 637 | – | 157 |
| | Bangiaceae | *Pyropia yezoensis* | – | 158 |
| | Cyanidiaceae | *Cyanidioschizon merolae* | – | 159 |
| ***Microalgae*** | Phaeodactylaceae | *Phaeodactylum tricornutum* | – | 160 |
| | Monodopsidaceae | *Nannochloropsis oceanic* | – | 161 |
| | Isochrysidaceae | *Tisochrysis lutea* | – | 162 |
| **Liverworts** | Marchantiaceae | *Marchantia polymorpha* | Umbrella liverwort | 163 |
| **Moss** | Funariaceae | *Physcomitrella patens* | Moss | 164 |
| ***Monocots*** | Poaceae | *Oryza sativa* var. Hwa-Chung and Japonica line 19 | Rice | 132,139 |
| ***Dicots*** | Solanaceae | *Nicotiana sylvestris*, *Nicotiana benthamiana* and *Nicotiana tabacum* var. Petit Havana | Tobacco | 165–168 |

**TABLE 9.1** *(Continued)*

| Plants | Family | Species | Name | Ref. |
|---|---|---|---|---|
| | | *Petunia hybrida* var. Pink Wave | Petunia | 169 |
| | | *Solanum tuberosum* cv. Desirée and line FL1607 | Potato | 170,171 |
| | | *Capsicum annuum* var. G4 | Pepper | 172 |
| | | *Solanum melongena* | Eggplant | 173 |
| | | *Solanum lycopersicum* | Tomato | 150 |
| | Brassicaceae | *Lesquerella fendleri* | Lesquerella | 174 |
| | | *Brassica napus* | Oilseed rape | 175 |
| | | *Brassica napus* cv. FY-4 | Rapeseed | 176 |
| | | *Arabidopsis thaliana* | Arabidopsis | 177 |
| | | *Brassica oleracea* var. capitata | Cabbage | 178 |
| | | *Brassica oleracea* var. botrytis | Cauliflower | 179 |
| | Apiaceae | *Daucus carota* cv. Half long | Carrot | 127 |
| | Malvaceae | *Gossypium hirsutum* cv. Coker310FR | Cotton | 180 |
| | | *Medicago sativa* cv. Longmu 803 | Alfalfa | 181 |
| | Asteraceae | *Lactuca sativa* cv. Verônica, cv. Cisco and cv. Flora | Lettuce | 182–184 |
| | | *Artemisia annua* | Artemisa | 185 |
| | Plantaginaceae | *Scoparia dulcis* | Sweet broom | 186 |
| | Fabaceae | *Glycine max* | Soybean | 187 |
| | Amaranthaceae | *Beta vulgaris* | Sugar beet | 188 |
| | Salicaceae | *Populus alba* | Poplar | 189 |
| | Cucurbitaceae | *Momordica charantia* | Bitter melon | 190 |

### 9.3.1 PEST AND DISEASE RESISTANCE

Due to the capacity of the chloroplast to overexpress foreign proteins and the observation of pest resistance to Cry proteins obtained by nuclear transformation, the Cry proteins were also plastid expressed (Table 9.2). These attempts began with the Cry1Ac protein expressed in tobacco protecting against the lepidopteran *Helicoverpa zea, Heliothis virescens,* and *Spodoptera exigua*[15] and opened the door for other proteins like Cry1Ia5, Cry2Aa2, Cry9Aa2, Cry1Ab, and Cry1C to be expressed in crops such as tobacco, soybean, cabbage, and poplar to control *Hyphantria cunea, Lymantria dispar, Phthorimaea operculella, P. xylostella, H. armigera,* and *Anticarsia gemmatalis.*[103–108] The results allowed us to notice that the insects did not develop resistance to the proteins expressed in plastids, so the idea of expressing the proteins in the plastids or redirecting the cytoplasmic Cry proteins already expressed toward the chloroplastid compartments to compartmentalize them and increase their accumulation has been gaining attention. This was tested with the expression and posterior redirectionalization to the chloroplast of Tvip3A*, Cry1Ac, Cry1Ah, and Cry2A,[48–50,109] confirming the potential of compartmentalization.

**TABLE 9.2** Agronomic Functional Traits Provided by Using the Plastid Genetic Transformation.

| Gene | Product | Host | Ref. |
|---|---|---|---|
| | **Resistance to pest and diseases** | | |
| *cry1Ab* | Cry1Ab | Cabbage | 103 |
| *cry1Ia5* | Cry1Ia5 | Tobacco | 107 |
| *cry1Ab* | Cry1Ab | Soybean | 106 |
| *cry1C* | Cry1C | Tobacco | 191 |
| *cry2Aa2* | Cry2Aa2 | Tobacco | 104 |
| *MSI-99* | Magainin-type antimicrobial peptide | Tobacco | 111 |
| *Pta* | *Pinellia ternata* agglutinin | Tobacco | 112 |
| *Cpo* | Chloroperoxidase | Tobacco | 113 |
| *cry1C* | Cry1C | Poplar | 108 |
| | **Abiotic stress tolerance** | | |
| *DHAR* | Dehydroascorbate reductase | | |
| *GST* | Glutathione-S-transferase | Tobacco | 121 |
| *Gor* | Glutathione reductase | | |
| *gor* | *Escherichia coli* glutathione reductase | Tobacco | 125 |
| *MnSOD* | Superoxide dismutase | | |
| *ArDH* | Arabitol dehydrogenase | Tobacco | 126 |
| *Fld* | Flavodoxin | Tobacco | 130 |
| *TCY* | Tocopherol cyclase | | |
| *HPT* | Homogentisate phytyltransfera | Tobacco<br>Tomato | 128 |
| *TMT* | γ-tocopherol methyltransferase | | |
| | Sporamin | | |
| * | Cystatin | Tobacco | 192 |
| | Chitinase | | |
| *CMO* | Choline monooxygenase | Tobacco | 122 |
| *coda* | Choline oxidase | Potato | 123 |

Despite that Cry proteins are effective in pest control, other proteins with insecticidal activity have been tested by plastid expression. In 2001, DeGray et al.[110] reported a magainin-type antimicrobial peptide (MSI-99) effective in controlling *Fusarium moniliforme Pseudomonas syringae*, *Colletotrichum destructivum, Aspergillus,* and *Verticillium dahliae.* Also, the use of *MSI-99* gene, the *cpo* gene that encodes the chloroperoxidase from *Pseudomonas pyrrocinia* and the *pta* gene from *Pinellia ternata* agglutinin were reported against *E. coli, Alternaria alternate, Aspergillus flavus, Fusarium verticillioides,* and *V. dahliae.*[111–113]

Although the strategy of expressing genes with insecticidal capacity is a path that should be explored in depth since it can solve the resistance problems observed in transgenic crops, problems associated with the expression of genes with a detrimental effect on the plant phenotype still need to be solved.[111]

### 9.3.2 USE OF RNAI TO PEST RESISTANCE

Although the expression of genes in chloroplasts has focused mainly on producing and accumulating proteins in this compartment, due to the impact of RNAi on insect control, the ability to express RNAi in chloroplasts has also been evaluated, but the reports are still limited. In 2015, Jin et al.[75] reported the efficient expression of RNAi in the chloroplast of tobacco using dsRNAs to silence cytochrome P450 monooxygenase (*P450* gene), chitin synthase (*Chi* gene), and V-ATPase from *H. armigera,* decreasing the growth and weight of larvae. Using a hpRNA targeting *ACE* gene that encodes the acetylcholinesterase of *H. armigera* and dsRNA targeting *ACT* and *SHR* genes that encode the Vps32 and β-actin proteins from Colorado potato beetle, it was possible to obtain strong protection against *H. armigera* and reduced growth of larvae of Colorado potato beetle in transplastomic tobacco and potato plants.[77,114] Recently, Ren et al.[115] reported the use of dsRNA *β-Actin* to control the beetle *Henosepilachna vigintioctopunctata* in potato plants.

Although reports of the use of RNAi in chloroplasts are encouraging, as with all technology, there are aspects that still need to be considered, such as the effective length of the RNAi, because it has been informed that dsRNA with ~200 nt shows greater protection than those dsRNA with length of 60 nt or less than 200 nt. However, this must be considered carefully because dsRNA accumulation has been reported to be inversely correlated with the length of the dsRNA itself.[116,117] It has also been reported that although the chloroplast synthesizes and accumulates large amounts of dsRNA and hpRNA, RNAi produced by nuclear expression shows more effective insecticidal activity.[118]

Despite the drawbacks that have been reported in this area, the production of RNAi in the chloroplast could be a correct bet because no RNAi processing machinery exists in the chloroplast, and RNAi produced in the chloroplast is not exported to the cytoplasm,[114] also, the chloroplast acts as a biocapsule protecting the RNAi when consumed by insects.[57,119] Therefore, the chloroplast is a viable solution to ensure the stability, accumulation, and protection of RNAi.

### *9.3.3 PLANTS WITH ABIOTIC STRESS TOLERANCE*

As with nuclear transformation, the decrease of ROS has become a target to allow plants to tolerate the stress to which they are subjected. Considering the photosynthetic processes that occur in the chloroplast and that these are an important source of ROS,[120] based on the results obtained with nuclear transformation, it has been sought to reduce ROS through the expression of foreign proteins in the chloroplast involved in metabolic pathways, such as glycine betaine (GB), to increase the antioxidant pathway.[121] Using potato plants, Zhang et al. and You et al.[122,123] analyzed the expression of the *codA* gene from *Arthrobacter globiformis* that encodes choline monooxygenase to increase GB, obtaining plants with a certain level of protection against different forms of abiotic stress, showing that the assumption is feasible. The expression of proteins such as glutathione-S-transferase (*GST* gene), dehydroascorbate reductase (*DHAR* gene),[121,124] glutathione reductase (*gor* gene), superoxide dismutase (*MnSOD* gene),[125] arabitol dehydrogenase (*ArDH*),[126] betaine aldehyde dehydrogenase (*badh* gene),[127] tocopherol cyclase (*TCY* gene), γ-tocopherol methyltransferase (*TMT* gene),[128] *otsB-A* operon (trehalose phosphate synthase/phosphatase),[129] homogentisate phytyltransferase (*HPT* gene), and flavodoxin (*fld* gene),[130] has also been analyzed and found to confer tolerance to heavy metal, salt, UV-B radiation, osmotolerance, and *freezing*,[131] indicating that protection against stress can be achieved by manipulating different metabolic pathways.

Because many types of stress activate similar metabolic pathways at different levels, detailed studies of the type of gene to use, the metabolic pathway to be modified, and the effect that the expression of different genes may have on the metabolism of plants are still required.

### 9.3.4 MONOCOTS, THE BARRIER OF PLASTID TRANSFORMATION

Although chloroplast engineering has made considerable progress in dicots, establishing the transformation of plastids in monocots has been complicated.[132] Cereals provide essential nutrients and energy in the daily human diet, contributing more than 50% of the energy supply. Global food security depends heavily on cereal production, which reached production of 2,793 million tons in 2021.[133–136] Efficiently establishing the transformation of plastids in monocots does not limit this technology. However, although the transformation of monocots has not yet been successfully established, there are some reports with promising results.

The attempts to transform monocots are not recent; they began in 1991 with the studies of Daniell et al.[137] on wheat and allowed us to know that it is possible to use dicot chloroplast promoters in the chloroplasts of monocots. In 1999, Khan et al.[138] inserted the FLARE-S gene into *rrn16-rps12/7* chloroplast sequences from *Oryza sativa* cv. Taipei 309, obtaining heteroplasmic plants, but without confirming whether the genes could be transmitted to the following generations. Since then, attempts to transform monocots have offered better results. Lee et al.[139] using the japonica rice as model plant, inserted the *sgfp* and *aadA* genes into the *trnI-trnA* chloroplast regions, producing fertile heteroplasmic plants with transmission of the transgenes to the progeny. In 2011, Cui et al.[140] reported the insertion of *nptII* and *gfp* genes into the *atpB-rbcL* intergenic region from wheat (*Triticum aestivum* L. cv. Bobwhite). The plants maintained a heteroplasmic state when immature inflorescences were used and homoplasmic plants using scutella. Unfortunately the results could not be validated and were retracted, until all experiments are confirmed.[141]

In 2018, a major advance was made in the transformation of monocots; the *smGFP* gene and the *hpt* gene (encoding soluble-modified Green Fluorescent Protein and hygromycin phosphotransferase, respectively) were inserted into the *trnI-trnA* chloroplast region from japonica rice line 19, resulting in homoplasmic plants.[132] However, it was not possible to determine if the plants were sterile because the plants did not reach the stage of seed production. Therefore, the transformation of chloroplast in monocots still remains incipient.

*Zea mays* is one of the main crops produced worldwide due to its nutritional quality, so much so that in 2020 its production was greater than 1162 million tons worldwide,[142] but it has never been possible to transform their chloroplasts. Although in 2020, a vector was developed by Arévalo-Gallegos et al.[143] to transform plastids from maize using immature embryos resulting in green fluorescent sections in scutellum, the capacity to express genes directly in the chloroplast compartments is not yet confirmed.

Chloroplast transformation in monocots has been highly challenging, mainly because of several hurdles such as the development of plant regeneration media, efficient transformation methods, and increase in the DNA content per plastid in seedlings on lighting condition to improve the recovery of transformed plants.[144] Also, efforts should focus on finding strategies for the transformation of non-green plastids, since cereals start from callus, meristems, or scutella and in these non-green tissues, genes are highly downregulated.[145] On the other hand, contrary to nuclear transformation, in chloroplast transformation, it is necessary to develop vectors for each species of plant because there are no universal vectors. Although the idea has been proposed,[16,146] the problem is that the chloroplast regions are not well conserved,[4] and the loss of homology between target regions of the chloroplast genome and homologous recombination sequences of transformation vectors decrease the frequencies of transformation, the functionality of transgenes,[4,147,148] and therefore, the recovery of transformed plants,[149–151] although the use of TALENs (Transcription Activator-Like Effector Nucleases) has given encouraging results to improve the transformation efficiency.[152]

Despite the challenges in the transformation of chloroplasts into monocots, it is a line of research that should be explored in depth because it represents an opportunity to reduce the unwanted effects that have been observed with nuclear transformation, mainly biosafety problems, and since cereals represent an important food group in agriculture, their genetic improvement is clearly supported.

## 9.4 CONCLUSION AND FUTURE REMARKS

The constant growth of the world population keeps increasing the need for food production, which maintains pressure on biotech crops with particular agronomic traits. Both nuclear and plastid genetic engineering offer the opportunity to make the protection of agricultural crops more efficient, and although these technologies are still under development and some challenges must be solved, they have demonstrated their potential in crop protection. Although nuclear transformation is well advanced with respect to plastid transformation in the development of crops approved for commercialization, plastid transformation can solve the ecological problems associated with nuclear transformation, but tasks such as the formulation of efficient culture media, optimization of regulatory sequences, elimination of pleiotropic effects in transformed crops, and gene expression in non-green plastids still need to be made more efficient to extend plastid engineering to a larger number of plant species, including cereals. The future of nuclear and plastid transformation genetic engineering still has some hurdles to overcome, but these technologies have the potential to reduce food production costs and reduce biotic and abiotic stress problems currently associated with agricultural crops.

## KEYWORDS

- **nuclear and plastid transformations**
- **agronomic traits**
- **advances and challenges**

## REFERENCES

1. Xu, J.; Towler, M.; Weathers, P. J. Platforms for Plant-Based Protein Production. In *Bioprocessing of Plant In Vitro Systems*; Pavlov, A., Bley, T., Trans. Eds.; Springer International Publishing: Cham, 2018; pp 509–548.
2. Dively, G. P.; Huang, F.; Oyediran, I.; Burd, T.; Morsello, S. Evaluation of Gene Flow in Structured and Seed Blend Refuge Systems of Non-Bt and Bt Corn. *J. Pest. Sci.* **2020,** *93* (1), 439–447.
3. Bock, R. Engineering Plastid Genomes: Methods, Tools, and Applications in Basic Research and Biotechnology. *Annu. Rev. Plant Biol.* **2015,** *66,* 211–241.
4. Ruhlman, T.; Verma, D.; Samson, N.; Daniell, H. The Role of Heterologous Chloroplast Sequence Elements in Transgene Integration and Expression. *Plant Physiol.* **2010,** *152* (4), 2088–2104.
5. Corigliano, M. G.; Albarracín, R. M.; Vilas, J. M.; López, E. F. S.; Luoni, S. A. B.; Deng, B.; Farran, I.; Veramendi, J.; Maiale, S. J.; Sander, V. A.; Clemente, M. Heat Treatment Alleviates the Growth and Photosynthetic Impairment of Transplastomic Plants Expressing *Leishmania Infantum* Hsp83-*Toxoplasma Gondii* SAG1 Fusion Protein. *Plant Sci.* **2019,** *284,* 117–126.
6. Morgenfeld, M. M.; Vater, C. F.; Alfano, E. F.; Boccardo, N. A.; Bravo-Almonacid, F. F. Translocation from the Chloroplast Stroma into the Thylakoid Lumen Allows Expression of Recombinant Epidermal Growth Factor in Transplastomic Tobacco Plants. *Transgenic Res.* **2020,** *29,* 295–305.
7. Espinoza-Sánchez, E. A.; Torres-Castillo, J. A.; Rascón-Cruz, Q.; Zavala-García, F.; Sinagawa-García, S. R. Production and Characterization of Fungal β-Glucosidase and Bacterial Cellulases by Tobacco Chloroplast Transformation. *Plant Biotechnol. Rep.* **2016,** *10* (2), 61–73.
8. Ruiz, O. N.; Daniell, H. Engineering Cytoplasmic Male Sterility via the Chloroplast Genome by Expression of β-Ketothiolase. *Plant Physiol.* **2005,** *138* (3), 1232–1246.

9. Brookes, G.; Barfoot, P. Farm Income and Production Impacts of Using GM Crop Technology 1996–2016. *GM Crops Food* **2018,** *9* (2), 59–89.
10. Tschofen, M.; Knopp, D.; Hood, E.; Stöger, E. Plant Molecular Farming: Much More than Medicines. *J. Annu. Rev. Anal. Chem.* **2016,** *9,* 271–294.
11. Finer, J. J. Plant Nuclear Transformation. In: *Genetic Modification of Plants*, Vol. 64; Kempken, F., Jung, C., Eds.; Springer: Berlin, Heidelberg, 2010; pp 3–21.
12. Matsumoto, T. K.; Gonsalves, D. Biolistic and Other Non-*Agrobacterium* Technologies of Plant Transformation. In: *Plant Biotechnology and Agriculture*; Altman, A.; Hasegawa, P. M., Trans, Eds.; Academic Press, **2012,** pp 117–129.
13. Bu, Y.; Sun, B.; Zhou, A.; Zhang, X.; Takano, T.; Liu, S. Overexpression of *AtOxR* Gene Improves Abiotic Stresses Tolerance and Vitamin C Content in *Arabidopsis Thaliana. J. BMC Biotechnol.* **2016,** *16* (1), 69.
14. Vaeck, M.; Reynaerts, A.; Höfte, H.; Jansens, S.; De Beuckeleer, M.; Dean, C.; Zabeau, M.; Montagu, M. V.; Leemans, J. Transgenic Plants Protected from Insect Attack. *Nature* **1987,** *328* (6125), 33–37.
15. McBride, K. E.; Svab, Z.; Schaaf, D. J.; Hogan, P. S.; Stalker, D. M.; Maliga, P. Amplification of a Chimeric *Bacillus* Gene in Chloroplasts Leads to an Extraordinary Level of an Insecticidal Protein in Tobacco. *Biotechnology* **1995,** *13* (4), 362–365.
16. Daniell, H.; Datta, R.; Varma, S.; Gray, S.; Lee, S.-B. Containment of Herbicide Resistance Through Genetic Engineering of the Chloroplast Genome. *Nat. Biotechnol.* **1998,** *16* (4), 345.
17. ISAAA. GM Approval Database [Online], 2022. https://www.isaaa.org/gmapprovaldatabase/ (accessed Mar 12, 2022).
18. FAO. Plant Health 2020 [Online], 2020. https://www.fao.org/plant-health-2020/en/ (accessed Mar 12, 2022).
19. Pimentel, D. *Techniques for Reducing Pesticide Use: Economic and Environmental Benefits*; John Wiley and Sons: New York, 1997.
20. Paoletti, M. G.; Pimentel, D. Environmental Risks of Pesticides Versus Genetic Engineering for Agricultural Pest Control. *J. Agric. Environ. Ethics* **2000,** *12* (3), 279–303.
21. FAO. Pesticides Indicators [Online], 2019. https://www.fao.org/faostat/en/#data/EP (11/03/2022).
22. Boedeker, W.; Watts, M.; Clausing, P.; Marquez, E. The Global Distribution of Acute Unintentional Pesticide Poisoning: Estimations Based on a Systematic Review. *BMC Public Health* **2020,** *20* (1), 1875.
23. Fauzi, N. I. M.; Fen, Y. W.; Omar, N. A. S.; Hashim, H. S. Recent Advances on Detection of Insecticides Using Optical Sensors. *Sensors* **2021,** *21* (11), 3856.
24. Olisah, C.; Rubidge, G.; Human, L. R. D.; Adams, J. B. A Translocation Analysis of Organophosphate Pesticides between Surface Water, Sediments and Tissues of Common Reed *Phragmites Australis*. *Chemosphere* **2021,** *284,* 131380.
25. Sapbamrer, R.; Hongsibsong, S. Organophosphorus Pesticide Residues in Vegetables from Farms, Markets, and a Supermarket Around Kwan Phayao Lake of Northern Thailand. *Arch. Environ. Contam. Toxicol.* **2014,** *67* (1), 60–67.
26. APRD. Arthropod Pesticide Resistance Database (APRD) [Online], 2022. https://www.pesticideresistance.org/ (accessed).
27. Ishiwatari, S. On a New Type of Severe flacherie (Sotto Disease). *Dainihon Sansi Kaiho* **1901,** *114,* 1–5.

28. Brar, S. K.; Verma, M.; Tyagi, R. D.; Valéro, J. R. Recent Advances in Downstream Processing and Formulations of *Bacillus Thuringiensis* based Biopesticides. *Proc. Biochem.* **2006,** *41* (2), 323–342.
29. Ye, W.; Zhu, L.; Liu, Y.; Crickmore, N.; Peng, D.; Ruan, L.; Sun, M. Mining New Crystal Protein Genes from *Bacillus Thuringiensis* on the basis of Mixed Plasmid-Enriched Genome Sequencing and a Computational Pipeline. *Appl. Environ. Microbiol.* **2012,** *78* (14), 4795–4801.
30. Guo, Y.; Weng, M.; Sun, Y.; Carballar-Lejarazú, R.; Wu, S.; Lian, C. *Bacillus thuringiensis* Toxins with Nematocidal Activity Against the Pinewood Nematode *Bursaphelenchus xylophilus*. *J. Invertebr. Pathol.* **2022,** *189,* 107726.
31. Singsit, C.; Adang, M. J.; Lynch, R. E.; Anderson, W. F.; Wang, A.; Cardineau, G.; Ozias-Akins, P. Expression of a *Bacillus thuringiensis cryIA (c)* Gene in Transgenic Peanut Plants and Its Efficacy Against Lesser Cornstalk Borer. *Transgenic Res.* **1997,** *6* (2), 169–176.
32. Shu, Q.; Ye, G.; Cui, H.; Cheng, X.; Xiang, Y.; Wu, D.; Gao, M.; Xia, Y.; Hu, C.; Sardana, R.; Altosaar, I. Transgenic Rice Plants with a Synthetic *cry1Ab* Gene from *Bacillus thuringiensis* were Highly Resistant to Eight Lepidopteran Rice Pest Species. *Mol. Breed.* **2000,** *6* (4), 433–439.
33. Cao, J.; Zhao, J. Z.; Tang, J.; Shelton, A.; Earle, E. Broccoli Plants with Pyramided *cry1Ac* and *cry1C* Bt Genes Control Diamondback Moths Resistant to *Cry1A* and *Cry1C* Proteins. *Theor. Appl. Genet.* **2002,** *105* (2), 258–264.
34. Cao, J.; Shelton, A. M.; Earle, E. D. Development of Transgenic Collards (*Brassica oleracea* L., var. Acephala) Expressing a *cry1Ac or cry1C* Bt Gene for Control of the Diamondback Moth. *Crop Prot.* **2005,** *24* (9), 804–813.
35. Cao, J.; Shelton, A. M.; Earle, E. D. Sequential Transformation to Pyramid Two Bt Genes in Vegetable Indian Mustard (*Brassica Juncea* L.) and Its Potential for Control of Diamondback Moth Larvae. *Plant Cell Rep.* **2007,** *27* (3), 479.
36. USDA. Economic Research Service U.S. Department of Agriculture, Recent Trends in GE Adoption [Online], 2020. https://www.usda.gov/ (accessed Mar 15, 2022).
37. Huang, F.; Buschman, L. L.; Higgins, R. A.; McGaughey, W. H. Inheritance of Resistance to *Bacillus Thuringiensis* Toxin (Dipel ES) in the European Corn Borer. *Science* **1999,** *284* (5416), 965–967.
38. Akhurst, R. J.; James, W.; Bird, L. J.; Beard, C. Resistance to the Cry1Ac Delta-Endotoxin of *Bacillus thuringiensis* in the Cotton Bollworm, *Helicoverpa armigera* (Lepidoptera: Noctuidae). *J. Econ. Entomol.* **2003,** *96* (4), 1290–1299.
39. Tabashnik, B. E.; Van Rensburg, J. B.; Carrière, Y. Field-Evolved Insect Resistance to Bt Crops: Definition, Theory, and Data. *J. Econ. Entomol.* **2009,** *102* (6), 2011–2025.
40. Wright, D. J.; Iqbal, M.; Granero, F.; Ferre, J. A Change in a Single Midgut Receptor in the diamondback Moth (*Plutella xylostella*) is Only in Part Responsible for Field Resistance to *Bacillus Thuringiensis* Subsp. Kurstaki and *B. Thuringiensis* subsp. Aizawai. *Appl. Environ. Microbiol.* **1997,** *63* (5), 1814–1819.
41. Welch, K. L.; Unnithan, G. C.; Degain, B. A.; Wei, J.; Zhang, J.; Li, X.; Tabashnik, B. E.; Carrière, Y. Cross-Resistance to Toxins Used in Pyramided Bt Crops and Resistance to Bt Sprays in *Helicoverpa zea*. *J. Inverteb. Pathol.* **2015,** *132,* 149–156.
42. Tabashnik, B. E.; Carrière, Y. Surge in Insect Resistance to Transgenic Crops and Prospects for Sustainability. *Nat. Biotechnol.* **2017,** *35* (10), 926–935.

43. Zukoff, S. N.; Ostlie, K. R.; Potter, B.; Meihls, L. N.; Zukoff, A. L.; French, L.; Ellersieck, M. R.; Wade French, B.; Hibbard, B. E. Multiple Assays Indicate Varying Levels of Cross Resistance in Cry3Bb1-Selected Field Populations of the Western Corn Rootworm to mCry3A, eCry3.1Ab, and Cry34/35Ab1. *J. Econ. Entomol.* **2016,** *109* (3), 1387–1398.
44. Bernardi, D.; Salmeron, E.; Horikoshi, R. J.; Bernardi, O.; Dourado, P. M.; Carvalho, R. A.; Martinelli, S.; Head, G. P.; Omoto, C. Cross-Resistance Between Cry1 Proteins in Fall Armyworm (*Spodoptera frugiperda*) may Affect the Durability of Current Pyramided bt Maize Hybrids in Brazil. *PLoS One* **2015,** *10* (10), e0140130–e0140130.
45. Bakhsh, A.; Dinç, T.; Hussain, T.; Demirel, U.; Aasim, M.; Çalişkan, M. E. Development of Transgenic Tobacco Lines with Pyramided Insect Resistant Genes. *Turk. J. Biol.* **2018,** *42* (2), 174–186.
46. Mandal, C. C.; Gayen, S.; Basu, A.; Ghosh, K. S.; Dasgupta, S.; Maiti, M. K.; Sen, S. K. Prediction-Based Protein Engineering of Domain I of Cry2A Entomocidal Toxin of *Bacillus thuringiensis* for the Enhancement of Toxicity Against Lepidopteran Insects. *Protein Eng. Des. Sel.* **2007,** *20* (12), 599–606.
47. Rajamohan, F.; Alzate, O.; Cotrill, J. A.; Curtiss, A.; Dean, D. H. Protein Engineering of *Bacillus thuringiensis* δ-Endotoxin: Mutations at Domain II of CryIAb Enhance Receptor Affinity and Toxicity Toward Gypsy Moth Larvae. *J. Proceed. Natl Acad. Sci.* **1996,** *93* (25), 14338–14343.
48. Wu, J.; Luo, X.; Zhang, X.; Shi, Y.; Tian, Y. Development of Insect-Resistant Transgenic Cotton with Chimeric TVip3A* Accumulating in Chloroplasts. *Transgenic Res.* **2011,** *20* (5), 963–973.
49. Kiani, S.; Mohamed, B. B.; Shehzad, K.; Jamal, A.; Shahid, M. N.; Shahid, A. A.; Husnain, T. Chloroplast-Targeted Expression of Recombinant Crystal-Protein Gene in Cotton: An Unconventional Combat with Resistant Pests. *J. Biotechnol.* **2013,** *166* (3), 88–96.
50. Li, X.; Li, S.; Lang, Z.; Zhang, J.; Zhu, L.; Huang, D. Chloroplast-Targeted Expression of the Codon-Optimized Truncated *cry1Ah* Gene in Transgenic Tobacco Confers a High Level of Protection Against Insects. *Plant Cell Rep.* **2013,** *32* (8), 1299–1308.
51. Tabashnik, B. E.; Brévault, T.; Carrière, Y. Insect Resistance to Bt Crops: Lessons from the First Billion Acres. *Nat. Biotech.* **2013,** *31* (6), 510–521.
52. Sauka, D. H.; Peralta, C.; Pérez, M. P.; Onco, M. I.; Fiodor, A.; Caballero, J.; Caballero, P.; Berry, C.; Del Valle, E. E.; Palma, L. *Bacillus toyonensis* Biovar Thuringiensis: A Novel Entomopathogen with Insecticidal Activity Against Lepidopteran and Coleopteran Pests. *Biol. Control* **2022,** *167,* 104838.
53. Aasen, S. S.; Hågvar, E. B. Effect of Potato Plants Expressing Snowdrop Lectin (GNA) on the Performance and Colonization Behaviour of the Peach-Potato Aphid *Myzus Persicae*. *J. Acta Agric. Scand. B Soil Plant Sci.* **2012,** *62* (4), 352–361.
54. Down, R. E.; Gatehouse, A. M. R.; Hamilton, W. D. O.; Gatehouse, J. A. Snowdrop Lectin Inhibits Development and Decreases Fecundity of the Glasshouse Potato Aphid (*Aulacorthum Solani*) when Administered In Vitro and via Transgenic Plants both in Laboratory and Glasshouse Trials. *J. Insect Physiol.* **1996,** *42* (11), 1035–1045.
55. Guo, P.; Wang, Y.; Zhou, X.; Xie, Y.; Wu, H.; Gao, X. Expression of Soybean Lectin in Transgenic Tobacco Results in Enhanced Resistance to Pathogens and Pests. *Plant Sci.* **2013,** *211,* 17–22.

56. Narva, K. E.; Storer, N. P.; Meade, T. Discovery and Development of Insect-Resistant Crops Using Genes from *Bacillus thuringiensis*. In: *Advances in Insect Physiology*, Vol. 47; Dhadialla, T. S.; Gill, S. S., Eds. Trans.; Academic Press, 2014; pp 177–247.
57. Baum, J. A.; Bogaert, T.; Clinton, W.; Heck, G. R.; Feldmann, P.; Ilagan, O.; Johnson, S.; Plaetinck, G.; Munyikwa, T.; Pleau, M. Control of Coleopteran Insect Pests Through RNA Interference. *Nat. Biotechnol.* **2007,** *25* (11), 1322–1326.
58. Mao, Y. B.; Cai, W. J.; Wang, J. W.; Hong, G. J.; Tao, X. Y.; Wang, L. J.; Huang, Y. P.; Chen, X. Y. Silencing a Cotton Bollworm P450 Monooxygenase Gene by Plant-Mediated RNAi Impairs Larval Tolerance of Gossypol. *Nat. Biotechnol.* **2007,** *25* (11), 1307–1313.
59. Guo, H.; Song, X.; Wang, G.; Yang, K.; Wang, Y.; Niu, L.; Chen, X.; Fang, R. Plant-Generated Artificial Small RNAs Mediated Aphid Resistance. *PLoS One* **2014,** *9* (5), e97410.
60. Bhatia, V.; Bhattacharya, R. Host-Mediated RNA Interference Targeting a Cuticular Protein Gene Impaired Fecundity in the Green Peach Aphid *Myzus Persicae*. *Pest Manag. Sci.* **2018,** *74* (9), 2059–2068.
61. Rauf, I.; Asif, M.; Amin, I.; Naqvi, R. Z.; Umer, N.; Mansoor, S.; Jander, G. Silencing Cathepsin L Expression Reduces *Myzus Persicae* Protein Content and the Nutritional Value as Prey for *Coccinella septempunctata*. *Insect Mol. Biol.* **2019,** *28* (6), 785–797.
62. Agrawal, A.; Rajamani, V.; Reddy, V. S.; Mukherjee, S. K.; Bhatnagar, R. K. Transgenic Plants Over-Expressing Insect-Specific microRNA Acquire Insecticidal Activity Against *Helicoverpa armigera*: An Alternative to Bt-Toxin Technology. *Transgenic Res.* **2015,** *24* (5), 791–801.
63. Chen, C.-Y.; Liu, Y.-Q.; Song, W.-M.; Chen, D.-Y.; Chen, F.-Y.; Chen, X.-Y.; Chen, Z.-W.; Ge, S.-X.; Wang, C.-Z.; Zhan, S.; Chen, X.-Y.; Mao, Y.-B. An Effector from Cotton Bollworm Oral Secretion Impairs Host Plant Defense Signaling. *Proc. Natl Acad. Sci. U S A.* **2019,** *116* (28), 14331.
64. Saini, R. P.; Raman, V.; Dhandapani, G.; Malhotra, E. V.; Sreevathsa, R.; Kumar, P. A.; Sharma, T. R.; Pattanayak, D. Silencing of *HaAce1* Gene by Host-Delivered Artificial microRNA Disrupts Growth and Development of *Helicoverpa Armigera*. *PLoS One* **2018,** *13* (3), e0194150.
65. Eakteiman, G.; Moses-Koch, R.; Moshitzky, P.; Mestre-Rincon, N.; Vassão, D. G.; Luck, K.; Sertchook, R.; Malka, O.; Morin, S. Targeting Detoxification Genes by Phloem-Mediated RNAi: A New Approach for Controlling Phloem-Feeding Insect Pests. *Insect Biochem. Mol. Biol.* **2018,** *100,* 10–21.
66. Xu, H.-X.; Qian, L.-X.; Wang, X.-W.; Shao, R.-X.; Hong, Y.; Liu, S.-S.; Wang, X.-W. A Salivary Effector Enables Whitefly to Feed on Host Plants by Eliciting Salicylic Acid-Signaling Pathway. *J. Proceed. Natl Acad. Sci.* **2019,** *116* (2), 490–495.
67. Hussain, T.; Aksoy, E.; Çalışkan, M. E.; Bakhsh, A. Transgenic Potato Lines Expressing Hairpin RNAi Construct of Molting-Associated *EcR* Gene Exhibit Enhanced Resistance Against Colorado Potato Beetle (*Leptinotarsa Decemlineata*, Say). *Transgenic Res.* **2019,** *28* (1), 151–164.
68. Sun, Y.; Sparks, C.; Jones, H.; Riley, M.; Francis, F.; Du, W.; Xia, L. Silencing an Essential Gene Involved in Infestation and Digestion in Grain Aphid Through Plant-Mediated RNA Interference Generates Aphid-Resistant Wheat Plants. *Plant Biotechnol. J.* **2019,** *17* (5), 852–854.

69. Liu, X.; Zou, Z.; Zhang, C.; Liu, X.; Wang, J.; Xin, T.; Xia, B. Knockdown of the Trehalose-6-Phosphate Synthase Gene Using RNA Interference Inhibits Synthesis of Trehalose and Increases Lethality Rate in Asian Citrus Psyllid, *Diaphorina Citri* (Hemiptera: Psyllidae). *Insects* **2020,** *11* (9), 605.
70. Fu, S.; Liu, Z.; Chen, J.; Sun, G.; Jiang, Y.; Li, M.; Xiong, L.; Chen, S.; Zhou, Y.; Asad, M.; Yang, G. Silencing Arginine Kinase/Integrin β1 Subunit by Transgenic Plant Expressing dsRNA Inhibits the Development and Survival of *Plutella xylostella. Pest Manag. Sci.* **2020,** *76* (5), 1761–1771.
71. Jiang, S.; Wu, H.; Liu, H.; Zheng, J.; Lin, Y.; Chen, H. The Overexpression of Insect Endogenous Small RNAs in Transgenic Rice Inhibits Growth and Delays Pupation of Striped Stem Borer (*Chilo suppressalis*). *Pest Manag. Sci.* **2017,** *73* (7), 1453–1461.
72. He, K.; Xiao, H.; Sun, Y.; Ding, S.; Situ, G.; Li, F. Transgenic microRNA-14 Rice Shows High Resistance to Rice Stem Borer. *Plant Biotechnol. J.* **2019,** *17* (2), 461–471.
73. EPA. U.S. Environmental Protection Agency, Registers Innovative Tool to Control Corn Rootworm [Online], 2017. https://www.epa.gov/
74. Head, G. P.; Carroll, M. W.; Evans, S. P.; Rule, D. M.; Willse, A. R.; Clark, T. L.; Storer, N. P.; Flannagan, R. D.; Samuel, L. W.; Meinke, L. J. Evaluation of SmartStax and SmartStax PRO Maize Against Western Corn Rootworm and Northern Corn Rootworm: Efficacy and Resistance Management. *Pest Manag. Sci.* **2017,** *73* (9), 1883–1899.
75. Jin, S.; Singh, N. D.; Li, L.; Zhang, X.; Daniell, H. Engineered Chloroplast dsRNA Silences Cytochrome p450 Monooxygenase, V-ATPase and Chitin Synthase Genes in the Insect Gut and Disrupts *Helicoverpa armigera* Larval Development and Pupation. *Plant Biotechnol. J.* **2015,** *13* (3), 435–446.
76. Gordon, K. H. J.; Waterhouse, P. M. RNAi for Insect-Proof Plants. *Nat. Biotechnol.* **2007,** *25* (11), 1231–1232.
77. Zhang, J.; Khan, S. A.; Hasse, C.; Ruf, S.; Heckel, D. G.; Bock, R. Full Crop Protection from an Insect Pest by Expression of Long Double-Stranded RNAs in Plastids. *Science* **2015,** *347* (6225), 991–994.
78. Niu, J.; Yang, W.-J.; Tian, Y.; Fan, J.-Y.; Ye, C.; Shang, F.; Ding, B.-Y.; Zhang, J.; An, X.; Yang, L.; Chang, T.-Y.; Christiaens, O.; Smagghe, G.; Wang, J.-J. Topical dsRNA Delivery Induces Gene Silencing and Mortality in the Pea Aphid. *Pest Manag. Sci.* **2019,** *75* (11), 2873–2881.
79. Goel, D.; Singh, A. K.; Yadav, V.; Babbar, S. B.; Murata, N.; Bansal, K. C. Transformation of Tomato with a Bacterial *codA* Gene Enhances Tolerance to Salt and Water Stresses. *J. Plant Physiol.* **2011,** *168* (11), 1286–1294.
80. Li, H.; Wang, Z.; Ke, Q.; Ji, C. Y.; Jeong, J. C.; Lee, H.-S.; Lim, Y. P.; Xu, B.; Deng, X.-P.; Kwak, S.-S. Overexpression of *codA* Gene Confers Enhanced Tolerance to Abiotic Stresses in Alfalfa. *Plant Physiol. Biochem.* **2014,** *85,* 31–40.
81. Ke, Q.; Wang, Z.; Ji, C. Y.; Jeong, J. C.; Lee, H.-S.; Li, H.; Xu, B.; Deng, X.; Kwak, S.-S. Transgenic Poplar Expressing *codA* Exhibits Enhanced Growth and Abiotic Stress Tolerance. *Plant Physiol. Biochem.* **2016,** *100,* 75–84.
82. Wang, M.; Ren, T.; Huang, R.; Li, Y.; Zhang, C.; Xu, Z. Overexpression of an *Apocynum venetum* Flavonols Synthetase Gene Confers Salinity Stress Tolerance to Transgenic Tobacco Plants. *Plant Physiol. Biochem.* **2021,** *162,* 667–676.
83. Gautam, R.; Meena, R. K.; Woch, N.; Kirti, P. B. Ectopic Expression of *BrALDH7B2* Gene Encoding an Antiquitin from *Brassica rapa* Confers Tolerance to Abiotic Stresses

and Improves Photosynthetic Performance Under Salt Stress in Tobacco. *Environ. Exp. Bot.* **2020,** *180,* 104223.

84. Jiang, D.; Zhang, H.; Cai, H.; Gao, Z.; Chen, G. Overexpression of *ZmPCK2*, A Phosphoenolpyruvate Carboxykinase Gene from Maize Confers Enhanced Tolerance to Water Deficit Stress in Rice. *Plant Sci.* **2022,** *317,* 111195.
85. Li, X.; Hou, Y.; Li, M.; Zhang, F.; Yi, F.; Kang, J.; Yang, Q.; Long, R. Overexpression of an ABA-Inducible Homeodomain-Leucine Zipper I Gene MsHB7 Confers Salt Stress Sensitivity to Alfalfa. *Ind. Crops Prod.* **2022,** *177,* 114463.
86. Ye, Y.; Jia, X.; Xue, M.; Gao, Y.; Yue, H.; Ma, F.; Gong, X. MpSnRK2.10 confers Salt Stress Tolerance in Apple via the ABA Signaling Pathway. *Sci. Hortic.* **2022,** *298,* 110998.
87. Yang, J.; Wang, M.; Zhou, S.; Xu, B.; Chen, P.; Ma, F.; Mao, K. The ABA Receptor Gene *MdPYL9* Confers Tolerance to Drought Stress in Transgenic Apple (*Malus domestica*). *Environ. Exp. Bot.* **2022,** *194,* 104695.
88. Miao, R.; Zang, W.; Yuan, Y.; Zhang, Y.; Zhang, A.; Pang, Q. The Halophyte Gene *ScVTC2* Confers Resistance to Oxidative Stress via AsA-Mediated Photosynthetic Enhancement. *Plant Physiol. Biochem.* **2021,** *169,* 138–148.
89. Torti, P.; Raineri, J.; Mencia, R.; Campi, M.; Gonzalez, D. H.; Welchen, E. The Sunflower TLDc-Containing Protein HaOXR2 Confers Tolerance to Oxidative Stress and Waterlogging when Expressed in Maize Plants. *Plant Sci.* **2020,** *300,* 110626.
90. Fang, X.; Li, W.; Yuan, H.; Chen, H.; Bo, C.; Ma, Q.; Cai, R. Mutation of ZmWRKY86 Confers Enhanced Salt Stress Tolerance in Maize. *Plant Physiol. Biochem.* **2021,** *167,* 840–850.
91. Luo, P.; Chen, Y.; Rong, K.; Lu, Y.; Wang, N.; Xu, Z.; Pang, B.; Zhou, D.; Weng, J.; Li, M.; Zhang, D.; Yong, H.; Han, J.; Zhou, Z.; Gao, W.; Hao, Z.; Li, X. *ZmSNAC13*, a Maize NAC Transcription Factor Conferring Enhanced Resistance to Multiple Abiotic Stresses in Transgenic *Arabidopsis*. *Plant Physiol. Biochem.* **2022,** *170,* 160–170.
92. Patel, M. K.; Pandey, S.; Patel, J.; Mishra, A. A Type 2 Metallothionein (*SbMT-2*) Gene Cloned from *Salicornia brachiata* Confers Enhanced Zn Stress-Tolerance in Transgenic Tobacco by Transporting $Zn^{2+}$ and Maintaining Photosynthesis Efficacy. *Environ. Exp. Bot.* **2021,** *191,* 104626.
93. Tiwari, P.; Indoliya, Y.; Chauhan, A. S.; Pande, V.; Chakrabarty, D. Over-Expression of Rice R1-type MYB Transcription Factor Confers Different Abiotic Stress Tolerance in Transgenic *Arabidopsis*. *Ecotoxicol. Environ. Saf.* **2020,** *206,* 111361.
94. Meng, X.; Li, W.; Shen, R.; Lan, P. Ectopic Expression of IMA Small Peptide Genes Confers Tolerance to Cadmium Stress in Arabidopsis Through Activating the Iron Deficiency Response. *J. Hazard. Mater.* **2022,** *422,* 126913.
95. Zhang, Y.; Yang, L.; Zhang, M.; Yang, J.; Cui, J.; Hu, H.; Xu, J. CfAPX, a Cytosolic Ascorbate Peroxidase Gene from Cryptomeria Fortunei, Confers Tolerance to Abiotic Stress in Transgenic Arabidopsis. *Plant Physiol. Biochem.* **2022,** *172,* 167–179.
96. Zhang, Y.; Yu, H.; Yang, X.; Li, Q.; Ling, J.; Wang, H.; Gu, X.; Huang, S.; Jiang, W. CsWRKY46, a WRKY Transcription Factor from Cucumber, Confers Cold Resistance in Transgenic-Plant by Regulating a Set of Cold-Stress Responsive Genes in an ABA-Dependent Manner. *Plant Physiol. Biochem.* **2016,** *108,* 478–487.
97. Songstad, D.; Petolino, J.; Voytas, D. F.; Reichert, N. Genome Editing of Plants. *J. Crit. Rev. Plant Sci.* **2017,** *36* (1), 1–23.

98. Rajeevkumar, S.; Anunanthini, P.; Ramalingam, S. Epigenetic Silencing in Transgenic Plants. *J. Front. Plant Sci.* **2015,** *6,* 693.
99. Pérez-González, A.; Caro, E. Benefits of Using Genomic Insulators Flanking Transgenes to Increase Expression and Avoid Positional Effects. *J. Sci. Rep.* **2019,** *9* (1), 1–11.
100. Han, S. M.; Lee, B.; Won, O. J.; Hwang, K. S.; Suh, S. J.; Kim, C.; Park, K. W.; Environment. Gene Flow from Herbicide Resistant Genetically Modified Rice to Conventional Rice (*Oryza sativa* L.) Cultivars. *J. Ecol.* **2015,** *38* (4), 397–403.
101. Kim, D.-S.; Song, I.; Ko, K. Low Risk of Pollen-Mediated Gene Flow in Transgenic Plants Under Greenhouse Conditions. *Hortic. Environ. Biotechnol.* **2018,** *59* (5), 723–728.
102. Farquhar, J.; Zerkle, A. L.; Bekker, A. Geological Constraints on the Origin of Oxygenic Photosynthesis. *Photosynth. Res.* **2011,** *107* (1), 11–36.
103. Liu, C.-W.; Lin, C.-C.; Yiu, J.-C.; Chen, J. J. W.; Tseng, M.-J. Expression of a *Bacillus Thuringiensis* Toxin (*cry1Ab*) Gene in Cabbage (*Brassica Oleracea* L. var. Capitata L.) Chloroplasts Confers High Insecticidal Efficacy Against *Plutella Xylostella. Theor. Appl. Genet.* **2008,** *117* (1), 75–88.
104. De Cosa, B.; Moar, W.; Lee, S. B.; Miller, M.; Daniell, H. Overexpression of the Bt cry2Aa2 Operon in Chloroplasts Leads to Formation of Insecticidal Crystals. *Nat. Biotechnol.* **2001,** *19* (1), 71–74.
105. Chakrabarti, S. K.; Lutz, K. A.; Lertwiriyawong, B.; Svab, Z.; Maliga, P. Expression of the *cry9Aa2* B.t. Gene in Tobacco Chloroplasts Confers Resistance to Potato Tuber Moth. *Transgenic Res.* **2006,** *15* (4), 481–488.
106. Dufourmantel, N.; Tissot, G.; Goutorbe, F.; Garçon, F.; Muhr, C.; Jansens, S.; Pelissier, B.; Peltier, G.; Dubald, M. Generation and Analysis of Soybean Plastid Transformants Expressing *Bacillus thuringiensis* Cry1Ab Protoxin. *Plant Mol. Biol.* **2005,** *58* (5), 659–668.
107. Reddy, V. S.; Leelavathi, S.; Selvapandiyan, A.; Raman, R.; Giovanni, F.; Shukla, V.; Bhatnagar, R. K. Analysis of Chloroplast Transformed Tobacco Plants with cry1Ia5 under Rice psbA Transcriptional Elements Reveal High Level Expression of Bt Toxin Without Imposing Yield Penalty and Stable Inheritance of Transplastome. *Mol. Breed* **2002,** *9* (4), 259–269.
108. Wu, Y.; Xu, L.; Chang, L.; Ma, M.; You, L.; Jiang, C.; Li, S.; Zhang, J. *Bacillus thuringiensis cry1C* Expression from the Plastid Genome of Poplar Leads to High Mortality of Leaf-Eating Caterpillars. *Tree Physiol.* **2019,** *39* (9), 1525–1532.
109. Muzaffar, A.; Kiani, S.; Khan, M. A. U.; Rao, A. Q.; Ali, A.; Awan, M. F.; Iqbal, A.; Nasir, I. A.; Shahid, A. A.; Husnain, T. Chloroplast Localization of Cry1Ac and Cry2A Protein—An Alternative Way of Insect Control in Cotton. *Biol. Res.* **2015,** *48* (1), 1–11.
110. DeGray, G.; Rajasekaran, K.; Smith, F.; Sanford, J.; Daniell, H. Expression of An Antimicrobial Peptide via the Chloroplast Genome to Control Phytopathogenic Bacteria and Fungi. *Plant Physiol.* **2001,** *127* (3), 852–862.
111. Wang, Y.-P.; Wei, Z.-Y.; Zhang, Y.-Y.; Lin, C.-J.; Zhong, X.-F.; Wang, Y.-L.; Ma, J.-Y.; Ma, J.; Xing, S.-C. Chloroplast-Expressed MSI-99 in Tobacco Improves Disease Resistance and Displays Inhibitory Effect Against Rice Blast Fungus. *Int. J. Mol. Sci.* **2015,** *16* (3), 4628–4641.
112. Jin, S.; Zhang, X.; Daniell, H. *Pinellia Ternata* Agglutinin Expression in Chloroplasts Confers Broad Spectrum Resistance Against Aphid, Whitefly, Lepidopteran Insects, Bacterial and Viral Pathogens. *Plant Biotechnol. J.* **2011,** *10* (3), 313–327.

113. Ruhlman, T. A.; Rajasekaran, K.; Cary, J. W. Expression of Chloroperoxidase from *Pseudomonas pyrrocinia* in Tobacco Plastids for Fungal Resistance. *Plant Sci.* **2014,** *228,* 98–106.
114. Bally, J.; McIntyre, G. J.; Doran, R. L.; Lee, K.; Perez, A.; Jung, H.; Naim, F.; Larrinua, I. M.; Narva, K. E.; Waterhouse, P. M. In-Plant Protection Against *Helicoverpa armigera* by Production of Long hpRNA in Chloroplasts. *Front. Plant Sci.* **2016,** 7, 1453.
115. Ren, B.; Cao, J.; He, Y.; Yang, S.; Zhang, J. Assessment on Effects of Transplastomic Potato Plants Expressing Colorado Potato Beetle β-Actin Double-Stranded RNAs for Three Non-Target Pests. *Pestic. Biochem. Phys.* **2021,** *178,* 104909.
116. Burke, W.; Kaplanoglu, E.; Kolotilin, I.; Menassa, R.; Donly, C. RNA Interference in the Tobacco Hornworm, *Manduca Sexta*, using Plastid-Encoded Long Double-Stranded RNA. *Front. Plant Sci.* **2019,** *10,* 313.
117. He, W.; Xu, W.; Xu, L.; Fu, K.; Guo, W.; Bock, R.; Zhang, J. Length-Dependent Accumulation of Double-Stranded RNAs in Plastids Affects RNA Interference Efficiency in the Colorado Potato Beetle. *J. Exp. Bot.* **2020,** *71* (9), 2670–2677.
118. Dong, Y.; Yang, Y.; Wang, Z.; Wu, M.; Fu, J.; Guo, J.; Chang, L.; Zhang, J. Inaccessibility to Double-Stranded RNAs in Plastids Restricts RNA Interference in *Bemisia tabaci* (whitefly). *Pest Manag. Sci.* **2020.** 10.1002/ps.5871
119. Turner, C.; Davy, M.; MacDiarmid, R.; Plummer, K.; Birch, N.; Newcomb, R. RNA interference in the Light Brown Apple Moth, *Epiphyas postvittana* (Walker) Induced by Double-Stranded RNA Feeding. *Insect Mol. Biol.* **2006,** *15* (3), 383–391.
120. Mittler, R. Oxidative Stress, Antioxidants and Stress Tolerance. *Trends Plant Sci.* **2002,** *7* (9), 405–410.
121. Le Martret, B.; Poage, M.; Shiel, K.; Nugent, G. D.; Dix, P. J. Tobacco Chloroplast Transformants Expressing Genes Encoding Dehydroascorbate Reductase, Glutathione Reductase, and Glutathione-S-Transferase, Exhibit Altered Anti-Oxidant Metabolism and Improved Abiotic Stress Tolerance. *Plant Biotechnol. J.* **2011,** *9* (6), 661–673.
122. Zhang, J.; Tan, W.; Yang, X.-H.; Zhang, H.-X. Plastid-Expressed Choline Monooxygenase Gene Improves Salt and Drought Tolerance Through Accumulation of Glycine Betaine in Tobacco. *Plant Cell Rep.* **2008,** *27* (6), 1113.
123. You, L.; Song, Q.; Wu, Y.; Li, S.; Jiang, C.; Chang, L.; Yang, X.; Zhang, J. Accumulation of Glycine Betaine in transplastomic potato plants expressing choline oxidase confers improved drought tolerance. *Planta* **2019,** *249* (6), 1963–1975.
124. Stavridou, E.; Michailidis, M.; Gedeon, S.; Ioakeim, A.; Kostas, S.; Chronopoulou, E.; Labrou, N. E.; Edwards, R.; Day, A.; Nianiou-Obeidat, I. Tolerance of Transplastomic Tobacco Plants Overexpressing a Theta Class Glutathione Transferase to Abiotic and Oxidative Stresses. *Front. Plant Sci.* **2019,** *9,* 1861.
125. Poage, M.; Le Martret, B.; Jansen, M. A. K.; Nugent, G. D.; Dix, P. J. Modification of Reactive Oxygen Species Scavenging Capacity of Chloroplasts Through Plastid Transformation. *Plant Mol. Biol.* **2011,** *76* (3), 371–384.
126. Khan, M. S.; Kanwal, B.; Nazir, S. Metabolic Engineering of the Chloroplast Genome Reveals that the Yeast *ArDH* Gene Confers Enhanced Tolerance to Salinity and Drought in Plants. *Front. Plant Sci.* **2015,** *6,* 725.
127. Kumar, S.; Dhingra, A.; Daniell, H. Plastid-Expressed Betaine Aldehyde Dehydrogenase Gene in Carrot Cultured Cells, Roots, and Leaves Confers Enhanced Salt Tolerance. *Plant Physiol.* **2004,** *136* (1), 2843–2854.

128. 128.Lu, Y.; Rijzaani, H.; Karcher, D.; Ruf, S.; Bock, R. Efficient Metabolic Pathway Engineering in Transgenic Tobacco and Tomato Plastids with Synthetic Multigene Operons. *Proc. Natl Acad. Sci. U S A.* **2013,** *110* (8), E623–E632.
129. Bansal, K.; Singh, A.; Wani, S. Plastid Transformation for Abiotic Stress Tolerance in Plants. In: *Plant Salt Tolerance*, Vol. 913; Shabala, S.; Cuin, T., Eds.; Humana Press: Totowa, NJ, 2012; pp 351–358.
130. Ceccoli, R. D.; Blanco, N. E.; Segretin, M. E.; Melzer, M.; Hanke, G. T.; Scheibe, R.; Hajirezaei, M.-R.; Bravo-Almonacid, F. F.; Carrillo, N. Flavodoxin Displays Dose-Dependent Effects on Photosynthesis and Stress Tolerance when Expressed in Transgenic Tobacco Plants. *Planta* **2012,** *236* (5), 1447–1458.
131. Grant, O. M.; Brennan, D. P.; Mellisho Salas, C. D.; Dix, P. J. Impact of Enhanced Capacity to Scavenge Reactive Oxygen Species on Cold Tolerance of Tobacco. *Int. J. Plant Sci.* **2014,** *175* (5), 544–554.
132. Wang, Y.; Wei, Z.; Xing, S. Stable Plastid Transformation of Rice, a Monocot Cereal Crop. *Biochem. Biophys. Res. Commun.* **2018,** *503* (4), 2376–2379.
133. FAO. World Food Situation, Cereal Production, Utilization, and Trade Reaching Record Levels in 2021/22 [Online], 2022. https://www.fao.org/worldfoodsituation/csdb/en/ (Mar 17, 2022).
134. Papageorgiou, M.; Skendi, A. 1 - Introduction to Cereal Processing and by-Products. *Sustainable Recovery and Reutilization of Cereal Processing By-Products*; Galanakis, C. M., Ed.; Woodhead Publishing, 2018; pp 1–25.
135. Laskowski, W.; Górska-Warsewicz, H.; Rejman, K.; Czeczotko, M.; Zwolińska, J. How Important are Cereals and Cereal Products in the Average Polish Diet? *Nutrients* **2019,** *11* (3), 679.
136. Labuschagne, M. A Review of Cereal Grain Proteomics and Its Potential for Sorghum Improvement. *J. Cereal. Sci.* **2018,** *84,* 151–158.
137. Daniell, H.; Krishnan, M.; McFadden, B. Transient Expression of β-Glucuronidase in Different Cellular Compartments Following Biolistic Delivery of Foreign DNA into Wheat Leaves and Calli. *J. Plant Cell Rep.* **1991,** *9* (11), 615–619.
138. Khan, M. S.; Maliga, P. Fluorescent Antibiotic Resistance Marker for Tracking Plastid Transformation in Higher Plants. *Nat. Biotechnol.* **1999,** *17* (9), 910.
139. Lee, S. M.; Kang, K.; Chung, H.; Yoo, S. H.; Xu, X. M.; Lee, S. B.; Cheong, J. J.; Daniell, H.; Kim, M. Plastid Transformation in the Monocotyledonous Cereal Crop, Rice (*Oryza sativa*) and Transmission of Transgenes to their Progeny. *Mol. Cells* **2006,** *21* (3), 401–410.
140. Cui, C.; Song, F.; Tan, Y.; Zhou, X.; Zhao, W.; Ma, F.; Liu, Y.; Hussain, J.; Wang, Y.; Yang, G. Stable Chloroplast Transformation of Immature Scutella and Inflorescences in Wheat (*Triticum aestivum* L.). *Acta Biochim. Biophys. Sin.* **2011,** *43* (4), 284–291.
141. He, G. Retraction of Article: Stable Chloroplast Transformation of Immature Scutella and Inflorescences in Wheat (*Triticum aestivum* L.). *Acta Biochim. Biophys. Sin.* **2012,** *44* (4), 373–373.
142. FAO. Crops and Livestock Products [Online], 2020. https://www.fao.org/faostat/en/#data/QCL (accessed Mar 24, 2022).
143. Arévalo-Gallegos, S.; Varela-Rodríguez, H.; Lugo-Aguilar, H.; Siqueiros-Cendón, T. S.; Iglesias-Figueroa, B. F.; Espinoza-Sánchez, E. A.; Aguado-Santacruz, G. A.; Rascón-Cruz, Q. Transient Expression of a Green Fluorescent Protein in Tobacco and Maize Chloroplast. *Electron J. Biotechnol.* **2020,** *45,* 1–20.

144. Zheng, Q.; Oldenburg, D. J.; Bendich, A. J. Independent Effects of Leaf Growth and Light on the Development of the Plastid and its DNA Content in *Zea* Species. *J. Exp. Bot.* **2011,** *62* (8), 2715–2730.
145. Valkov, V. T.; Scotti, N.; Kahlau, S.; MacLean, D.; Grillo, S.; Gray, J. C.; Bock, R.; Cardi, T. Genome-Wide Analysis of Plastid Gene Expression in Potato Leaf Chloroplasts and Tuber Amyloplasts: Transcriptional and Posttranscriptional Control. *Plant Physiol.* **2009,** *150* (4), 2030–2044.
146. Madanala, R.; Gupta, V.; Singh, P. K.; Tuli, R. Development of Chloroplast Transformation Vectors, and a New Target Region in the Tobacco Plastid Genome. *Plant Biotechnol. Rep.* **2012,** *6* (1), 77–87.
147. Kuroda, H.; Maliga, P. Complementarity of the 16S rRNA Penultimate Stem with Sequences Downstream of the AUG Destabilizes the Plastid mRNAs. *Nucleic Acids Res.* **2001,** *29* (4), 970–975.
148. Rigano, M. M.; Scotti, N.; Cardi, T. Unsolved Problems in Plastid Transformation. *Bioengineered* **2012,** *3* (6), 329–333.
149. Scotti, N.; Valkov, V. T.; Cardi, T. Improvement of Plastid Transformation Efficiency in Potato by Using Vectors with homologous flanking sequences. *GM Crops* **2011,** *2* (2), 89–91.
150. Ruf, S.; Hermann, M.; Berger, I. J.; Carrer, H.; Bock, R. Stable Genetic Transformation of Tomato Plastids and Expression of a Foreign Protein in Fruit. *Nat. Biotechnol.* **2001,** *19* (9), 870.
151. Verma, D.; Daniell, H. Chloroplast Vector Systems for Biotechnology Applications. *Plant Physiol.* **2007,** *145* (4), 1129–1143.
152. Li, D.; Tang, N.; Fang, Z.; Xia, Y.; Cao, M. Co-transfer of TALENs Construct Targeted for Chloroplast Genome and Chloroplast Transformation Vector into Rice Using Particle Bombardment. *J. Nanosci. Nanotechnol.* **2016,** *16* (12), 12194–12201.
153. Gutierrez, C. L.; Gimpel, J.; Escobar, C.; Marshall, S. H.; Henríquez, V. Chloroplast Genetic Tool for the Green Microalgae *Haematococcus pluvialis* (Chlorophyceae, Volvocales). *J. Phycol.* **2012,** *48* (4), 976–983.
154. Georgianna, D. R.; Hannon, M. J.; Marcuschi, M.; Wu, S.; Botsch, K.; Lewis, A. J.; Hyun, J.; Mendez, M.; Mayfield, S. P. Production of Recombinant Enzymes in the Marine Alga *Dunaliella tertiolecta*. *Algal Res.* **2013,** *2* (1), 2–9.
155. Goldschmidt-Clermont, M. Transgenic Expression of Aminoglycoside Adenine Transferase in the Chloroplast: A Selectable Marker for Site-Directed Transformation of *Chlamydomonas*. *Nucleic Acids Res.* **1991,** *19* (15), 4083–4089.
156. Doetsch, N. A.; Favreau, M. R.; Kuscuoglu, N.; Thompson, M. D.; Hallick, R. B. Chloroplast Transformation in *Euglena gracilis*: Splicing of a Group III Twintron Transcribed from a Transgenic psbK Operon. *Curr. Genet.* **2001,** *39* (1), 49–60.
157. Lapidot, M.; Raveh, D.; Sivan, A.; Arad, S. M.; Shapira, M. Stable Chloroplast Transformation of the Unicellular Red Alga *Porphyridium* Species. *Plant Physiol.* **2002,** *129* (1), 7–12.
158. Kong, F.; Zhao, H.; Liu, W.; Li, N.; Mao, Y. Construction of Plastid Expression Vector and Development of Genetic Transformation System for the Seaweed *Pyropia yezoensis*. *Mar. Biotechnol.* **2017,** *19* (2), 147–156.
159. Zienkiewicz, M.; Krupnik, T.; Drożak, A.; Golke, A.; Romanowska, E. Transformation of the *Cyanidioschyzon merolae* Chloroplast Genome: Prospects for Understanding Chloroplast Function in Extreme Environments. *Plant Mol. Biol.* **2017,** *93* (1), 171–183.

160. Xie, W.-H.; Zhu, C.-C.; Zhang, N.-S.; Li, D.-W.; Yang, W.-D.; Liu, J.-S.; Sathishkumar, R.; Li, H.-Y. Construction of Novel Chloroplast Expression Vector and Development of an Efficient Transformation System for the Diatom *Phaeodactylum tricornutum*. *Mar. Biotechnol.* **2014,** *16* (5), 538–546.
161. Gan, Q.; Jiang, J.; Han, X.; Wang, S.; Lu, Y. Engineering the Chloroplast Genome of Oleaginous Marine Microalga *Nannochloropsis oceanica*. *Front. Plant Sci.* **2018,** *9,* 439.
162. Bo, Y.; Wang, K.; Wu, Y.; Cao, H.; Cui, Y.; Wang, L. Establishment of a Chloroplast Transformation System in *Tisochrysis lutea*. *J. Appl. Phycol.* **2020,** *32* (5), 2959–2965.
163. Chiyoda, S.; Linley, P. J.; Yamato, K. T.; Fukuzawa, H.; Yokota, A.; Kohchi, T. Simple and Efficient Plastid Transformation System for the Liverwort *Marchantia polymorpha* L. Suspension-Culture Cells. *Transgenic Res.* **2007,** *16* (1), 41–49.
164. Sugiura, C.; Sugita, M. Plastid Transformation Reveals that Moss tRNAArg-CCG is Not Essential for Plastid Function. *Plant J.* **2004,** *40* (2), 314–321.
165. Svab, Z.; Hajdukiewicz, P.; Maliga, P. Stable Transformation of Plastids in Higher Plants. *Proc. Natl Acad. Sci. U S A.* **1990,** *87* (21), 8526–8530.
166. Svab, Z.; Maliga, P. High-Frequency Plastid Transformation in Tobacco by Selection for a Chimeric *aadA* Gene. *Proc. Natl Acad. Sci. U S A.* **1993,** *90* (3), 913–917.
167. Davarpanah, S. J.; Jung, S. H.; Kim, Y. J.; Park, Y.-I.; Min, S. R.; Liu, J. R.; Jeong, W. J. Stable Plastid Transformation in *Nicotiana benthamiana*. *J. Plant Biol.* **2009,** *52* (3), 244–250.
168. Maliga, P.; Svab, Z. Engineering the Plastid Genome of *Nicotiana sylvestris*, a Diploid Model Species for Plastid Genetics. In: *Plant Chromosome Engineering: Methods and Protocols*, Vol. 701; Birchler, J., Eds. Trans.; Humana Press: Totowa, NJ, 2010, pp. 10.1007/978-1-61737-957-4_2
169. Zubko, M. K.; Zubko, E. I.; Zuilen, K. v.; Meyer, P.; Day, A. Stable Transformation of Petunia Plastids. *Transgenic Res.* **2004,** *13* (6), 523–530.
170. Sidorov, V. A.; Kasten, D.; Pang, S. Z.; Hajdukiewicz, P. T.; Staub, J. M.; Nehra, N. S. Stable Chloroplast Transformation in Potato: Use of Green Fluorescent Protein as a Plastid Marker. *Plant J.* **1999,** *19* (2), 209–216.
171. Valkov, V. T.; Gargano, D.; Manna, C.; Formisano, G.; Dix, P. J.; Gray, J. C.; Scotti, N.; Cardi, T. High Efficiency Plastid Transformation in Potato and Regulation of Transgene Expression in Leaves and Tubers by Alternative 5′ and 3′ Regulatory Sequences. *Transgenic Res.* **2011,** *20* (1), 137–151.
172. Kota, S.; Lakkam, R.; Kasula, K.; Narra, M.; Qiang, H.; Rao Allini, V.; Zanmin, H.; Abbagani, S. Construction of a Species-Specific Vector for Improved Plastid Transformation Efficiency in *Capsicum annuum* L. *3 Biotech.* **2019,** *9* (6), 226.
173. Singh, A. K.; Verma, S. S.; Bansal, K. C. Plastid Transformation in Eggplant *(Solanum melongena* L.). *Transgenic Res.* **2010,** *19* (1), 113–119.
174. Skarjinskaia, M.; Svab, Z.; Maliga, P. Plastid Transformation in *Lesquerella Fendleri,* an Oilseed *Brassicacea*. *Transgenic Res.* **2003,** *12* (1), 115–122.
175. Hou, B.-K.; Zhou, Y.-H.; Wan, L.-H.; Zhang, Z.-L.; Shen, G.-F.; Chen, Z.-H.; Hu, Z.-M. Chloroplast Transformation in Oilseed Rape. *Transgenic Res.* **2003,** *12* (1), 111–114.
176. Cheng, L.; Li, H.-P.; Qu, B.; Huang, T.; Tu, J.-X.; Fu, T.-D.; Liao, Y.-C. Chloroplast Transformation of Rapeseed (*Brassica napus*) by Particle Bombardment of Cotyledons. *J. Plant Cell. Rep.* **2010,** *29* (4), 371–381.

177. Sikdar, S.; Serino, G.; Chaudhuri, S.; Maliga, P. Plastid Transformation in *Arabidopsis thaliana*. *Plant Cell Rep.* **1998,** *18* (1–2), 20–24.
178. Liu, C. W.; Lin, C. C.; Chen, J. J.; Tseng, M. J. Stable Chloroplast Transformation in Cabbage (*Brassica oleracea* L. var. capitata L.) by Particle Bombardment. *Plant Cell Rep.* **2007,** *26* (10), 1733–1744.
179. Nugent, G. D.; Coyne, S.; Nguyen, T. T.; Kavanagh, T. A.; Dix, P. J. Nuclear and Plastid Transformation of *Brassica Oleracea* var. Botrytis (Cauliflower) Using PEG-Mediated Uptake of DNA into Protoplasts. *Plant Sci.* **2006,** *170* (1), 135–142.
180. Kumar, S.; Dhingra, A.; Daniell, H. Stable Transformation of the Cotton Plastid Genome and Maternal Inheritance of Transgenes. *Plant Mol. Biol.* **2004,** *56* (2), 203–216.
181. Wei, Z.; Liu, Y.; Lin, C.; Wang, Y.; Cai, Q. A.; Dong, Y.; Xing, S. Transformation of Alfalfa Chloroplasts and Expression of Green Fluorescent Protein in a Forage Crop. *Biotechnol. Lett.* **2011,** *33* (12), 2487–2494.
182. Queiroz, L. N.; Maldaner, F. R.; Mendes, É. A.; Sousa, A. R.; D'Allastta, R. C.; Mendonça, G.; Mendonça, D. B.; Aragão, F. J. Evaluation of Lettuce Chloroplast and Soybean Cotyledon as Platforms for Production of Functional Bone Morphogenetic Protein 2. *Transgenic Res.* **2019,** *28* (2), 213–224.
183. Lelivelt, C. L.; McCabe, M. S.; Newell, C. A.; Bastiaan deSnoo, C.; Van Dun, K. M.; Birch-Machin, I.; Gray, J. C.; Mills, K. H.; Nugent, J. M. Stable Plastid Transformation in Lettuce (*Lactuca sativa* L.). *Plant Mol. Biol.* **2005,** *58* (6), 763–774.
184. Kanamoto, H.; Yamashita, A.; Asao, H.; Okumura, S.; Takase, H.; Hattori, M.; Yokota, A.; Tomizawa, K.-I. Efficient and Stable Transformation of *Lactuca sativa* L. cv. Cisco (Lettuce) Plastids. *Transgenic Res.* **2006,** *15* (2), 205–217.
185. Kaushal, C.; Abdin, M. Z.; Kumar, S. Chloroplast Genome Transformation of Medicinal Plant *Artemisia annua*. *Plant Biotechnol. J.* **2020**.
186. Muralikrishna, N.; Srinivas, K.; Kumar, K. B.; Sadanandam, A. Stable Plastid Transformation in *Scoparia dulcis* L. *Physiol. Mol. Biol. Plants* **2016,** *22* (4), 575–581.
187. Dufourmantel, N.; Pelissier, B.; Garcon, F.; Peltier, G.; Ferullo, J. M.; Tissot, G. Generation of Fertile Transplastomic Soybean. *Plant Mol. Biol.* **2004,** *55* (4), 479–489.
188. De Marchis, F.; Wang, Y.; Stevanato, P.; Arcioni, S.; Bellucci, M. Genetic Transformation of the Sugar Beet Plastome. *Transgenic Res.* **2008,** *18* (1), 17.
189. Okumura, S.; Sawada, M.; Park, Y. W.; Hayashi, T.; Shimamura, M.; Takase, H.; Tomizawa, K.-I. Transformation of Poplar (*Populus alba*) Plastids and Expression of Foreign Proteins in Tree Chloroplasts. *Transgenic Res.* **2006,** *15* (5), 637–646.
190. Narra, M.; Kota, S.; Velivela, Y.; Ellendula, R.; Allini, V. R.; Abbagani, S. Construction of Chloroplast Transformation Vector and Its Functional Evaluation in *Momordica charantia* L. *3 Biotech.* **2018,** *8* (3), 140.
191. Lin, C.-H.; Chen, Y.-Y.; Tzeng, C.-C.; Tsay, H.-S.; Chen, L.-J. Expression of a *Bacillus Thuringiensis cry1C* Gene in Plastid Confers High Insecticidal Efficacy Against Tobacco Cutworm a Spodoptera Insect. *Bot. Bull. Acad. Sinica* **2003,** *44,* 199–210.
192. Chen, P. J.; Senthilkumar, R.; Jane, W. N.; He, Y.; Tian, Z.; Yeh, K. W. Transplastomic *Nicotiana benthamiana* Plants Expressing Multiple Defence Genes Encoding Protease Inhibitors and Chitinase Display Broad-Spectrum Resistance Against Insects, Pathogens and Abiotic Stresses. *Plant Biotechnol. J.* **2014,** *12* (4), 503–515.

CHAPTER 10

# Impact of *Moringa oleifera* in Agronomic Sciences: A Review

ROMEO ROJAS, DARYTZA MABEL TREVIÑO-GÓMEZ, ISRAEL BAUTISTA-HERNÁNDEZ, MARÍA DOLORES GONZÁLEZ-HERNÁNDEZ, and GUILLERMO CRISTIAN GUADALUPE MARTÍNEZ-ÁVILA

*Universidad Autónoma de Nuevo León. School of Agronomy, General Escobedo, Nuevo León, México*

## ABSTRACT

The Moringa plant (Moringa oleifera) is native to India and Africa. It is characterized by growing in tropical and arid climates. It has been extensively studied because it is a potential source of bioactive compounds with applications in agriculture and food additives. The compounds are present in leaves, stems, and roots and today have a wide application in modern agriculture since they have been used successfully in bioremediation for their ability to remove heavy metals, promote the germination of seeds of important crops such as corn and wheat. Also, it has allelopathic properties and in animal nutrition it has been used as feed for cattle due to its high protein content. In addition, the oil that is extracted from its seeds has been used to produce biodiesel since it has a high content of oleic acid, being an ideal substitute for petrodiesel. All this, without considering that it also has medicinal properties. Therefore, Moringa is a viable alternative for its use and application in modern agriculture, bioremediation, and livestock feeding. Likewise, the bioactive compounds that are extracted can be applied in other areas as a food additive (for humans) and in the medical area due to their antioxidant properties.

Sustainable Agricultural Production Systems: Traditional and Modern Perspectives.
Romeo Rojas, Guillermo Cristian Guadalupe Martínez-Ávila, Julia Mariana Márquez-Reyes, and Ma del Carmen Ojeda-Zacarías (Eds.)

## 10.1 INTRODUCTION

Moringa belongs to a *Moringaceae* family, includes 13 species of trees and shrubs that originate in India and Africa; likewise, it is distributed in various tropical and arid countries.[44] Due to is chemical composition, it has been demonstrated that Moringa has many benefits and applications, which include several antimicrobial activities (bacteria, fungi, and yeasts);[3] modulation of the expression of fertility;[11] antioxidant activity;[49] water waste management;[18] and many others for human health and nutrition.[1,44] Moreover, moringa extracts have also been applied to improve the quality and growth of some crops with agronomic importance in different ways.[2,29] A wide number of bioactive compounds (Figure 10.1), moringa plant has been recognized in the plant including carbohydrates, functional proteins, and peptides, among others.[44] According to some authors, the bioactive compounds attributed to this plant can be found in all the organs, but most of the researches focus on the leaves. For example, some studies have reported the presence and importance of polyphenolic compounds extracted from various parts of Moringa, as well as their activity in vitro and in vivo. Torres-Castillo et al.[49] reported the content of phenolic compounds of leaves, roots, and stems of Moringa, being the leaves the best source of phenolic compounds and with a higher antioxidant activity. Among the compounds found, the main ones include 4 lignans, 26 flavonoids, and 11 phenolic acids and derivatives.[42] Thus, due to their wide chemical composition, *M. oleifera* extracts have several potential applications in agronomic sciences.

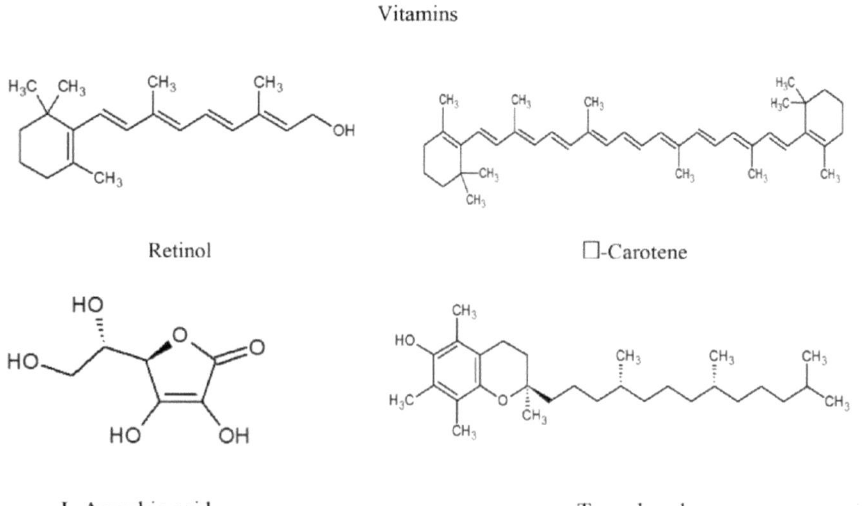


**FIGURE 10.1** *(Continued)*

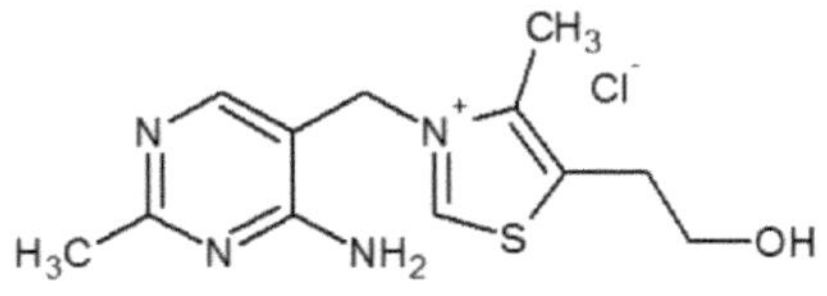

Thiamine

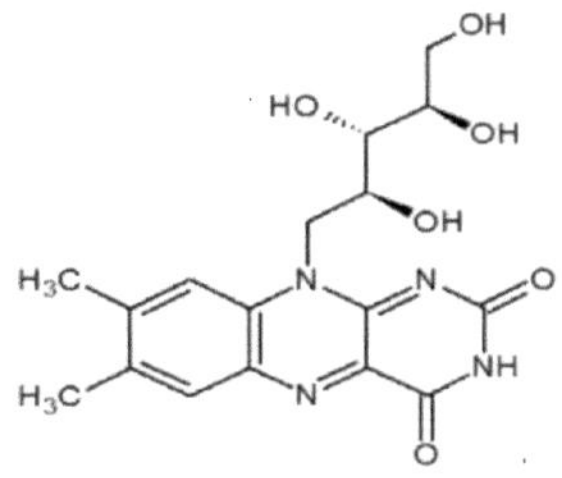

Riboflavin

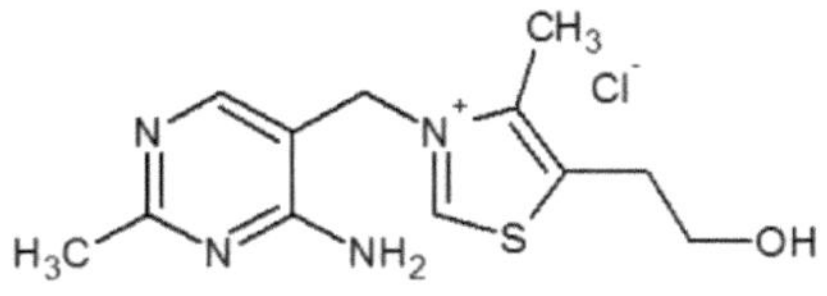

Thiamine

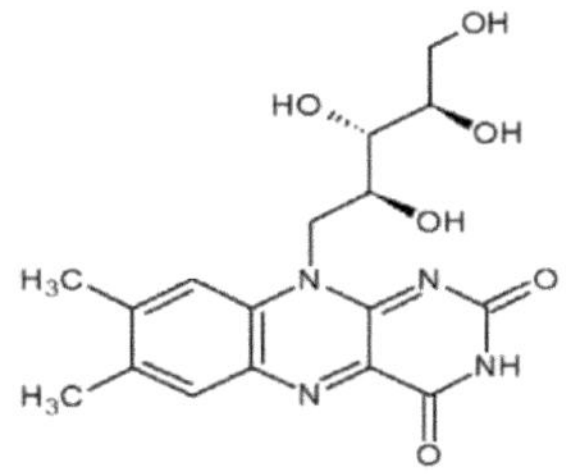

Riboflavin

**Flavonoids**

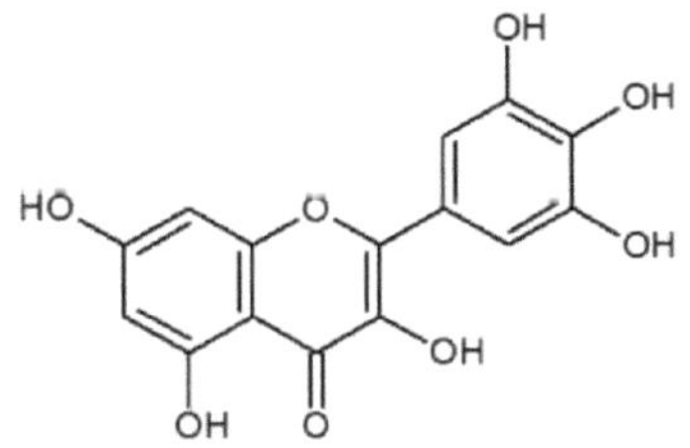

Myricetin

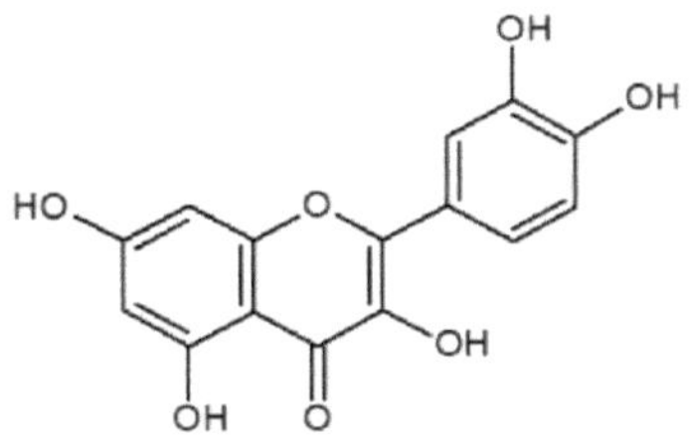

Quercetin

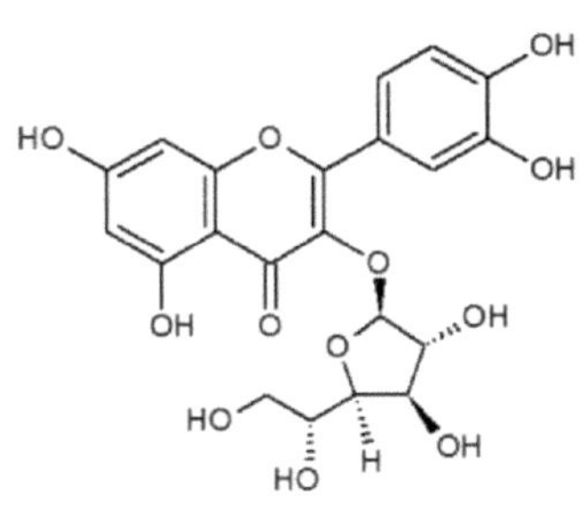

Isoquercetin

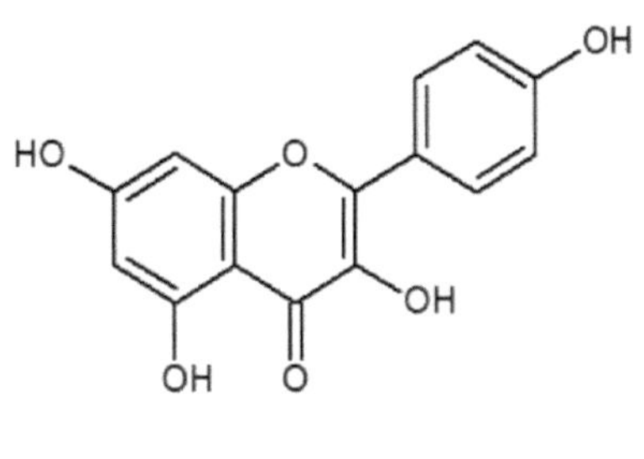

Kaempferol

**FIGURE 10.1** *(Continued)*

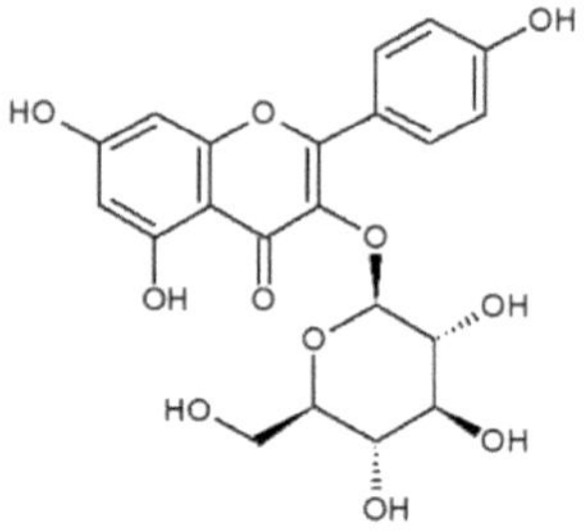

Astragalin

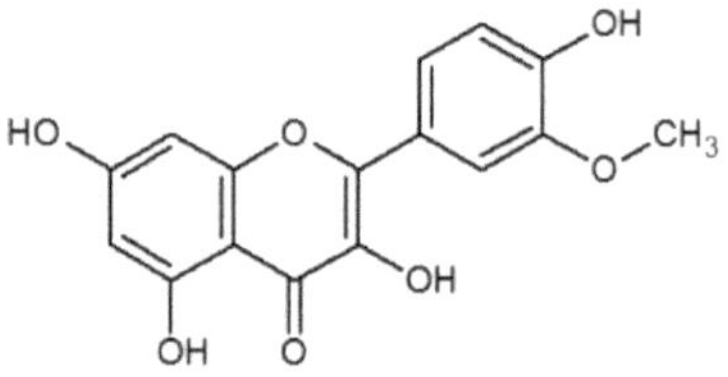

Isorhamnetin

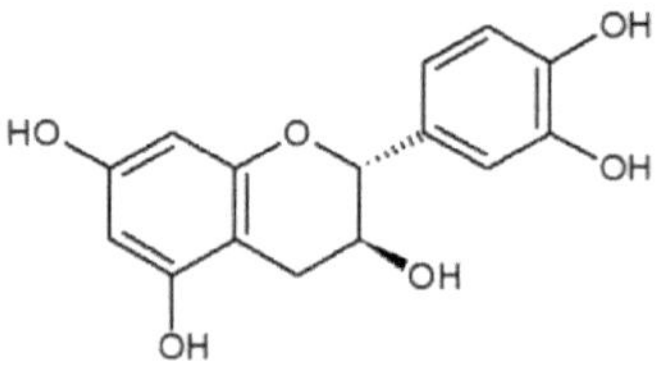

(-) Epicatechin

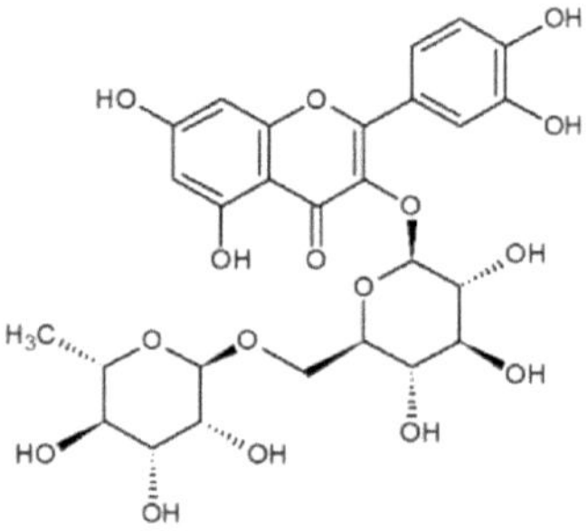

Rutin

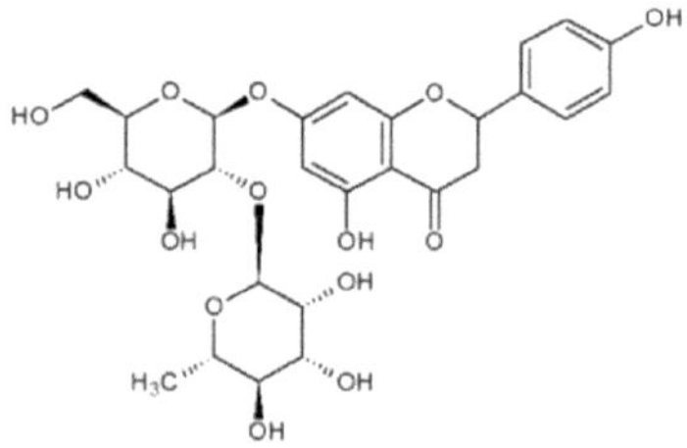

Naringin

**Phenolic acids**

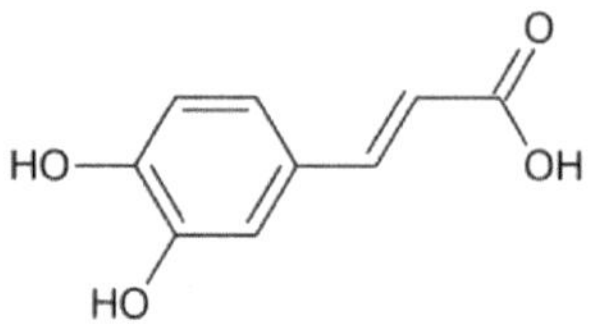

Caffeic acid

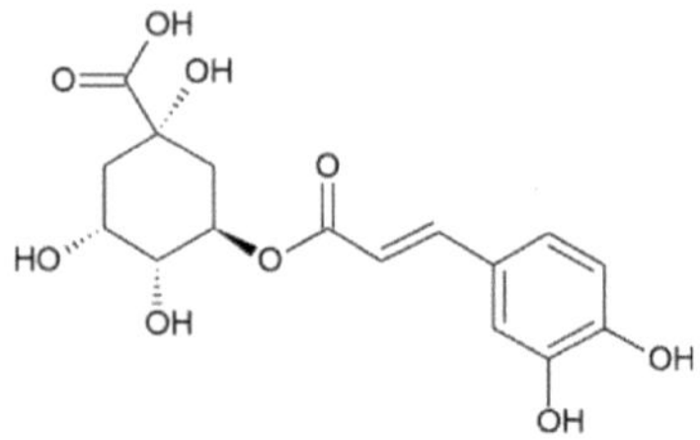

Chlorogenic acid

**FIGURE 10.1** *(Continued)*

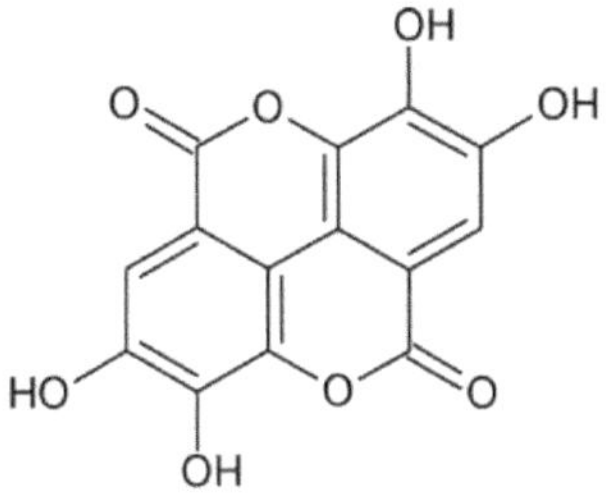
Ellagic acid

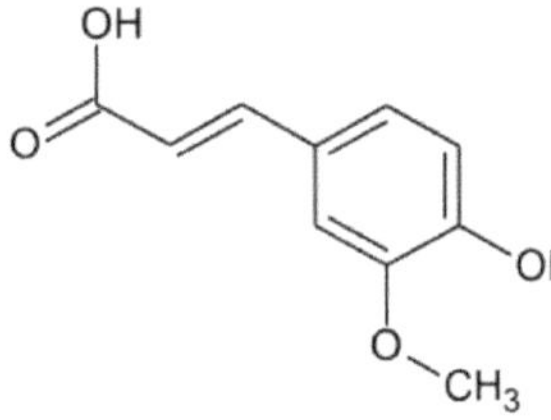
Ferulic acid

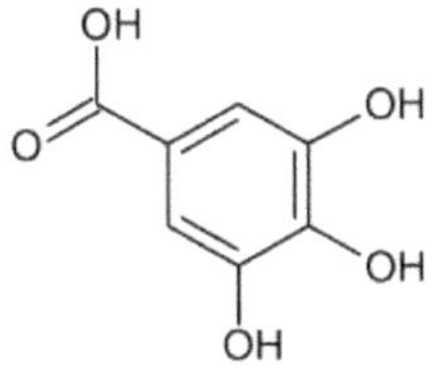
Gallic acid

**Glucosinolates**

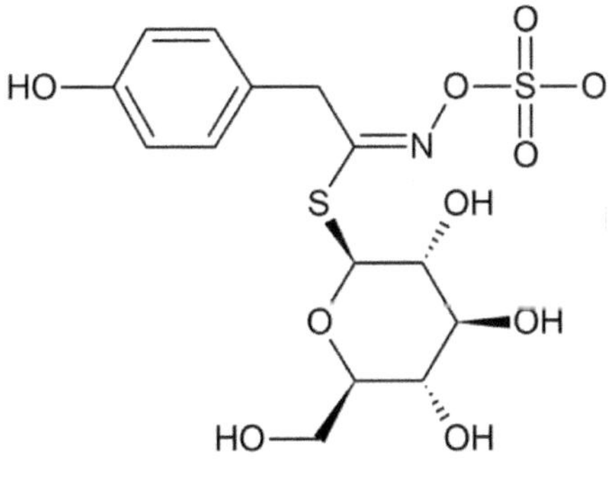
Sinalbin

**FIGURE 10.1** Main bioactive compounds in *Moringa oleifera*.

## 10.2 MAIN CHARACTERISTICS OF *MORINGA OLEIFERA* PLANT

### *10.2.1 ECOLOGY OF MORINGA PLANT*

Moringa is a versatile plant that grows as a tree or as perennial under intensive cultivation; it can reach a height of up to 12 m,[47] it grows in countries that have a dry to humid tropical or subtropical climate with annual rainfall of 760–2500 mm (due to the fact that only less than 800 mm of irrigation is necessary) and with a temperature between 18–28°C.[17] Moreover, this plant grows on any soil type except those with heavy clays, some growing conditions including slightly acidic and alkaline soils, preferring well-drained sandy or loamy soil with a pH of 4.5–8.0 at an altitude of up to 2000 m.[4,31,35] This specie tolerates water with electrical conductivities (EC) of 3 dS $m^{-1}$ at germination, and subsequently increases its resistance to saline water.[23] Once the moringa crop is established, its antioxidant system allows its development in moderate saline conditions (EC: 8 dS $m^{-1}$), with a slight decrease in its nutritional quality.[32]

### *10.2.2 PRODUCTION AND CULTIVATION*

India is the main producer of *Moringa oleifera* with a production of 1.1–1.3 million tons in an area of 380 km.[39] Moringa is a cross-pollinated tree with a high heterogeneity in form and yield;[22] it grows very fast in optimal conditions reaching a height of 1–2 m/year in the first 3–4 years. Its life span is unknown when the crop (tree) is properly cared for (providing nutrients, taking care of erosion, drought, and high temperatures).[9,27] This species propagates in two ways (sexual and asexual). For forage production, the most used is sexual reproduction.[9] *M. oleifera* is easily propagated by cuttings, but it is difficult to propagate by air layering.[45] Despite seedling growth being a fast process;[21] seeds should not be pretreated because scarification limits germination.[45] The optimal sowing depth is 1–2 cm.[27,45] The percentage of germination of fresh seeds is 60–90%, between 7 and 30 days after sowing.[21,30,36,45]

Flowering phenology varies among varieties and with geographic location. Meanwhile in some countries, trees may flower once between April and June, however, in south India up to twice a year. Flowering also may occur continuously throughout the year.[16,40] Flowering begins periodically within the first year (6 months after sowing). The trees produce seeds for 12–16 years. Pollination is carried out by bees, insects, and/or birds.[5,27] After 6 months the production of fruits begins; this is in the case of trees from stem and branch cuttings.[40] It has been demonstrated that Moringa has many benefits and applications, which include antimicrobial activity against bacteria, fungi, and yeasts;[3] modulation of the expression of fertility;[11] antioxidant activity;[49] water waste management[18] and many others for human health and nutrition.[1,44] Moreover, moringa extracts have also been applied to improve the quality and growth of some crops with agronomic importance in different ways.[2,29] *M. oleifera* has significant amounts of nutrients, which could play an important role in plant growth; these nutrients include bioactive compounds such as polyphenols, flavonoids, saponins; and other important constituents such as vitamins, protein, minerals, fibers, phytohormones, among others, whose amount can vary depending on the part of the plant used to obtain them.[1,49,50]

## 10.3 IMPACT OF *M. OLEIFERA* IN AGRONOMIC SCIENCES

In some reports, *M. oleifera* has been described as a medicinal plant having various medicinal or therapeutic properties. However, Feglie (1999) extended its uses and applications to biomass production, domestic cleaning (crushed leaves), forage (leaves and treated seed-cake), biogas production (leaves), blue dye (wood), fencing, foliar nutrient, fertilizer, water purification, among others. There are several studies regarding the effect of moringa extracts from leaves, roots, and seeds on physiology and physicochemical characteristics of some crops with agronomic importance. Those studies include effects on the reproductive growth, nutrimental composition, and functional properties of different plants and crops.

### *10.3.1 GROWTH ENHANCER*

One of the alternatives for progress in the agriculture, which is commercially available, is the use of growth promoters that have successfully been used to improve the growth and development under adverse conditions for a variety of crops, but in some cases, it turns out to be costly and its availability is low for farmers. Moringa is a potential source of growth regulators (zeatin), ascorbate, phenolic compounds, and minerals such as Ca, K, Mg, Mn, P, B, Zn, and Fe.[6,12,16,24,25] According to (Hussain et al. (2013),[19] the isolates of this plant such as secondary metabolites promote plant growth and defense mechanisms against biotic and abiotic stresses, due to a particular feature present in this plant; this effect has been recognized as an allelopathy effect.

In crops science, the mechanism of action of allelopathy is determined by the exogenous application of plant extracts used as crops fertilizers, and thus fits into crop rotation for suppressing weeds, insects, diseases and improving growth.[12] Moringa leaf extracts (MLE) have been successfully tested as plant growth promoters and have been shown to be an excellent source of compounds that enable plant growth.[6,25,48] The MLE have been used as a foliar spray or seed preparation agent to promote growth.[32,48] Among the advantages generated using MLE is accelerated growth, firm stems, resistance to biotic and abiotic stresses, increasing the number of roots, stems, and leaves, lengthening life, generating additional fruits obtaining a higher yield (20–35%). Earlier the authors in Ref. [38] reported increased growth and yield of crops using Moringa as it has zeatin (growth hormone from the cytokinin group). Likewise, several authors indicate the improvement in the early establishment of stands, depending on the effect of the concentration of MLE applied. Another study reported that the use of MLE (1:30) increases the number of leaves, roots and tillers, shoots, length, weight of roots applied in three grasses (yard grass, blue grass, and buffle grass).[32] Likewise, Iqbal et al.[20] reported that the foliar application of 2% Moringa + 2% brassica extract increased the yield of canola (Brassica napus) by up to 35% compared to the control. Also, the exogenous application of MLE in the spike stage of late-sown wheat (*Triticum aestivum*) improved 1000-grain weight, biological yield, grain yield, and harvest index (6.84, 3.17, 6.80, and 3.51%, respectively) compared to control. Finally, Yasmeen et al.[50] reported improvements in grain weight, biological yield, and harvest index by foliar application of MLE in late wheat. On the other hand, Hussain et al.[19] reported the use of canola and moringa (*M. oleifera*) extracts as foliar sprays on maize. In this study, the application of two sprays of moringa/canola mixture sprayed at 30 and 40 DAS promotes an increase of 83% of the corn grain.

### *10.3.2 BIODIESEL PRODUCTION*

The environmental impact and the social consequences of production and mass marketing, the supply of first-world countries, and the development of some countries generate the increase of deforestation, misuse of natural resources, and the destruction of ecosystems and biodiversity. One of the alternatives proposed has been the biodiesel production, the use of natural fats such as vegetable oils to produce a new generation of fuels. Biodiesel is an alternative to replace petrodiesel since it does not contain sulfur or aromatic compounds and allows lower emission of monoxides and particles. Moringa seeds have oil in percentages between 30–40% with a high content of oleic acid (>70%). In addition, it presents resistance to oxidative degradation. According to Rashid et al.,[41] biodiesel from Moringa seed oil is an acceptable substitute for petrodiesel, also when compared to biodiesel from oils from other vegetable sources, due to the high cetane number and the oxidative stability; while Salaheldeen et al.[43] reported similar physicochemical results in other specie of the family: *Moringa peregrina*, which was considered a potential source for sustainable biodiesel production.

## 10.3.3 BIOREMEDIATION

Water is a vital resource for a multiple number of activities, such as agriculture, livestock, industrial, home, environmental, among others. However, its availability is becoming scarcer every day. The contamination of soil and water resources with environmentally harmful chemicals represents a problem of great concern not only in relation to the biota in the receiving environment, but also to humans. Growth and urbanization have led to a greater amount of pollutants being discharged into the environment, mainly heavy metals.[7] It is estimated that 10% of the waste generated in each country contains heavy metals. Due to this, it has been reported that the presence of heavy metals in the environment is one of the factors responsible for diseases of the central nervous system (Mn, Hg, Pb, As), kidneys or liver (Hg, Pb, Cd, Cu) and skin, bones, or teeth (Ni, Cd, Cu, Cr).[51] Due to this, it is necessary to search for alternatives that allow the removal of the environment since even if they are at low concentrations they do not decompose and are toxic. Two of the most used methods for the removal of heavy metals from liquid wastes are chemical precipitation and ion exchange. However, they are not economical methods, so they become economically unviable; likewise, they do not eliminate heavy metals in their entirety, hence the need to search for alternative methods for the elimination of heavy metals.[10,23,37] Biosorption is defined as the ability of some biomolecules or biomass to bind and concentrate ions or other molecules from aqueous solutions. The use of biomass of vegetable origin is an innovative and alternative technology for the removal of contaminants from contaminated water. It presents a series of advantages, among which its low cost, high efficiency, reduction of chemical and/or biological sludge, as well as the reuse of the bio-absorbent stand out. Moringa is a material that can be used for this purpose.[13,34,46] The best-known application is for the treatment of cloudy water and the purification of water using Moringa seeds and fruits. This is because the seeds contain cationic polyelectrolytes and can be used as a substitute for aluminum sulfate. Likewise, they can eliminate turbidity and have antimicrobial properties.[44] It is known that the most used part of Moringa is the leaves; however, the seeds and roots contain glucosinolates (contains a central carbon atom that is linked to the thioglucose group (forming a sulfated ketoxime) through a sulfur atom and to the sulfate group through a nitrogen atom). These functional groups containing S and N can sequester metals from a solution.[7] Due to these characteristics, Moringa seeds are an alternative for the elimination of metallic ions from contaminated water with a percentage of elimination of 60–90% of Cd (II), Cu (II), Ni (II), Pb (II), As(III), As(V), Cr(III), and Zn(II). Likewise, the metal ion removal process is improved if leaves, bark, and pods of the plant are used.[14]

### 10.3.4 ANIMAL NUTRITION

Moringa has excellent nutritional properties, which is why it is used as fresh fodder to feed cattle. Its leaves are a source of carotene, protein, ascorbic acid, and iron. However, the pod is rich in lysine. The antinutritional compounds present in *Moringa oleifera* do not have negative effect when consumed by animals.[40] Mendieta-Araica et al.[26] studied the forms of cultivation of Moringa, as well as its use as fodder for livestock and food for fish, resulting in a potential alternative for its use as food. Also, Moyo et al.[28] concluded that the results are a clear indication that the plant has the potential as a feed additive as a source of fatty acids, proteins, vitamins, and minerals and that it can be incorporated into food matrices for humans and animals. Moreover, Asaolu et al.[8] recommended that the moringa forage is mixed with other forages for better use in feeding goats. In sheep, there are also positive results with Moringa supplementation in the diet.[9] In the feeding of dairy cows, they have also conducted studies in which it is concluded that Moringa leaves are a potential source of protein to be incorporated as a supplement in low-protein foods such as elephant grass. In this way it can replace commercial concentrates.[36] According to the report by Foidl et al.,[15] another advantage of moringa is the high productivity of fresh material per hectare compared to other forage crops. By this reason, for the cultivation of moringa to forage production suggested planting at high densities and make cuts when the plant reaches 120–150 cm in height, reporting that they can give up to eight cuts per year.

## 10.4 CONCLUSION

Moringa oleifera is a plant that can provide many alternatives within the agronomic sciences since it has many bioactive compounds with various properties as growth regulators applied exogenously on the foliage of crops. Likewise, the extracted oil allows the production of biodiesel as a substitute for petrodiesel. Finally, the seeds are considered a natural bio-absorbent against heavy metals, which is why it has applications in bioremediation and the leaves are a potential source for animal nutrition.

## KEYWORDS


- **petrodiesel**
- **animal nutrition**
- **moringa**
- **modern agriculture**
- **bioremediation**


## REFERENCES


1. Abdulkadir, A. R.; Zawawi, D. D.; Jahan, M. S. Proximate and Phytochemical Screening of Different Parts of Moringa Oleifera. *Russ. Agric. Sci.* **2016,** *42* (1), 34–36.
2. Adandonon, A.; Aveling, T. A. S.; Labuschagne, N.; Tamo, M. Biocontrol Agents in Combination with Moringa Oleifera Extract for Integrated Control of Sclerotium-Caused Cowpea Damping-off and Stem Rot. *Eur. J. Plant Pathol.* **2006,** *115* (4), 409–418.
3. Al_husnan, L. A.; Alkahtani, M. D. F. Impact of Moringa Aqueous Extract on Pathogenic Bacteria and Fungi In Vitro. *Ann. Agric. Sci.* **2016,** *61* (2), 247–250.
4. Anjorin, T. S.; Ikokoh, P.; Okolo, S. Mineral Composition of Moringa Oleifera Leaves, Pods and Seeds from Two Regions in Abuja, Nigeria. *Int. J. Agric. Biol.* **2010,** *12* (3), 431–434.
5. Anwar, F.; Ashraf, M.; Bhanger, M. I. Interprovenance Variation in the Composition of Moringa Oleifera Oilseeds from Pakistan. *J. Am. Oil Chem. Soc.* **2005,** *82* (1), 45–51.
6. Anwar, F.; Latif, S.; Ashraf, M.; Gilani, A. H. (2007). Moringa Oleifera: A Food Plant with Multiple Medicinal Uses. *Phytother. Res. 21* (1), 17–25.
7. Araújo, C. S. T.; Carvalho, D. C.; Rezende, H. C.; S. Almeida, I. L.; Coelho, L. M.; Coelho, N. M. M.; Alves, T. L. M. and V. N.; Araújo, C. S. T.; Carvalho, D. C.; Rezende, H. C.; S. Almeida, I. L.; Coelho, L. M.; Coelho, N. M. M.; Alves, T. L. M. and V. N. Bioremediation of Waters Contaminated with Heavy Metals Using Moringa oleifera Seeds as Biosorbent. In *Applied Bioremediation—Active and Passive Approaches*, 2013. DOI: 10.5772/56157
8. Asaolu, V. O.; Binuomote, R. T.; Akinlade, J. A.; Oyelami, O. S.; Kolapo, K. O. Utilization of Moringa Oleifera Fodder Combinations with Leucaena Leucocephala and Gliricidia Sepium Fodders by West African Dwarf Goats. *Int. J. Agric. Res.* **2011,** *6* (8), 607–619.
9. Aslam, M.; Anwar, F.; Nadeem, R.; Rashid, U.; Kazi, T. G.; Nadeem, M. Mineral Composition of Moringa oleifera Leaves and Pods from Different Regions of Punjab, Pakistan. *Asian J. Plant Sci.* **2005,** *4* (4), 417–421.
10. Babel, S.; Kurniawan, T. A. Cr(VI) Removal from Synthetic Wastewater using Coconut Shell Charcoal and Commercial Activated Carbon Modified with Oxidizing Agents and/or Chitosan. *Chemosphere* **2004,** *54* (7), 951–967.

11. Barakat, I. A. H.; Khalil, W. K. B.; Al-Himaidi, A. R. Moringa Oleifera Extract Modulates the Expression of Fertility Related Genes and Elevation of Calcium Ions in Sheep Oocytes. *Small Rumin. Res.* **2015,** *130,* 67–75.
12. Basra, S. M. A.; Iftikhar, N.; Afzal, I. Potential of Moringa (Moringa oleifera) Leaf Extract as Priming Agent for Hybrid Maize Seeds. *Int. J. Agric. Biol.* **2011,** *13* (6), 1006–1010.
13. Bhatti, H. N.; Mumtaz, B.; Hanif, M. A.; Nadeem, R. Removal of Zn(II) Ions from Aqueous Solution using Moringa Oleifera Lam. (horseradish tree) Biomass. *Process Biochem.* **2007,** *42* (4), 547–553.
14. Demirbas, A. Heavy Metal Adsorption onto Agro-Based Waste Materials: A Review. *J. Hazard. Mat.* **2008,** *157* (2–3), 220–229.
15. Foidl, N.; Makkar, H. P. S.; Becker, K. The Potential of Moringa Oleifera for Agricultural and Industrial Uses. In *The Miracle Tree: The Multiple Attributes of Moringa*, 1st ed.; Vol. 1; 2001; pp 45–76.
16. Fuglie, L. J. *The Miracle Tree: Moringa Oleifera, Natural Nutrition for the Tropics*, 1st ed.; Vol. 1, Issue 1; Dakar (Senegal) CWS, 1999. DOI: 10.3/JQUERY-UI.JS
17. García Roa, M. *Producción de Semillas Forestales de Especies Forrajeras Enfatizadas en Sistemas Silvopastoriles* [Online], 2003. INAFOR. https://www.academia.edu/21154492/PRODUCCI%C3%93N_DE_SEMILLAS_FORESTALES_DE_ESPECIES_FORRAJERAS_ENFATIZADOS_EN_SISTEMAS_SILVOPASTORILES_POR_MARIO_GARCIA_ROA
18. Garde, W. K.; Buchberger, S. G.; Wendell, D.; Kupferle, M. J. Application of Moringa Oleifera Seed Extract to Treat Coffee Fermentation Wastewater. *J. Hazard. Mater.* **2017,** *329*, 102–109.
19. Hussain, M.; Farooq, M.; Basra, S. M. A.; Lee, D. J. Application of Moringa Allelopathy in Crop Sciences. In *Allelopathy: Current Trends and Future Applications*; 2013, pp 469–483.DOI: 10.1007/978-3-642-30595-5_20/TABLES/3
20. Iqbal, M. A.; Iqbal, A.; Akbar, N.; Abbas, R. N.; Khan, H. Z.; Maqsood, Q. Response of Canola to Foliar Application of Moringa (Moringaolifera L.) and Brassica (Brassica napus L.) Water Extracts. *Int. J. Agric. Crop Sci.* **2014,** *7* (14), 1431–1433.
21. Jahn, S.; Musnad, H. A.; Burgstaller, H. The Tree that Purifies Water: Cultivating Multipurpose Moringaceae in the Sudan. *Unasylva*, 2012.
22. Leone, A.; Spada, A.; Battezzati, A.; Schiraldi, A.; Aristil, J.; Bertoli, S. Cultivation, Genetic, Ethnopharmacology, Phytochemistry and Pharmacology of Moringa oleifera Leaves: An Overview. *Int. J. Mol. Sci.* 2015, *16* (6), 12791–12835.
23. Li, H.; Liu, T.; Li, Z.; Deng, L. Low-Cost Supports Used to Immobilize Fungi and Reliable Technique for Removal Hexavalent Chromium in Wastewater. *Bioresour. Technol.* **2008,** *99* (7), 2234–2241.
24. Mahmood, K.; Mugal, T.; Haq, I. Moringa Oleifera: A Natural Gift—A Review. *J. Pharm. Sci. Res.* **2010,** *2* (11), 775–781.
25. Makkar, H. P. S.; Becker, K. Nutrional Value and Antinutritional Components of Whole and Ethanol Extracted Moringa Oleifera Leaves. *Anim. Feed Sci. Technol.* **1996,** *63* (1–4), 211–228.
26. Mendieta-Araica, B.; Spörndly, R.; Reyes-Sánchez, N.; Spörndly, E. Moringa (Moringa oleifera) Leaf Meal as a Source of Protein in Locally Produced Concentrates for Dairy Cows Fed Low Protein Diets in Tropical Areas. *Livestock Sci.* **2011,** *137* (1–3), 10–17.

27. Morton, J. F. The Horseradish Tree, Moringa Pterygosperma (Moringaceae)—A Boon to Arid Lands? *Econ. Bot.* **1991,** *45* (3), 318–333.
28. Moyo, B.; Masika, P. J.; Hugo, A.; Muchenje, V. Nutritional Characterization of Moringa (Moringa oleifera Lam.) Leaves. *Afr. J. Biotechnol.* **2011,** *10* (60), 12925–12933.
29. Nasir, M.; Khan, A. S.; Basra, S. M. A.; Malik, A. U. Foliar Application of Moringa Leaf Extract, Potassium and Zinc Influence Yield and Fruit Quality of 'Kinnow' Mandarin. *Sci. Hortic.* **2016,** *210,* 227–235.
30. Nautiyal, B. P.; Venhataraman, K. G. Moringa (Drumstick)—An Ideal Tree For Social Forestry: Growing Conditions and Uses—Part I. *Myforest* **1987,** *23* (1), 53–58.
31. Nouman, W.; Basra, S. M. A.; Siddiqui, M. T.; Yasmeen, A.; Gull, T.; Alcayde, M. A. C. Potential of Moringa oleifera L. as Livestock Fodder Crop: A Review. *Turk. J. Agric. Forest.* **2014,** *38,* 1–14.
32. Nouman, W.; Siddiqui, M. T.; Basra, S. M. A.; Khan, R. A.; Gull, T.; Olson, M. E.; Munir, H. Response of Moringa Oleifera to Saline Conditions. *Int. J. Agric. Biol.* **2012,** *14,* 757–762.
33. de Oliveira, F. R. A.; de Oliveira, F. de A.; Guimarães, I. P.; de Medeiros, J. F.; de Oliveira, M. K. T.; de Freitas, A. V. L.; de Medeiros, M. A. Emergency of Seedlings of Moringa Oleifera Lam Irrigated with Water of Different Levels of Salinity. *Biosci. J.* **2009,** *25* (5), 66–74.
34. Olsen, A. Low Technology Water Purification by Bentonite Clay and Moringa Oleifera Seed Flocculation as Performed in Sudanese Villages: Effects on Schistosoma Mansoni Cercariae. *Water Res.* **1987,** *21* (5), 517–522.
35. Palada, M. C. Moringa (Moringa oleifera Lam.): A Versatile Tree Crop with Horticultural Potential in the Subtropical United States. *HortSci.* **1996,** *31* (5), 794–797.
36. Parrotta, J. A. Moringa Oleifera Lam. Reseda, Horseradish Tree. Moringaceae. Horseradish Tree Family. *USDA Forest Service* [Online], 1993. International Institute of Tropical Forestry (SO-ITF-SM; 61). http://www.fs.usda.gov/research/treesearch/30357
37. Petroni, S. L. G.; Pires, M. A. F.; Munita, C. S. Adsorção de Zinco e Cádmio em Colunas de Turfa. *Química Nova* **2000,** *23* (4), 477–481.
38. Price, M. L. The Moringa tree. *ECHO Technical Note* [Online], 1985. http://echonetwork.org/files_pdf/Moringa.pdf
39. Rajangam, J.; Azahakia Manavalan, R. S.; Thangaraj, T.; Vijayakumar, A.; Muthukrishan, N. Status of Production and Utilisation of Moringa in Souther India. *Development Potential for Moringa Products: International Workshop, 29th October - 2nd November 2001*, Dar Es Salaam, Tanzania, 2002.
40. Ramachandran, C.; Peter, K. V.; Gopalakrishnan, P. K. Drumstick (Moringa oleifera): A Multipurpose Indian Vegetable. *Econ. Bot.* **1980,** *34* (1), 276–283.
41. Rashid, U.; Anwar, F.; Moser, B. R.; Knothe, G. Moringa Oleifera Oil: A Possible Source of Biodiesel. *Bioresour. Technol.* **2008,** *99* (17), 8175–8179.
42. Rodríguez-Pérez, C.; Quirantes-Piné, R.; Fernández-Gutiérrez, A.; Segura-Carretero, A. Optimization of Extraction Method to Obtain a Phenolic Compounds-Rich Extract from Moringa Oleifera Lam Leaves. *Ind. Crops Prod.* **2015,** *66,* 246–254.
43. Salaheldeen, M.; Aroua, M. K.; Mariod, A. A.; Cheng, S. F.; Abdelrahman, M. A.; Atabani, A. E. Physicochemical Characterization and Thermal Behavior of Biodiesel and Biodiesel–Diesel Blends Derived from Crude Moringa Peregrina Seed Oil. *Energy Convers. Manag.* **2015,** *92,* 535–542.

44. Saucedo-Pompa, S.; Torres-Castillo, J. A.; Castro-López, C.; Rojas, R.; Sánchez-Alejo, E. J.; Ngangyo-Heya, M.; Martínez-Ávila, G. C. G. Moringa Plants: Bioactive Compounds and Promising Applications in Food Products. *Food Res. Int.* **2018,** *111*, 438–450.
45. Sharma, G. K.; Raina, V. Propagation Techniques of Moringa oleifera Lam. In *Improvement of Forest Biomass: Symposium Proceedings*,Khosla, P. K., Ed.; 1982. DOI: 10.3/JQUERY-UI.JS
46. Sharma, P.; Kumari, P.; Srivastava, M. M.; Srivastava, S. Removal of Cadmium from Aqueous System by Shelled Moringa Oleifera Lam. Seed Powder. *Bioresour. Technol.* **2006,** *97* (2), 299–305.
47. Sharma, V.; Ritu Paliwal, P.; Sharma, S. Phytochemical Analysis and Evaluation of Antioxidant Activities of Hydro-Ethanolic Extract of Moringa Oleifera Lam. Pods. *J. Pharm. Res.* **2011,** *4* (2), 554–557.
48. Shrestha, A.; Pradhan, S.; Shrestha, J.; Subedi, M. Role of Seed Priming in Improving Seed Germination and Seedling Growth of Maize (Zea mays L.) Under Rain fed Condition. *J. Agric. Nat. Res.* **2019,** *2* (1), 265–273.
49. Torres-Castillo, J. A.; Sinagawa-García, S. R.; Martínez-Ávila, G. C. G.; López-Flores, A. B.; Sánchez-González, E. I.; Aguirre-Arzola, V. E.; Torres-Acosta, R. I.; Olivares-Sáenz, E.; Osorio-Hernández, E.; Gutiérrez-Díez, A. Moringa Oleifera: Phytochemical Detection, Antioxidants, Enzymes and Antifugal Properties. *Phyton (Buenos Aires)* **2013,** *82,* 193–202.
50. Yasmeen, A.; Basra, S. M. A.; Ahmad, R.; Wahid, A. Performance of Late Sown Wheat in Response to Foliar Application of Moringa oleifera Lam. Leaf Extract. *Chil. J. Agric. Res.* **2012,** *72* (1), 92–97.
51. Zevenhoven, R.; Kilpinen, P. Control of Pollutants in Flue Gases and Fuel Gases. *Control of Pollutants in Flue Gases and Fuel Gases* [Online], 2001. 2.1-2.12. http://users.abo.fi/rzevenho/gases.PDF

CHAPTER 11

# Allelopathic Responses and Mechanisms of Action of Essential Oils on Weeds

BRENDA LUNA-SOSA[1], CRISTIAN MARTÍNEZ ÁVILA[1], NESTOR ARANDA-LEDEZMA[1], NANCY MEDINA-HERRERA[1], JULIO C. TAFOLLA ARELLANO[2], and ROMEO ROJAS[1]

[1]*Universidad Autonoma de Nuevo Leon, School of Agronomy, Francisco I. Madero S/N, Ex Hacienda el Cañada, Gral. Escobedo, N.L.*

[2]*Laboratorio de Biotecnología y Biología Molecular, Departamento de Ciencias Básicas, Universidad Autónoma Agraria Antonio Narro, Calzada Antonio Narro 1923, Buenavista, Saltillo, Coahuila, Mexico*

## ABSTRACT

Weeds in crops are the cause of 30% of potential economic losses recorded in agricultural production. In turn, herbicides have allowed a large number of farmers worldwide to eliminate weeds and thus favor agricultural production. However, to provide quality and safety to the consumer of the basic food products, it is important to address the problems that agriculture faces due to the use of agrochemicals that are harmful to health. An alternative application is the essential oils (EOs) of aromatic plants that could naturally have various allelopathic effects. These are considered a good alternative to replace synthetic herbicides. Some constituents of interest that we can find in EOs inhibit the germination of weed seeds, in addition to reducing their growth.

The metabolites extracted from essential oils are quite friendly to the environment due to their high volatilization and degradation. In this review we will focus on the identifiable phytotoxic effects of the *Lamiaceae* and *Umbelliferae* families, *Rutaceaes* and *Poaceas* classes for the development of an oil-encapsulated polymer through emulsions on various crops.

Sustainable Agricultural Production Systems: Traditional and Modern Perspectives.
Romeo Rojas, Guillermo Cristian Guadalupe Martínez-Ávila, Julia Mariana Márquez-Reyes, and Ma del Carmen Ojeda-Zacarías (Eds.)

## 11.1 INTRODUCTION

Farmers have been using herbicides to eradicate weeds, some shrubs, and generally all unwanted vegetation for decades. Herbicide mechanisms of action are highly efficient, cost-effective, and affordable compared to mechanical weed removal. Today, Mexico and the world are facing the secondary effects that herbicides cause due to their excessive use in crops. Some synthetic substances generate substantial damage to crops, the environment, and human health. How is it that these herbicides cause damage to the growth of the plant? One of the collateral effects generated by the inadequate application of the indications for use is the activation of resistance in the plant. Regarding the environment, there have been losses of endemic species in terms of flora and fauna due to the exposure of organisms that have contact with the treated plant and this causes toxicity to occur in the environment and does not allow degradation.[11]

In general, herbicides have provided significant benefits to farmers and producers to protect yields and manage weeds without ruling out that the consumer demand for products remains balanced. Unfortunately, the pros of herbicides are not enough to balance sustainability and food safety. Despite the negative consequences, many growers continue to choose to apply herbicides because they are cheaper and easier to use.[5] Research has shown herbicides have some kind of selective characteristics, meaning that they only kill some particular plant species but not others; some other weeds are broad-spectrum, meaning they work effectively in most crops.

The most widely used synthetic herbicides in agriculture are 2, 4-D, paraquat, and glyphosate; it is worth mentioning that the mechanism of action that they exert on weeds has been extensively studied and therefore, their negative effects are well known. These have an allelopathic or phytotoxic potential that involves the photosynthesis and respiration processes of plants, which in turn are strongly related to the inhibition of enzymatic activities. Biochemical, physiological, molecular, and chemical parameters are also strongly linked to the inhibition effect on seed germination. Weeds have developed resistance to 23–26 sites of action to known herbicides and to 167 unknown ones; there is a record of at least 94 primary crops from various countries that suffer resistance due to constant rotation and soil erosion.

A condition that limits the farmer or producer is the cost of controlling the elimination of weeds by means of herbicides, since it is much lower than the implementation of manual weeding in organic and sustainable agricultural productions. Considering the above, governmental institutions and organizations; Greenpeace, WHO, Sader, Senasica to name a few, seek alternatives to reduce the use of herbicides and even eliminate their complete use in crops. Nowadays, increasing agricultural production in a sustainable way is one of agriculture's main purposes, that is the importance of minimizing the use of herbicides as much as possible. The use and exploitation of renewable resources of medicinal and aromatic plants arise as a replacement for herbicides due to the allelopathic properties that they provide.[8]

Taking into account the previous points, essential oils from aromatic plants open a wide portal for the generation of environmentally friendly bioherbicides thanks to allelopathy. Each component of each extract offers positive effects of inhibition in the germination of weeds without impairing the conservation of endemic species. Essential oils (EO) are defined as metabolites from natural sources that are highly volatile, are characterized by having a penetrating odor, and are synthesized in plant organs. Ecologically and environmentally, these compounds help the interaction with pathogens, herbivores, and symbiotic insects, as well as for pollination and seed dissemination.[2,9]

EOs chemical composition is diverse and depends on factors of the micro and macro environment (origin, climate, species, and recovery extraction). The classification of EOs is described as terpenes, terpenoids, ethers, and nitrogenous. In this review we will present a compilation of reports based on the challenges and trends of the substitution of synthetic herbicides to biodegradable ones. Similarly, allelopathy studies the components of various aromatic species and their mechanisms of action in inhibiting germination in weeds at the plant-seed level.

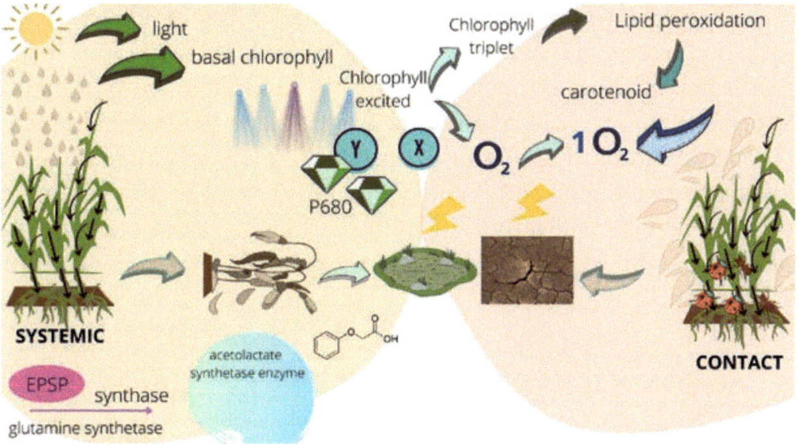


**FIGURE 11.1** Mechanisms of action of herbicides according to their application.

## 11.2 HERBICIDE MECHANISMS OF ACTION

It is known that, to eliminate any active weed in primary crops, a herbicide carries out rigorous biological and chemical processes that will depend on the product's label formulation.[29]

The selectivity and action of an herbicide are completely related to the active ingredients of the formulation. A systemic herbicide can have destructive effects on weeds in a dry environment and in the presence of sunlight. In the case of the contact product, it will work in moist substrates, so the plant must absorb the chemical through the stomata, that is, through the leaves, and distribute it throughout its structure until it penetrates the roots and settles in the soil.[17]

The active sites of herbicides that are involved in the mechanism of action are: amino acid synthesis (participation of the enzyme EPSP synthase and acetohydroxyacid synthase), photosynthesis inhibitors (presence of glutamine synthase and quinone band proteins) that can be mobile and immobile, discoloration (photosystem 1 and the enzyme protoporphyrinogen oxidase), lipid synthesis (enzyme Acetyl Co-enzyme A Carboxylase), synthesis of carotenoids (enzyme phyto desaturase, lycopene ß-cyclase, Ž-carotene desaturase, isopentenyl-diphosphate delta isomerase and prenyltransferase), cell division (ß-tubulin), synthesis of cellulose (cellulose synthase) and folate (dihydropteroate synthase).

Figure 11.1 shows the process of a systemic herbicide that acts on a plant that begins with the penetration through the leaves with the intention of inhibiting the passage of bacteria, viruses, or other pests. The plant begins to show symptoms of imbalance and kills its leaves and root system. Other symptoms that plants present when they receive penetration or contact with the herbicide are marginal necrosis (darkening of leaves and loss of rigidity) and interveinal chlorosis (yellowing of leaf tissue). For its part, the contact herbicide does not penetrate directly into the plant but rather keeps it protected with pest control and acts as a barrier against certain pathogens. The broad spectrum of action of the product affects and kills a wide variety of organisms (beneficial and nonbeneficial insects). The absorption, translocation, metabolism, and active site of the herbicide participate in the death of a plant and mainly require starting to attack the growing points of the plant, the meristems, or the nodes.[19]

## 11.3 ACTION MECHANISMS OF OILS

The phytotoxicity and allelopathy of EOs, through their effects on cellular and biochemical dysfunctions, are strongly related to the inhibition of plant germination and growth. The etymological origin of the word allelopathy is explained from the roots aleon = each other and phatía = reciprocal. The objective of allelopathy is that, through the chemical, physiological, and biochemical study of its secondary metabolites of a plant, it serves to help it for its self-defense against pathogens or products foreign to its habitat. Every action has a reaction and in plants it works like that too; when they identify an antagonistic stimulus that could threaten their vegetative structure, they make use of allelochemicals (substances) to attack this phenomenon.

The stress to which we subject a plant will depend on the conditions we give it, since a plant will always try to adapt. The bioactive compounds of the EOs of the *Lamiaceae* and *Umbelliferae* families have an excellent biostimulation and defense profile against weeds in a particular crop of interest. The mechanisms of action of EOs from aromatic species in crops have not been fully exploited and understood. An assumption of the allelopathic activity of EOs in a plant is that, by themselves, EOs could only affect the surface of a host and not penetrate the tissues due to the hydrophobicity of phase separation in which they reach.[15]

However, studies carried out on OEs have shown in recent years that there are positive impacts and cellular, physiological, and biochemical effects on weeds, although their studies are limited on translocation. Next, Figure 11.2 shows the 10 steps that consider AE as an allelopathy response to weeds.[30]

**FIGURE 11.2** Steps that EOs consider in an allelopathic profile on weeds.

The first step of entry of an EO extract is through the leaves of a plant (leaf wax) in an aerial application. In the case of fixing the EO metabolites in a seed, the inhibition would start contrarily to what is known. The bioactive molecules (mainly monoterpenes) of the EOs enter through the cuticle and the stomata of the plant due to gas exchange. The second step consists of the separation of phases, in this case gaseous and liquid by the transpiration of the plant. The amount of gas dissolved in the liquid must be directly proportional to the partial pressure. In turn, the liquid phase allows the adherence of the EO bioactives through the cellular cytosol of the plant that generates the formulation and ratio of oil and water.[7] Limonene and thymol mainly allow the translocation of monoterpenes in the vegetative cells of plants at the time they begin to enter.

When water alteration state (WAS) occurs, mainly in the xylem, this organ interrupts the passage of water in the transport veins of the leaves of the plants, which generates the desiccation of the natural protective cells of the plants. The symptomatology of WAS in a plant with EO that presents is dryness in the tips of leaves and strongly brownish stains (brown-yellow). Figure 11.3 shows that by incorporating the bioactive compounds (1.8 cineol, geraniol, thymol, menthol, and camphor) of essential oils (BC-EO) in the cell membrane of the plant, these increase the amount of fatty acids unsaturated phospholipids. Two cases of active sites are also presented, one of them belongs to the modification of the organization of the lipid bilayer due to EO monoterpenes (OEMS) and the other case is the participation of BC-EO receptors that participate in the pathway. of phenylpropanoid (IRRMFE).

Phytoremediation enables antioxidant defense in plants by a burst of calcium that reduces molecular oxygenation and increases the vapor pressure on the leaf surface of the lipid membrane. The presence of secondary metabolites of the EO allows the active synthesis that is induced by BC-EO in the plant by subjecting it to oxidative stress with a high plant density, competition with other beings due to lack of nutrients or energy that generates tension in the plant. The IF and IERON steps are fundamental in the cellular signals and stimuli at the moment in which BC-EO are loaded on a plant, permitting macromolecules degradation (carbohydrates, lipids, proteins, and genetic material). The production of superoxide radicals that transform hydrogen peroxide into superoxide dismutase is one of the results of uncoupling of photosynthesis and cellular respiration.

Likewise, enzymes superoxide dismutase, guaiacol peroxidase, catalase, peroxidase, ascorbate, glutathione reductase permit growth inhibition in plant roots. Oxidative stress enables the lipid peroxidation of the membrane in which the secondary metabolites of the EOs enter, which in turn generate an immense leak of electrolytes and coloration and/or vital staining of the plant that is provided by the chloroplasts.

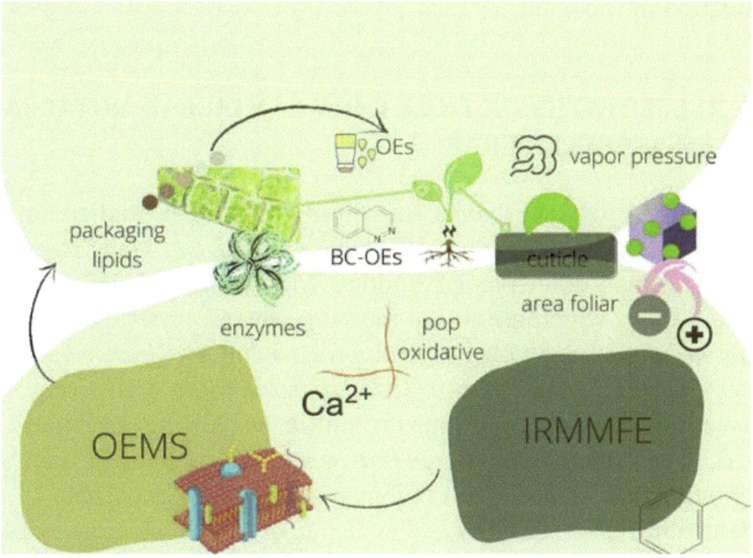


**FIGURE 11.3** MIP of BC-EO in plants.

MRI, DMG, and IZR determine that monoterpenes (mainly camphor and limonene) found in BC-EO decrease respiratory oxygen consumption in plant structures. Pinene is responsible for uncoupling oxidative phosphorylation and electron transfer inhibition. Due to the above, adenosine triphosphate (ATP) is interrupted by an alteration that the secondary metabolites of the EO generated in the mitochondrial membrane. In addition, BC-EOs present a mito-depressive activity in plants and weed seeds due to the inhibition of DNA synthesis and the genotoxic effect of the metabolites. The allelopathy and phytotoxicity of BC-EO occurs in weeds due to the inhibition of carbonic anhydrase that catalyzes the reversible conversion of $CO_2$ to bicarbonate. This enzyme participates in respiration and photosynthesis.

The FCDV and the MD are processes responsible for maintaining the balance of concentration and distribution in the cellular structure of auxins and ethylene of the plant. These promoters allow root growth and radicle elongation. A gene modified by OE metabolites allows the direct–indirect affectation of the PIN protein that limits meristem growth and plant phenotypes. A process that involves the last steps of inhibition of growth and germination of seeds in weeds is the sequestration and biotransformation of exogenous volatile organic compounds of BC-EOs such as the lack of protein reception, packaging, and shipments to endogenous growth hormones of the plants. Therefore, BC-EOs generate cell death by inhibiting growth and reproduction of the plant.[1]

## 11.4 ALLELOPATHY OF THE FAMILY LAMIACEAE, RUTACEAE AND THE CLASS UMBELLIFERAE

The allelopathy of the *Lamiaceae* family has been studied in different ways. This includes Oregano (*Origanum vulgare L.*), Pennyroyal (*Mentha pulegium L.*), and other plants. Another widely studied class is that of the *Umbelliferae*, in which are found the Hojasen (*Fluorensia cernua*), Laurel (*Laurus nobilis*), Eucalyptus (*Eucalyptus)*, Pepper (*Piperaceae*), Yerbanís (*Tagetes lucida*), Mango ataulfo (*Mangifera indica*), and coriander (*Coriandrum sativum*). In these groups or families we also find Rutaceae such as lemon (*Citrus x limon*) and orange (*citrus x sinensis*), to name a few. Finally, the lemongrass (*Cymbopogon*) belongs to the *Poaceaes*. Table 11.1 shows the allelopathic characteristics and evidence of work that various researchers have studied in the aforementioned species.

The *Lamiaceae* family has a broad profile of beneficial activities due to its chemical and biological composition of the aromatic plants from which the EOs are extracted. An activity that stands out in the field of agronomy is antimicrobial and allelopathic. The EOs that characterize this group are due to the volatility, lipophilicity, and odorifericity that are obtained from the BC-EOs. The factors that intervene so that the allelopathy of the *Lamiaceae* family is positive in weeds are synergy between the EO-plant interaction, EO species, geographical location, aromatic plant stage of EO extraction, and EO extraction methods that differ by the appearance of a hydrolate and oil phase. Although the allelopathic activity of the species that are part of the *Lamiaceae* family has been studied, the type of interaction at the cellular and enzymatic levels that responds as a defense mechanism against weeds produced by EO is still unknown.

In addition, it is a good field for research to learn about the interactions that occur between EO and weeds at different stages, application, and synergy of patterns of phytotoxic activity. In *Umbelliferae*, the inhibitory effect of photosynthetic rate (thylakoid) has been evaluated in contrast to

**TABLE 11.1** Allelopathic Characteristics and Phytotoxic Responses of BC-EOs of Different Aromatic Species of the *Lamiaceae*, *Rutaceae*, *Poaceaes* Families and the *Umbelliferae* Class.

| EOs | BC-OEs | Weed to be removed | Allelopathy | Reference |
|---|---|---|---|---|
| Origanum vulgare | Carvacrol, β-phenkyl alcohol, thymol and γ-terpinene | *Arabidopsis thaliana* | Photosynthesis inhibition | Araniti et al.,[3] Teixeira et al.,[27] |
| Mentha pulegium L. | Geraniol, β-citronellol, α-pinene, β-pinene, linalool, menthone, DL-menthol, and R-(+)-limonene | *Cyperus rotundus, E.crus-galli, Oryza sativa* | Exposure to steam and foliar spray | Singh et al.,[23] Mohammed et al.,[17] |
| Fluorensia cernua | – | – | – | – |
| Laurus nobilis | | | | |
| Eucalyptus | Citronellol | *E. crus-galli L.* | – | Vokou et al., 2003 |
| Piperaceae | – | – | – | – |
| Tagetes lucida | – | – | – | – |
| Coriandrum sativum | α-pinene, camphene, β-myrcene and linalool | *F. culmorum* | *In vitro* inhibition by FCDV | Raveau et al.,[18] |
| Citrus x limón | Limonene | *Echinochloa crus-galli, Phalaris minor, Avena fatua* | FS and IRM | Saroj et al., 2017 |
| Citrus sinensis | - | *Echinochloa crus-galli* | Chlorophyll removal | Duangporn et al., 2016 |
| Cymbopogon | L-limoneno, (Z) citral (neral), (E)-citral (geranial) y nerol | *Lysimachia arvensis* | Foliar spray | Sharma et al.,[21] |

atrazine, which is a chemical molecule more used in synthetic herbicides. One of the most studied BC-EOs is falcarinol, although its allelopathic activity is unknown due to other mechanisms at the cellular and structural levels of inhibition against weeds. This family has been less studied than *Lamiaceae* in terms of a potential bioherbicide due to the allelopathic profiles that various species have.[16] The *Rutaceae* and *Poaceae* have been studied for secondary metabolites recently by Sothearith *et al.*,[24] on lettuce radicles and their volatile allelochemicals that enabled inhibition.

The allelopathic effects caused by the BC-EO of the aforementioned aromatic plants vary in species, micro and macro environment of collection of the vegetative material. Sesquiterpenes are intended to inhibit seedling growth at an early stage. Along with this, soluble phenols have the effect of reducing plant growth (Kong et al., 2021). Glucosinolates and various products derived from hydrolysis generate interactions between the invasive plant and the recipient plant, the plant against a phytopathogenic organism and against nematodes. If it is necessary to eliminate the growth of legumes mainly and their rhizospheres, as well as some grasslands, the 7-8 benzoflavones and catechins generate a positive control effect. The group of isopropyl and sec-butyl glucosinolates allow inhibiting the germination of seeds and seedlings. Limonene, lineol, menthol, and thymol allow the reduction of biomass, decrease in germination, and increase in weed mortality. Monoterpenes in general decrease the aerial biomass of plants.[22] Polyacetylenes and diterpenes have positive effects against weeds during seedling growth. If reducing germination and at the same time the weight and growth of gram (-) phytopathogenic bacteria are required, cnicin is the active ingredient. Finally, onopordopicrin allows the root length of weeds to be reduced.

## 11.5 PENETRATION OF ESSENTIAL OILS ON THE VEGETABLE STRUCTURE OF WEEDY PLANTS

Currently, it is difficult to estimate the foliar volatilization of essential oils on plants. This is because all BC-EO's compounds are concentrated in leaf hairs and specifically in the meristem of a seedling.

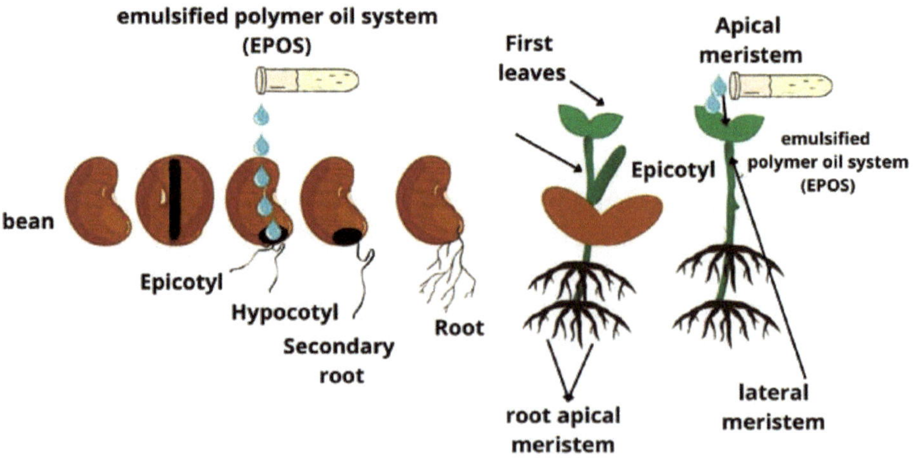


**FIGURE 11.4** Modes of EO application on weed seeds and seedlings.

The application of EPOS can be done by means of immersion in the seed as well as by foliar application in seedlings of certain weeds. The emergence of the embryo from the seed begins with the process of absorbing water. In this case, EPOS is incorporated into the seminal envelopes, which in turn perforates the seed coat to enter and distribute its molecules. The meristematic dome on the main root is the most affected by the presence of EPOS chemicals. At this point, the seed is intoxicated and what it generates is that the vascular aces take the EO molecules as a nutrient.[6] The seed will try to anchor itself in the substrate where it will germinate, it will have independent nutrient reserves in the cotyledons. Once the seed is anchored to the ground, its secondary roots and absorbent hairs will begin to grow. At this stage, a regressive death will be generated through the xylem and phloem, specifically from the apical root of the meristem.

In the case of foliar application of EPOS in seedlings, the lower epidermis, which has a fine cuticle, will allow the penetration of the apical meristem with the help of a surfactant, through which it will be damaged. The surfactant acts as a disguise so that the seedling can accept EPOS, mistaking it for minerals and nutrients necessary for its growth. Therefore, this process facilitates the maximum adsorption of the active compounds found in the oils and generates allelopathy and drought of their structures, specifically of the leaves and stem. Possibly, the incorporation of EPOS in weedy plants directly affects photosynthesis since the high volatility generates the mobility of monoterpenes and these are absorbed by the main and secondary roots and are translocated through the xylem toward the leaves during transpiration. Therefore, an energy block and limitation in electron transfer (-ATP) is generated and a decomposition of photosynthetic pigments occurs.[2]

Similarly, weeds generate changes in the micro and ultrastructure of their plant cells by various allelochemicals. Until now, the distribution of allelochemicals in the tissues of a plant, as well as the quantity of their active ingredients and the possible effects on the allelopathy process, are unknown. Monoterpenes: eucalyptol and camphor can widen and shorten the cells of the main root, as well as induce nuclear abnormalities and increase the number of vacuoles. Limonene, linalool, and camphor in turn affect the germination and radicle appearance of weed seeds. Some allelochemicals present in the pollen (hordenine and gramine) of sorghum allow mitotic activity to be reduced by more than 50% at the seedling level, in addition to generating pyknotic nuclear irregularities and inhibiting the growth of the radicle and the hypocotyl of the imbibition and vegetative stage of the plant undergrowth.

## 11.6 STABILITY AND THERMOABILITY OF EPOS

Essentially, various chemical and edaphic factors impact the degradation of EO, which can influence both the ability of the essential oil to oxidize and the course of the reaction. It is for this reason that external factors such as light, temperature, and oxygen must be carefully taken care of. In addition to this, the composition of the essential oil, structures, and the existence of impurities can also be decisive in its stability.[14] The use of ultraviolet light (UV) and visible light (Vis) is considered adequate to accelerate the auto-oxidation processes of EPOS, since these activate the hydrogen that results in the formation of alkyl radicals.[25]

However, there are considerable and rapid changes in composition when it comes to lighting, especially monoterpenes have been shown to undergo rapid degradation when exposed to the influence of light. Likewise, illumination plays another essential role in oxidation called photooxidation in which air triplet oxygen is converted into an excited singlet state in the presence of an organic sensitizer. This nonradical electrophilic molecule can directly react with a double bond, thus abstracting an allylic hydrogen atom leading to hydroperoxide structuring.

## 11.7 SCIENTIFIC CHALLENGE BEFORE THE USE OF MONOTERPENS AS ALLELOCHEMICAL IN AGRICULTURE

The EPOS monoterpenes released into the environment differ in their purity and dosage concentration. A good synergy, antagonism, and positive inhibition effects on weed seeds will allow them to be associated with the defense mechanisms of the seedlings' own metabolites to be eliminated in a crop.[28] Environmental and microenvironmental factors influencing allelopathy include species type, variety, growth stage, chemical profile, and monoterpene extraction. The high endogenous levels in the process of allelopathy allow resistance to fabiotic stress in weeds, while the exogenous ones increase the endogenous level of the receptors with a simultaneous increase in growth and resistance against abiotic stress.

It has been proven that when monoterpenes act on a weed in an abiotic stress, they cause an increase in allelopathy. EO monoterpenes can be degraded after being released either by contact or by foliar route to the seedling in substrate. The average life of EO chemicals varies from hours to months and this is associated with the volatility, type and enzymes of the soil, microbial population, and edaphological structure in which a study plant is kept.[13] Very little is known about the transport and biodegradation of EPOS in the soil for a genetic population of phallelopathic species, dosage, and volatility in weeds so that they allow their entry without being detected as foreign agents to their defense signal systems.

Numerous investigations have established that EO monoterpene allelopathy has many advantages in agricultural production. Allelopathic potentials of EPOS have been studied in crops in agricultural production, but applications are limited to regional and marginal areas. However, the dosage and implementation in agricultural crops of EO monoterpenes such as EPOS could provide control as effective as the base dose of traditional herbicides.

Due to their structural relationship within the same chemical group, it is known that essential oil components oxidize, isomerize, cyclize, or produce dehydrogenated actions via enzymatic or chemical means. After evaluating the stability of essential oils, it should be kept in mind that the chemical composition can vary in the starting material, being influenced by the health of the plant, the stage of growth, habitat, including climate, edaphic factors, and harvest time.[4]

Terpenoids tend to be both volatile and thermolabile and can be easily oxidized or hydrolyzed depending on their respective chemical structures. A good management of EPOS will allow the farmer to generate better control of various weeds through sustainable and effective practices for their eradication.

## 11.8 CONCLUSION

The problem of invasion and dissemination of seeds in a crop by non-native species leads to a severe problem of loss of endemic and/or native species. Therefore, a BC-EO interaction only in plants that compete with a crop of interest is defined and proposed. Allelopathy plays a fundamental role in the control and inhibition of weeds. The mechanisms of action at the cellular and structural levels of the BC-EOs that allow interaction among the weeds are completely related by environmental conditions and phytotoxicity that gradually compete with synthetic herbicides. An alternative to replace toxic and harmful chemicals for biodiversity with regulation and normative measures in weeds suggest the use of essential oils. EOs are promising metabolites in agriculture; however, their effectiveness is required in agricultural fields where primary crops are established and with greater economic demand.

## KEYWORDS


- **aromatic essences**
- **bioherbicide**
- **emulsification**
- **compound encapsulation**


## REFERENCES


1. El-Gawad, A. A.; Elshamy, A.; El Gendy, A. E.; Gaara, A.; Assaeed, A.. Volatiles Profiling, Allelopathic Activity, and Antioxidant Potentiality of Xanthium Strumarium Leaves Essential Oil from Egypt: Evidence from Chemometrics Analysis. *Molecules* **2019,** *24,* 584.
2. Ahuja, N.; Batish, D. R.; Singh, H. P.; Kohli, R. K. Herbicidal Activity of Eugenol Towards Some Grassy and Broad-Leaved Weeds. *J. Pest Sci.* **2015,** *88,* 209–218.
3. Araniti, F.; Landi, M.; Lupini, A.; Sunseri, F.; Guidi, L.; Abenavoli, M. R. Origanum Vulgare Essential Oils Inhibit Glutamate and Aspartate Metabolism Altering the Photorespiratory Pathway in Arabidopsis Thaliana Seedlings. *J. Plant. Physiol.* **2018,** *231,* 297–309.
4. Barra, A. Factors Affecting Chemical Variability of Essential Oils: A Review of Recent Developments. *Nat. Prod. Commun.* **2009,** *4,* 1147–1154.
5. Bordin, E. R.; Camargo, A. F.; Stefanski, F. S.; Scapini, T.; Bonatto, C.; Zanivan, J.; Preczeski, K.; Modkovski, T. A.; Júnior, F. R.; Mossi, A. J.; Fongaro, G.; Ramsdorf, W. A.; Treichel, H.. Current Production of Bioherbicides: Mechanisms of Action and Technical and Scientific Challenges to Improve Food and Environmental Security. *Biocatal. Biotransform.* **2020,** *39,* 1–14.
6. Cheng, F.; Cheng, Z. Research Progress on the use of Plant Allelopathy in Agriculture and the Physiological and Ecological Mechanisms of Allelopathy. *Front. Plant Sci.* **2015,** *6,* 1020.
7. Cloyd, R. A.; Galle, C. L.; Keith, S. R.; Kalscheur, N. A.; Kemp, K. E. Effect of Commercially Available Plant-Derived Essential Oil Products on Arthropod Pests. *J. Econ. Entomol.* **2009,** *102,* 1567–1579.
8. Dayan, F.; Barker, A.; Bough, R.; Ortiz, M.; Takano, H.; Duke, S. Herbicide Mechanisms of Action and Resistance. **2019**. DOI: 10.1016/B978-0-444-64046-8.00211-1
9. Ghasemi, H.; Mozaffari, S.; Mohammadghasemi, H.; Jemere, A.; Nazemifard, N. Microfluidic Platform for Characterization of Crude Oil Emulsion Stability. *Can. J. Chem.* **2022,** DOI: 10.1139/cjc-2021-0248.
10. Grana, E.; Sanchez-Moreiras, A. M.; Reigosa, M. J. Mode of Action of Monoterpenes in Plant-Plant Interactions. *Curr. Bioact. Compd.* **2012,** *8,* 80–89.
11. *Investigación: Uso y Regulación de Herbicidas en México.* Palacio Legislativo de San Lázaro, Ciudad de México. Mayo 2020. Recuperado de: http://www.cedrssa.gob.mx/files/b/13/76Herbicidas.pdf

12. Jiang, Z.; Akhtar, Y.; Bradbury, R.; Zhang, X.; Isman, M. Comparative Toxicity of Essential Oils of Litsea pungens and Litsea cubeba and Blends of Their Major Constituents against the Cabbage Looper, Trichoplusia ni. *J. Agric. Food Chem.* **2009,** *57,* 4833–4837.
13. Kaur, S.; Singh, H. P.; Batish, D.; Kohli, R. Chemical Characterization and Allelopathic Potential of Volatile Oil of Eucalyptus Tereticornis Against Amaranthus Viridis. *J. Plant Interact.* **2011,** *6,* 297–302.
14. Khare, P.; Srivastava, S.; Nigam, N.; Singh, A.; Singh, S. Impact of Essential Oils of E. Citriodora, O. Basilicum and M. Arvensis on Three Different Weeds and Soil Microbial Activities. *Environ. Technol. Innov.* **2019,** *14,* 100343.
15. Maes, C.; Bouquillon, S.; Fauconnier, M. L. Maes Encapsulation of Essential Oils for the Development of Biosourced Pesticides with Controlled Release: A Review. *Molecules* **2019,** *24,* 2539.
16. Merad, N.; Andreu, V.; Chaib, S.; de Carvalho Augusto, R.; Duval, D.; Bertrand, C.; Boumghar, Y.; Pichette, A.; Djabou, N. Essential Oils from Two Apiaceae Species as Potential Agents in Organic Crops Protection. *Antibiotics (Basel, Switzerland)* **2021,** *10* (6), 636.
17. Mohammed Naguib Abd El-Ghany Hasaneen. *Herbicides—Mechanisms and Mode of Action*; InTech: Croatia, 2011, pp 1–214.
18. Raveau, R.; Fontaine, J.; Soltani, A.; Ballesteros, E.; Pilkington, L.; Quek, S.-Y.; Mediouni Ben Jemâa, J. In Vitro Potential of Clary Sage and Coriander Essential Oils as Crop Protection and Post-Harvest Decay Control Products. *Foods* **2022,** *11*. DOI: 10.3390/foods11030312.
19. Yerasi, P. K. R. Herbicide Development and Registration of the Herbicide. **2022**. DOI: 10.22271/int.book.111.
20. Sawi, S.; Ibrahim, M.; El-Rokiek, K.; El-Din, S. Allelopathic Potential of Essential Oils Isolated from Peels of Three Citrus Species. *Ann. Agric. Sci.* **2019,** *64*. DOI: 10.1016/j.aoas.2019.04.003.
21. Sharma, A.; Singh, H. P.; Batish, D. R.; Kohli, R. K. Chemical Profiling, Cytotoxicity and Phytotoxicity of Foliar Volatiles of Hyptis Suaveolens. *Ecotoxicol. Environ. Saf.* **2019,** *171,* 863–870.
22. Silva, L.; Oliveira Ferreira, O.; Cruz, J.; Franco, C.; Anjos, T.; Cascaes, M.; Almeida, W.; Aguiar, E.; Oliveira, M. Lamiaceae Essential Oils, Phytochemical Profile, Antioxidant, and Biological Activities. *Evid. Based Complement Alternat Med.* **2021,** 1–18. DOI: 10.1155/2021/6748052.
23. Singh, N.; Singh, H. P.; Batish, D. R.; Kohli, R. K.; Yadav, S. S. Chemical Characterization, Phytotoxic, and Cytotoxic Activities of Essential Oil of Mentha Longifolia. *Environ. Sci. Pollut. Res.* **2020,** *27,* 13512–13523.
24. Sothearith, Y.; Appiah, K.; Mardani, H.; Motobayashi, T.; Yoko, S.; Hourt, K.; Sugiyama, A.; Fujii, Y. Determination of the Allelopathic Potential of Cambodia's Medicinal Plants Using the Dish Pack Method. *Sustainability* **2021,** *13,* 9062.
25. Subelzu, N.; Schöneich, C. Near UV and Visible Light Induce Iron-Dependent Photodegradation Reactions in Pharmaceutical Buffers: Mechanistic and Product Studies. *Mol. Pharm.* **2020,** *17* (11), 4163–4179.
26. Suwanagul, D.; Pitaksaringkarn, W.; Suwanagul, A. Allelopathic Potential of Citrus Reticulata Peel Extract for Weed Control. **2019,** 17–21. DOI: 10.14456/rjas.2016.2.

27. Teixeira, B.; Marques, A.; Ramos, C.; Serrano, M.; Matos, O.; Neng, N.; Nogueira, J.; Saraiva, J.; Nunes, M. Chemical Composition and Bioactivity of Different Oregano (Origanum Vulgare) Extracts and Essential Oil. *J. Sci. Food Agric.* **2013,** *93,* DOI: 10.1002/jsfa.6089.
28. Turek, C.; Stintzing, F. Stability of Essential Oils: A Review. *Compr. Rev. Food Sci. Food Saf.* **2013,** *12*. DOI: 10.1111/1541-4337.12006.
29. Universidad de las Fuerzas Armadas (ESPE). *Carlos Cárdenas Tello*; Las Plantas Alelopáticas: Recuperado depp; pp 1–204 : https://repositorio.espe.edu.ec/bitstream/21000/9218/3/Las%20Plantas%20Alelopaticas.pdf
30. Werrie, P. Y.; Durenne, B.; Delaplace, P.; Fauconnier, M. L. Phytotoxicity of Essential Oils: Opportunities and Constraints for the Development of Biopesticides. A Review. *Foods* **2020,** *9*. DOI: 10.3390/foods9091291.

CHAPTER 12

# Traditional and Modern Agricultural Practices in a Periurban Community Close to Mexico City

SILVANA ESCALANTE HIDALGO[1], VÍCTOR ÁVILA AKERBERG[2], and NATALIA DE GORTARI LUDLOW[3]

[1]*Licenciatura en ciencias de la Tierra, Facultad de Ciencias, UNAM*

[2]*Instituto de Ciencias Agropecuarias y Rurales, UAEMéx*

[3]*Instituto de Ciencias Sociales y Humanidades Alfonso Vélez Pliego, BUAP*

## ABSTRACT

Our research deals with soil quality, plant diversity, and social perception in Tlazala, a rural town vulnerable to changes in land use, technification, and loss of traditional knowledge due to its closeness to Mexico City. The main objective was to evaluate and compare the physical and chemical soil characteristics in Mexican traditional maize fields (milpas) with different agricultural practices ranging from traditional to intensive, according to their use of agrochemicals. Three treatments were identified and compared between them and with the surrounding forest vegetation and soil qualities with the purpose of outlining the contrasts between natural and altered lands and their implications in ecosystem services. The diversity of noncultivated plants present on the plots was collected and their traditional uses identified (edible, medicinal, forage, ornamental, ceremonial). Interviews were conducted with farmers to understand their perception of the impacts of their management and decisions on the environment, and to learn if they perceived any changes in Tlazala's natural landscape. Also surveys to kids were carried out to understand their relation to agriculture. We observed

Sustainable Agricultural Production Systems: Traditional and Modern Perspectives.
Romeo Rojas, Guillermo Cristian Guadalupe Martínez-Ávila, Julia Mariana Márquez-Reyes, and Ma del Carmen Ojeda-Zacarías (Eds.)

negative tendencies in some soil characteristics, which increased with the intensification of management. Plant diversity and earthworms were higher in traditionally managed fields.

## 12.1 INTRODUCTION

Throughout human history it has been observed how bad soil management generates poverty[9] and environmental degradation. Climate change, the even stronger connection between the social and commercial systems, pollinators decline, and the rise of plagues and diseases make unstabilities that can alter the provision of ecosystem services from the agricultural landscape, including food production.[33]

Agriculture is an activity that has the potential to deteriorate ecosystem services or, on the contrary, to promote their improvement through a rational design of agroecosystems.[51,55,56]. Provisioning ecosystem services from agricultural lands (food and fodder) have degraded other less direct services, but fundamental for ecosystems, such as climate and water regulation, biodiversity, and the protection against soil erosion.[51] The Millennium Ecosystem Assessment proposed that agriculture can be the human activity that represents the highest threat to biodiversity and ecosystem functioning.

Biodiversity is the variability that exists between species, and between levels of organization, ecosystems, and landscapes. Ecosystem function is closely linked to diversity at the species and population levels, and changes in it can influence the supply of ecosystem services.[5,14,63]. At the same time, this relationship is influenced by interactions between elements or species[34] that depend directly on the conditions and availability that a habitat provides.[41] Species richness (number of different species), equitability (equity of elements between systems), and heterogeneity (disparity between elements and their functions) are useful tools to know this environmental variability.[7]

Although traditional low-impact agriculture contributes to the maintenance of biodiversity,[60] Mexican fields tend toward intensification, the use of elements and products from the market, and the loss of traditional knowledge, and with it the loss of biological and cultural diversity.

The use of agrochemicals in intensive agriculture is the factor that most negatively impacts soil properties, even when used in a semi-intensive or controlled manner.[44] This management can also be related to pest resistance, fluctuations, and excess of nutrients (MEA, 2005). The sustainability of agriculture and environmental quality are determined by the quality and health of the soil.[1,50] Doran[21] defines soil quality or health as "the ability of a living soil to function, within natural or managed ecosystem barriers, to sustain biological productivity, maintain and improve air and water quality, and promote plant and animal health." Among the basic functions that a healthy soil must carry out are the promotion of the activity and productivity of biodiversity, filtration, buffering, degradation and detoxification of organic and inorganic materials, regulation and distribution of solutes and water flows, cycling of carbon and nutrients, and the proportion of physical stability for plants and animals.[61]

As a complex system, the soil is characterized by having quantifiable properties that can be used for its evaluation and indicate adverse changes that it has suffered due to degradation processes.[20] The main processes that negatively affect the health and quality of the soil in agricultural land are erosion, compaction, leaching of nutrients, reduction of fertility, loss of organic matter, acidification, pollution, and loss of biodiversity.

However, the consequences on sustainability and agrodiversity that alter services through management do not depend only on the biophysical conditions, but also on the perception of the people who do the management.[6,35,43,45] For the practice of environmental management, it is essential to know the social perception of the landscape, that is, the ways in which people think about nature and to know if the actors responsible for certain management perceive the failures and positive and negative impacts that their practices generate.

It is currently known that it is necessary to promote research that intersects quantitative scientific analysis with social analysis, both factors that make up a socio-environment.[11,32] That is why within this study, analytical tools from the natural and social sciences were used to generate a comprehensive theory of land degradation.

Finding the balance between increasing the production and yield of crops and maintaining the ecosystem services that make the land productive is a challenge, and even more so when considering the increase in population.[8,29,52,66]

Through the evaluation of soil quality and biodiversity we can know the state of health that an agroecosystem presents and use this knowledge to promote actions that allow the conservation of biodiversity and agricultural productivity and, with this, reduce the abandonment of land of cultivation and land use change from natural vegetation to agricultural land.

The development and results of a study on soil quality and plant diversity in agricultural lands in the town of Tlazala, State of Mexico, are presented below. The mosaic of landscapes observed in Tlazala, and its surroundings represents an opportunity to describe trends in soil degradation due to agricultural strategies in a peri-urban area (vulnerable to urbanization), allowing comparison with a reference soil, in this case of surrounding forest, which has not undergone anthropogenic transformations.[19]

## 12.2 THE MEXICAN CONTEXT

During the last three decades the processes of urbanization and globalization have configured a new panorama for the agricultural sector.[25,26] Two sectors of peasants have been generated: those that are associated with the export market, that receive government support and that carry out intensive management applying technologies, and those that carry out extensive traditional subsistence agriculture.[53] In the case of Mexico, this transformation has increased the levels of poverty, migration, and "de-agrarianization" of the rural environment, where nonagricultural activities represent more than 50% of the income of rural families,[3,24,58,] as is the case of the town of Tlazala.

The work of the milpa and the knowledge behind it are part of the biocultural heritage of Mexicans and humanity. However, farmers not only respond to their culture and traditions, but also are immersed in a globalized context and often find themselves in the need or desire to participate in the market, a situation that makes them dependent on external prices influenced by supply and demand. The need to adapt to changing circumstances in their social, political, and economic environment significantly alters the way they work their land and the impact they have on it.

On the other hand, the Mexican territory presents deforestation rates above the world average, the main causes being the increase in cultivation and grazing areas, and urban expansion.[46,47,54,]

The Organization for Economic Cooperation and Development (OECD) places Mexico among the first places for degraded soils (with almost 50% of severely and very severely degraded soils), while UNEP-SEMARNAT-INECC (2004) establish that more than half of this degradation is caused by agricultural activities.

Much of this degradation is due to a lack of knowledge about the ecological importance of the soil, the limits and aptitudes for its use, and the appropriate techniques for good management (PNUMA-SEMARNAT-INECC, 2004). This lack of information is sown from the institutional level, with the lack of land use policies and the promotion of practices that accelerate its degradation, without considering that its loss may be irreversible. This misinformation about the impacts generated by poor agricultural management does not only affect the soil, but also directly and indirectly affects water resources, puts biodiversity at risk, unbalances the environment, and puts the health of living beings at risk.

## 12.3 STUDY OBJECTIVES

The objective of this study was to evaluate soil quality and plant diversity in milpas (Mexican traditional multidiverse maize agriculture) under management systems with different intensities and to learn about the farmers' perceptions of the environment, resource management, and ecosystem services in Tlazala, Isidro Fabela municipality, State of Mexico.

## 12.4 EXPERIMENTAL METHODS AND MATERIALS

### *12.4.1 STUDY AREA*

The town of Tlazala is the municipal capital of Isidro Fabela, located northwest of Mexico City, in the northern part of the Sierra de las Cruces mountain range (INEGI, 2006), within the Guadalupe Dam Basin (Fig. 12.1). Despite its proximity to Mexico City and Toluca City, the territory still retains its rural essence and a low environmental degradation. The geographical location of its localities and its low population density have allowed the preservation of its natural resources.[62] The main land uses of the territory of Isidro Fabela are forestry (76%), rural (7%), seasonal agriculture (7%), pasture (7%), and scrubland (3%).[4] The dominant soils in Isidro Fabela are Andosol; however, Luvisol is the best represented soil type in the agricultural lands of Tlazala.

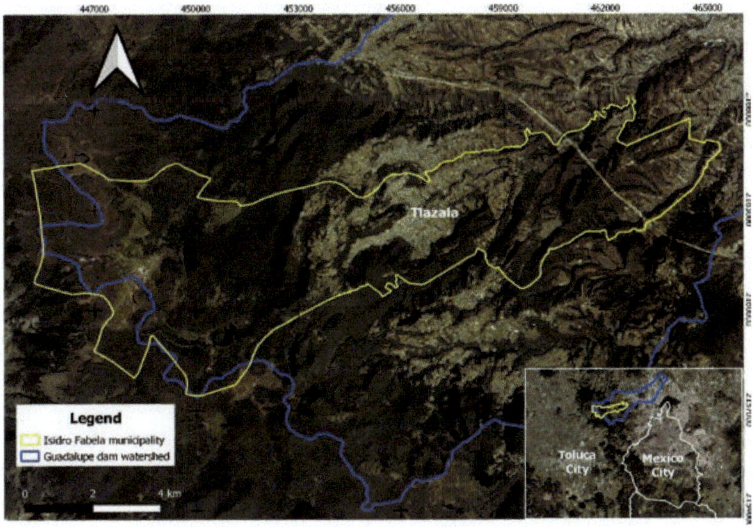


**FIGURE 12.1** Geographical location of Tlazala de Fabela, in relation to the Guadalupe Dam watershed and the two biggest cities in central Mexico.

*Source*: Adapted from Ávila-Akerberg & González-Martínez, 2023

#### *12.4.1.1 DELIMITATION OF MANAGEMENT SYSTEMS*

Three types of agricultural management systems used in Tlazala were identified: traditional, medium, and intensive. Intensive management is distinguished using pesticides (carbofuran and organophosphates) and artificial (ammonium sulfate and simple calcium super phosphate) and natural (animal manure) fertilizers; in the medium system artificial and natural fertilizers are used (urea, 18–46 fertilizer and chicken manure) and in the traditional system only animal manure. In the intensive treatment herbicides are applied systematically (atrazine, dicamba, and glyphosate), in the medium treatment they are used occasionally, and in the traditional treatment they are not used. The milpas worked in the traditional way are generally for personal consumption of food and fodder and occasionally participate in the local market, unlike medium and intensive management, which supply both the local market and that of other nearby regions.

#### *12.4.1.2 SELECTION OF AGRICULTURAL SAMPLING SITES*

The criteria for the selection were the following: (1) that the main crop of the plot be some variety of maize, (2) that the plots be found, as far as possible, in a similar environmental unit, encompassing the largest part of the town of Tlazala. Four replicates were obtained for each treatment. In addition, soil samples from the surrounding forest were taken to have reference values for nonagricultural soil in the area.[32]

#### *12.4.1.3 INDICATORS OF SOIL QUALITY AND PLANT DIVERSITY*

For the evaluation of soil quality and plant diversity, the parameters described in Table 12.1 were used.

#### *12.4.1.4 INTERVIEWS AND SURVEYS*

Semistructured interviews were conducted among farmers from the sample plots. A total of 15 farmers were randomly interviewed while they carried out their everyday activities. The aim of these interviews was to get to know the use of the maize fields, their history, the motivation to keep them, and the farmers' perception of their land use and its impact on the environment to

**TABLE 12.1** Indicators of Soil Quality and Plant Diversity Used in This Study (Own Elaboration from Refs. [22, 23, 39, 48]).

| Indicator | Indicator | Relationship with soil condition and its function |
|---|---|---|
| **Physical** | Texture | Retention and transport of water and minerals |
| | Bulk density and porosity | Potential for leaching and water transport, compaction, productivity, and erosivity |
| | Moisture percentage | Water availability for organisms, transport of nutrients and minerals |
| | Percent bare soil (soil without vegetation) | Indicates possible compaction, poor moisture retention, little or no organic matter contribution, poor soil structure, lack of diversity |
| **Chemicals** | Organic material | Determines soil fertility, nutrient availability, aggregate stability, stability, and erosion measurement |
| | pH | Determines the thresholds of biological and chemical activity, and limits for plant growth |
| | Electric conductivity | Determines plant and microbiological activity thresholds, limits plant growth |
| | Interchangeable bases ($Na^+$, $K^+$, $Mg^{2+}$ y $Ca^{2+}$) | Plant nutrient storage, contaminant retention, and pH buffering |
| **Biological** | Plant diversity and richness | Yield, regulating services, biological pest control, soil fertility, pollinator attraction |
| | Shannon-Wiener and Simpson diversity indexes | Uniformity of the relative abundance of a certain species and the degree of dominance between species |
| | Herbaceous functional diversity | Through the total richness and the monocotyledonous/dicotyledonous index, the potential of certain species to attract pollinators is discussed |
| | Sörensen Floristic Similarity Index | Allows you to compare communities with similar attributes |
| | Presence of worms | Indicates good soil structure, aeration, moisture retention, presence of organic matter, and diversity |

know whether they have observed changes in Tlazala's environment. A total of 1126 surveys were carried out among children in five elementary schools and three secondary schools in the Isidro Fabela municipality. The aim was to get to know more about their points of view and the local problems.

## 12.5 RESULTS

### *12.5.1 SOIL*

The quantitative texture analysis showed that the forest soil is silty-clay (RL), while 83% of the maize field soil is silty-clay and 17% is clay-loam (CR; they contain more sand than the others).

Moisture, bulk density, porosity, pH, electrical conductivity, and organic matter were significantly different between the maize fields and the forest soil. Differences are mainly due to the removal of the natural vegetation and land plowing (which produces loss of structure and compaction). Vegetation removal opens the system by generating a demand for external nutrients,[12] it exposes the soil to weathering, and it modifies the dynamic properties, affecting their capacity to provide ecosystem services.[42]

In forest environments, carbon and nitrogen provided by vegetal waste maintain soil health. Altering this natural supply of nutrients by removing forest vegetation to create agricultural fields results in a severe affectation of the biological indicators due to changes in the quantity and quality of vegetal waste.[10,28] On the other hand, diversity allows organisms to avoid an intense competition for food or space, it decreases invasion and maintains the ecosystem working constantly even in fluctuating environmental conditions.[17]

The moisture ratio in the forest is significantly higher than that in the maize fields due to differences in altitude and the presence of forest cover, which provides a higher quantity of organic matter with high moisture retention capacity in the soil.

Significant differences were expected between the values of the forest soil and the agricultural soil properties, as farmers mentioned that the removal of natural vegetation from the plots happened more than 150 years ago. All the interviewees stated that their lands have been used for agriculture for as long as they can remember, and that their ancestors worked on them.

In all treatments, pH values were moderately acidic (NOM-021-RECNAT-2000) with values between 5.1 and 6.5, which are adequate for growing corn. Given the generally low conductivity in the agricultural soils studied, it is inferred that they contain little dissolved salts and that they are nonsaline soils.

The bare soil that was observed in the maize fields is a product of the vegetal management used. Intensive management always makes use of herbicides (sometimes medium management does too), and mechanical removal of plants that are not part of the crops. This reduces soil moisture and the input of organic matter and nutrients that happen naturally, which impacts the soil's macro and microfauna and the transport of fluids and solutes. These actions are reflected in the presence of earthworms, which were found more frequently in the forest and in traditional maize fields (83%) than in medium-management (49.7%) or in intensive-management ones (29.7%).

A rising trend was observed regarding bulk density, whereas porosity decreased as management became more intense (Figs. 12.2 and 12.3). Bulk density is a dynamic property of the soil that easily changes due to management. Intensive plowing and the use of heavy machinery increase soil density, debilitate its structure, and decrease porosity and the soil's capacity to transport water, air, and solutes. The use of tractors, common in the intensive method, could be the cause of the increase in bulk density and the decrease in porosity in these plots, either because of the loss of structure or because of the collapse of the macropores. In time, this situation leads to an increase in surface runoff and soil erosion.[40] Despite these visibly negative trends, porosity and bulk density do not show significant differences among treatments. Growing conditions are adequate for vegetable growth in the three treatments (< 1.1).

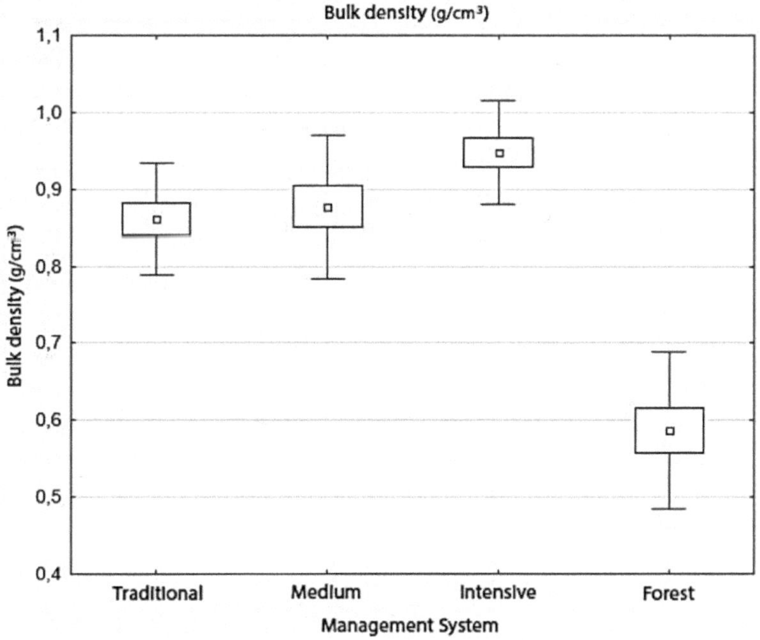


**FIGURE 12.2** Values obtained for bulk density in forest and agricultural soils under three types of management in Tlazala de Fabela, Estado de México. The smaller box represents the average, the bigger box represents the standard error (±), and the vertical lines represent the standard deviation (±).

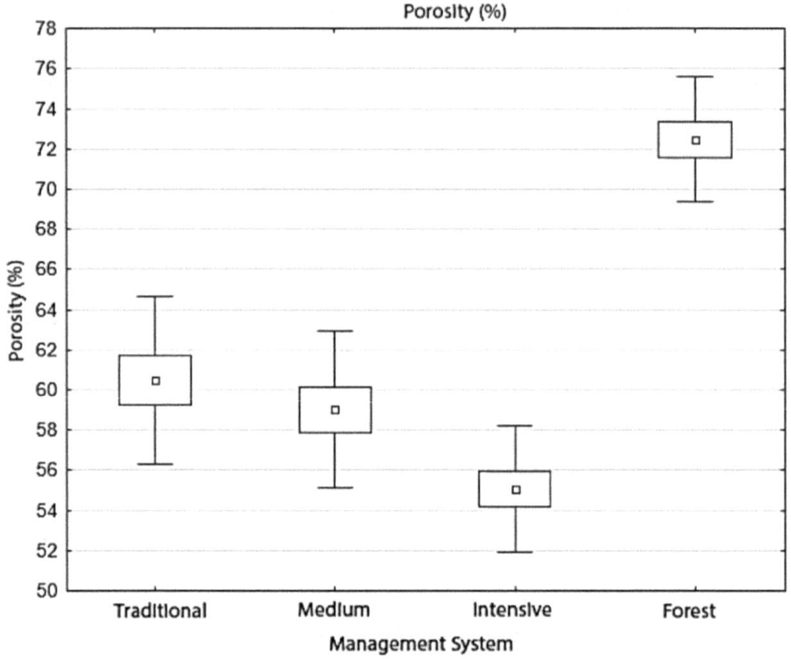


**FIGURE 12.3** Values obtained for porosity in forest and agricultural soils under three types of management in Tlazala de Fabela, Estado de México. The smaller box represents the average, the bigger box represents the standard error (±), and the vertical lines represent the standard deviation (±).

As management becomes more intensive, a downward trend in soil moisture can be observed. This is possibly because in intensive management, weeds are avoided and the soil is more compact, which allows it to retain less water. The available water can be a determining factor in microbiological activity in the soil. The more porosity present, the higher the moisture ratio. This factor and the fact that the organic matter contribution in the forest (which has not been affected) is higher in quantity and quality than that in the maize fields, lead us to infer that microbiological activity is higher in the forest than in the maize fields. This inference is highlighted when we compare the forest soil with the intensive maize fields, as pesticides and herbicides destroy part of their microfauna.[44]

The presence of organic matter is higher in traditional maize fields and electrical conductivity is higher in intensive maize fields. These values are related to salinity and the presence of more exchangeable bases produced by the addition of chemical inputs.

In plots with traditional management, earthworms were found as frequently as in the forest soil. The earthworms mix and decompose the soil components, which is beneficial for microbiological activity, root growth, soil structure, and water and air movement and retention. Their presence in these plots indicates better quality and better biodiversity. The low frequency of earthworms in the medium and intensive maize fields could be related to the addition of agrochemicals, especially pesticides. It is therefore believed that the fauna and flora in the soil have suffered the highest impact under intensive management.

Traditional and medium management show more similar characteristics between them than compared with intensive management. It could be argued that this is because of the occasional use of heavy machinery and pesticides, which cause a more significant impact than the use of chemical fertilizers. However, the physical and chemical properties in the maize field soils using the three types of treatment do not show significant differences. This could happen because, even though an intensive treatment is used in some parts, it does not cause a great impact, or it has been implemented for too short a time for changes to be observed.[49] This could in fact be one of the main reasons, as technical agriculture has been present in Tlazala for a short time, and those who use intensive management are still exploring the available products. Still, it is believed that if the intensive management continues, it will lead to a more severe degradation process.

The ecosystems and landscapes induced by human beings always involve changes in the soil.[64] If we take into consideration that a natural soil (without any management) should show better health and stability than a soil that receives human management, we can argue that the agricultural soil in Tlazala is slightly deteriorated, as some of its ecological functions have been slightly truncated or replaced, without affecting its general functionality.

### *12.5.2 PLANT DIVERSITY*

The outcomes of the parameters used to analyze plant diversity are shown in Table 12.2. The study registered a total of 55 species grouped into 21 families present in the maize fields. The most important families, in terms of number of species, were *Asteraceae* (18.2%), *Cucurbitaceae* (10.9%), *Fabaceae* (9.1%), and *Polygonaceae* (9.1%). Regarding life forms, some herbs and vines were found. There were no instances of plant species from the forest found in the maize fields.

**TABLE 12.2** Values of Species Richness and Diversity According to Treatment.

| | Richness | Simpson index | Shannon-Wiener (H') Index | Functional diversity (mono/dico) |
|---|---|---|---|---|
| **Traditional** | 41 | 0.20 | 2.03 | 0.24 |
| **Medium** | 34 | 0.18 | 1.93 | 0.22 |
| **Intensive** | 24 | 0.46 | 1.29 | 0.20 |

Richness was higher in the plots where traditional treatment is used (r=41): almost double compared with those where intensive treatment is used. This was expected, as in those spaces no herbicides or pesticides are used, and there is no constant mechanical removal of non-cultivated plants. Medium management (without pesticides) is found in second place (r=34). This could be due to the occasional use of herbicides. Lastly, the intensive treatment had a species richness of 24. Richness in each type of management is represented both by species shared among treatments (tolerant) and by specific and exclusive species, that is, species only found in maize fields that use a certain type of treatment. Most species found were tolerant to all types of management, but most exclusive species were found in traditional management.

The Simpson index is based on dominance, and it considers the abundance of the most important species. Plots under intensive management had a higher Simpson value, which shows dominance of certain species (mainly Oxalis pes-caprae L.; an exotic species; CONABIO, 2009) and displacement of specialized species due to the pressure exerted using agrochemicals, which allows generalist species to spread thanks to their resistance. It also causes the seed bank to grow richer with propagules at the expense of plant diversity. The Simpson index is the inverse value of equitability and diversity.[30] Therefore, diversity in the traditional and medium maize fields is higher. This diversity allows organisms to avoid intense competition for food or space. It also decreases invasion and allows the ecosystem to keep working even in fluctuating environmental conditions.[17]

The Shannon-Wiener diversity index was higher for traditional treatment (H'= 2.03), followed by medium management (H'= 1.93), and then intensive management (H'= 1.29). This result reveals that there is more heterogeneity in traditional and medium maize fields, as they show a fairer distribution of abundance than that in intensive maize fields.

Functional diversity, represented in this study by the monocotyledons/dicotyledons ratio, is based on the premise that dicotyledons are more attractive for pollinators than monocotyledons.[32] This ratio was relatively similar in the tree treatments. Traditional management was the one with a higher ratio

(0.24), followed by medium treatment (0.22) and intensive treatment (0.20). A higher value in this ratio indicates a larger monocotyledon representation. The values obtained for the three types of management are low, which shows a larger presence of dicotyledons in all cases and, therefore, higher pollinator attraction. It is believed that the functional diversity value was higher for the traditional maize fields because these plots generally present a larger quantity of noncultivated plants.

The current crop diversity is shown in Figure 12.4. We can observe that corn is not only the main element traditionally present in the maize fields, but also one of the main crops in lower Tlazala, followed by fava beans and beans. In the highest parts of the municipality, the main crops are potatoes and oats. Even though some elements are not actively planted by the farmers, they germinate in the land through their wild relatives (such as zucchini), as a product of the seed bank in the soil and as noncultivated plants.

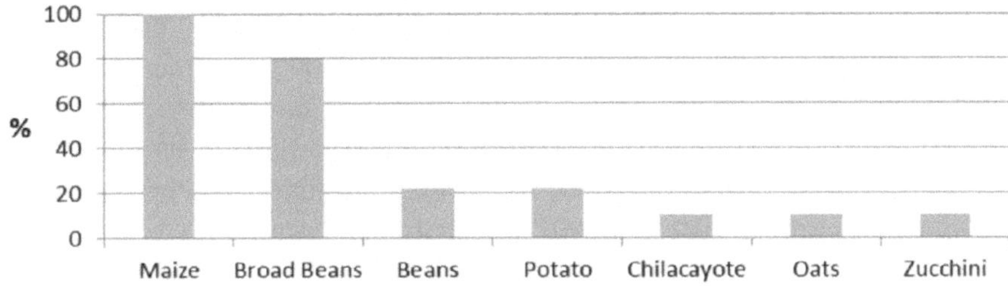


**FIGURE 12.4** Common crops in the interviewed farmers' maize fields in Tlazala de Fabela, Estado de México.

The "Noncultivated plants" refers to spontaneous or adventitious crop companions in places disturbed by agriculture .[15] These plants take part in the trophic webs as primary producers. They can be soil protectors, preventing erosion, providing organic matter, and sometimes supporting the presence of beneficial insects as natural enemies of plagues.[2,13,36] The inhabitants of Tlazala use wild herbs, ruderals, and noncultivated plants to cover their food and health needs and to raise livestock or domestic animals. Research on the use of the species found was done based on the sampling conducted in the maize fields and on the identification of the specimens (Fig. 12.5). The noncultivated plants are used mainly as forage for animals that will later provide organic manures and other by-products. Medicinal plants are the second most mentioned, followed by edible, ornamental, melliferous, and ceremonial plants, as well as those used as pigments.

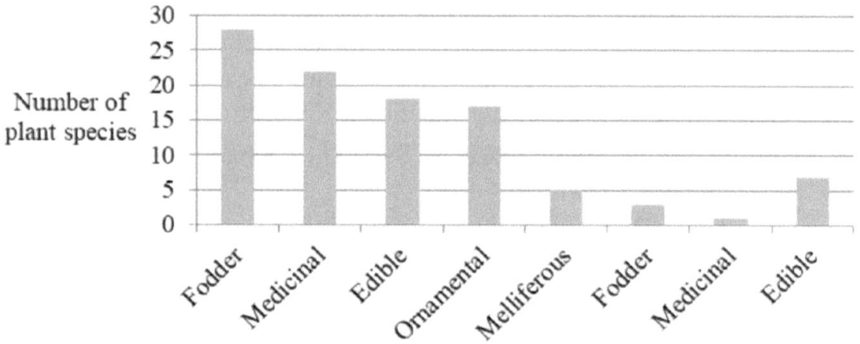


**FIGURE 12.5** Uses of the plant species found in the maize fields studied in Tlazala de Fabela, Estado de México.

The diversity of uses for noncultivated plants is part of the traditional knowledge of the region. They represent an important part of their biological culture. They provide important ecosystem services: provisioning (food, medicine, and genetic resources), supporting (pollinator attraction, biologic control, carbon accumulation), regulating (they reduce the soil's water erosion, promote moisture retention, retain sediments, and provide organic matter), and cultural (use in ceremonies and as pigments).

### *12.5.3 CANONICAL CORRESPONDENCE ANALYSIS*

The canonical correspondence analysis (CCA; Fig. 12.6) shows the general variation pattern in the composition of the vegetal community explained by edaphic variables (pH, electrical conductivity, moisture ratio, organic matter, porosity, real and bulk density, and exchangeable cations) and by species distribution along each variable.

The points represent the species and plots studied, and the colors represent the type of treatment (traditional = orange, medium = yellow, intensive = pink, and forest = green). The arrows show the variability in the soil's properties, their direction shows the correlation between factors, and their length is the same as the exchange ratio. This is therefore a measurement of how species distribution varies along each environmental variable.[59] The most important variables are represented by longer arrows.

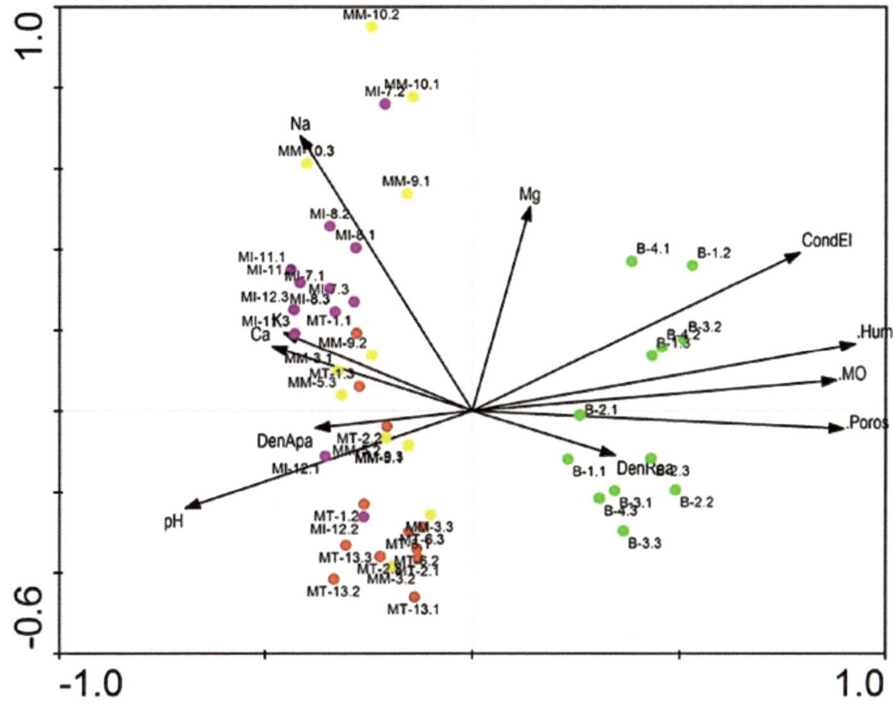


**FIGURE 12.6** Canonical correspondence analysis (CCA). Soil property values and treatment per sample plot, including the forest soil in Tlazala de Fabela, Estado de México. TM = Traditional Management, MM = Medium Management, IM = Intensive Management, and F = Forest. The numbers represent the sampling unit.

In Figure 12.6, we can clearly see the difference in composition between the agricultural management and the forest soils. The properties with a higher exchange ratio from one type of vegetation to another are electrical conductivity, moisture ratio, organic matter, porosity, and pH. These differences have a significant influence on the vegetation. As we mentioned before, these differences are mainly due to the removal of natural vegetation and plowing.

Electrical conductivity and pH are the most determinant factors in the composition. They are inversely related: the higher the pH, the lower the electrical conductivity is. Based on the CCA and the size of the vector corresponding to bulk density, it could be possible to believe that the bulk density is a secondary variant that seemingly does not affect the vegetation composition significantly.

### 12.5.4 SOCIAL PERCEPTION

#### 12.5.4.1 INTERVIEWS

##### 12.5.4.1.1 Agricultural Management

The interviewees were aged 17–81, with an average age of 52. They were 3 women and 12 men. When asked about their occupation, 52% replied that they were farmers.

Most (87%) of the agricultural lands were acquired through family inheritance and almost all of them are exclusively seasonal (69%). The work on the fields was learned from their parents or grandparents and it is taught in the same way to the children. Few question the management taught by their parents and make changes; most continue doing the same because of tradition.

Agricultural production in Tlazala is for household consumption (65%) and local (25%) or external (10%) sales. The farmers do not want to enter the market as they say that corn is not profitable. They claim that they cannot make a living out of it because the buying prices are very low, the aids are minimal and the (agrochemical) products are very expensive. This is the reason why they seek to get an income from other sectors such as trade, industry, or construction, which results in production being for household use only. This situation has benefitted the soils and the biodiversity in Tlazala, as no intensive management measures have been implemented significantly. However, families need to get their incomes in urban zones or by urbanizing Tlazala, a situation that causes migration and environmental decay because of land use change and plant removal.

As for management, plowing is generally done with a yoke. Few use tractors for this activity, and most commonly they are the ones who produce more and sell externally. It is mentioned that the land in Tlazala is not good for tractors, due to the steep terrain.

Almost half the farmers use manure to input nutrients into the land, as represented in Figure 12.7. Urea is the most common chemical fertilizer used, followed by diammonium phosphate, known as abono 18–46, and hydrated lime. In Tlazala, traditional management (with manure) is still the most common, due to the production size, as it is for household consumption and therefore non-remunerated. This is the reason why the farmers want to invest as little as possible in inputs. The plots where intensive management is used, and where two or more fertilizers from Figure 12.7 are used, sell their products inside and outside Tlazala.

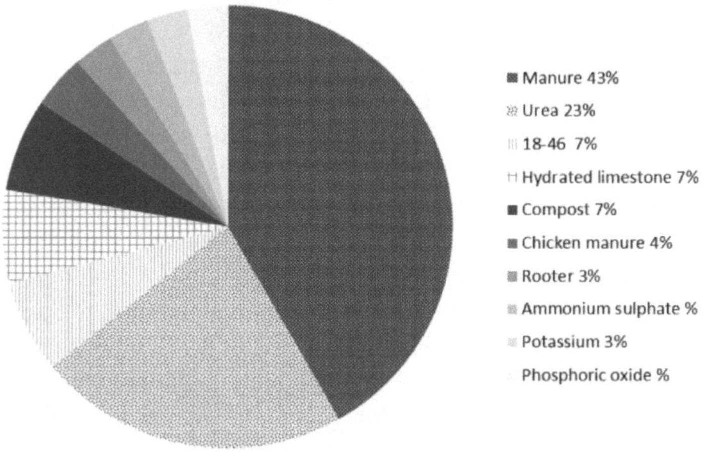


**FIGURE 12.7** Natural and synthetic elements used in Tlazala to fertilize the soil.

Regarding the use of pesticides, the inhabitants of Tlazala avoid their use because of the costs, and because they have noticed negative effects. Most inhabitants have used them in critical moments and not as part of a management system. Again, those who use them systematically are those who have bigger productions, and they represent a minority (20%). More than 50% of the population uses pesticides. Marvel (dicamba and atrazine) and Faena Fuerte (glyphosate) are the most used.

Corn seeds used in Tlazala are 70% creole. They are selected and stored by the farmers every year. If they do not have any in storage, they usually receive them from a family member or friend or they buy them. All the interviewees state that they know about the "improved seeds" since they are offered by the government as a form of aid. Most (66%) have used them at some point, but they think that they do not work (50%), as they are not designed for Tlazala's environment. Another part of the population (33%) claims not to be interested in them and a minority (17%) says "they're wrong" with no further explanation.

Not only does the work in the maize fields provide people with food, but it also produces intangible worth in the form of identity, satisfaction, security, and well-being. All the interviewees claim that they like sowing. The main reason for this answer was that it is their food (36%), followed by personal satisfaction, feeling safe with what they consume and duty (11%). However, as Jackson et al.[38] noted, farmers do not perceive the external benefits of preserving agro-diversity and environmental health at a larger scale. They do not recognize the role their agricultural lands play in supporting and regulating services, either.

When asked about the relationship they see between young people and the fields, the answers indicate that young people have no interest in continuing working on the maize field (43%), as the activity seems exhausting, they get earn too little, or "they lack love for the fields." The percentage of young people who still know how to work on the fields is 22%. Some of the interviewees' children and grandchildren have moved to the city and they work in the secondary and tertiary sectors.

The outlook is worrisome. As young people abandon the fields, tradition is at risk in this community, and consumption of foreign products is promoted. In rural areas, access to processed food is easy and sometimes seen as "progress," which makes the population more vulnerable to deficient nutrition. The low costs of processed food (associated with industrial agriculture) and changes in lifestyle (associated with a rising urbanization) have left Mexico in a public health crisis (ENES, s.f). Moreover, land abandonment and population growth promote land use change and decay in environmental quality.

#### *12.5.4.1.2 Perception of the Environment*

When interviewees were asked to give their definition of "environment," 37% of them referred to environmental factors such as the weather, rain, and plants. A similar percentage (36%) did not define it, as their first answer was talking about the relationship, they think they should have with it (taking care of it and tending to it). The rest of the interviewees stated that they did not know what it was (18%) or that they could not explain it (9%).

It was clear that all the interviewees have noticed changes in the watercourses. Tlazala is a town that uses water from the springs nearby for human consumption. They mentioned that the flow has been noticeably less recently, and that the quality of the water has deteriorated.

Regarding rainfall patterns, 38% claimed that the rain is now more intense and that it happens in shorter periods of time compared with previous years. A similar percentage stated that the patterns are irregular. The rest thought that rain brings less water (13%) or for a shorter time (13%). Those who have not observed changes in rainfall are 13%. Most (55%) have not observed changes in the climate regime, while 27% mentioned climate change, and 18% pointed out that heat has increased.

All the interviewees spoke about loss of biodiversity, as they remember seeing more medium-sized mammals in the past, and they blame this on the human population growth. Chameleons and axolotls (*Phrynosoma orbiculare* and *Ambystoma altamirani*) are among the species that used to be easily spotted. They also speak about loss of plants that used to be easily found in their wild form, and that were used for human consumption.

Finally, the interviewees were asked how they thought their agricultural management affects the environment. More than half of them (55%) consider that their management benefits the environment, 36% believes that it is a neutral relationship, and 9% does not see any relationship between their management and the environment. No interviewee believed that their management system affected the environment in a negative way. Among the answers obtained when asked why they considered that their management was positive were: "because the soil regenerates," "we do not pollute nor burn," "trees are planted," and "plants are life." Almost half the interviewees (45%) overlook the relationship between their plot and its surroundings, which entails a lack of awareness of the relationship that exists between their decisions and the environmental health, and their impact.

The interviewees were asked about the problems they can see in Tlazala: 27% mentioned environmental problems, 20% problems related to work, 20% political problems, 7% demographic ones, 7% agricultural ones, and 20% did not see any problem. When asked for more details, they talked about lack of jobs, lack of governmental support, illegal logging, population growth, loss of crops, soil erosion, and land use change. As for agricultural problems in Tlazala, they mentioned economic (50%), political (25%), environmental (13%), and social (13%) problems.

The biggest problem perceived by the interviewees is that sowing is not a good business. They cannot make a living out of it, as they invest more than they earn. That is the reason why people tend to work exclusively for their households. Among the perceived problems, they mention land misuse, lack of information about adequate crops, plot abandonment, and loss of agricultural activity. In general, these situations are caused by a lack of information about rational environmental management that seeks to be sustainable and to be a household income; a management that also prevents land abandonment and recognizes traditional knowledge as valuable cultural and environmental wisdom.

Thirty-three percent of the inhabitants know about the existence of organic agriculture, while the rest do not know the concept. From the interviews, it was apparent that "organic agriculture" is understood as traditional agriculture, which used to be practiced in Tlazala (and in the rest of the country) before agricultural technologies were implemented. Currently, organic agriculture is widely accepted, and its foundations and methods are being studied. Moreover, it represents a growing market.[65]

#### 12.5.4.1.3 Surveys to Children and Young People

The surveys were applied to a total of 1126 primary-school (32%), secondary (18%), and high school (18%) students. Most of the surveys were answered before 10 a.m., and by this time, 32% said they had consumed processed food, whereas 23% said they had eaten processed food and home-made food. This confirms the general trend of consuming processed food more than natural food, a situation that could get worse due to loss of family plots and work on the fields.

Seventy-two percent of the students have a family maize field, which is worked on mainly by their grandparents (23%), by people outside their families (19%), or by the family (15%). The rest of the students did not answer (14%).

Most of the students consider that they know how to sow (66%), while 2% estimate that they know how to sow at an intermediate level, 32% admit that they do not know how to do it, and 3% did not answer. However, 63% do not like sowing, 30% like it, and 4% like it moderately, while 3% did not answer. This confirms what we established during the interviews with the adult farmers, who pointed out that young people are disinterested in the activity. When asked why they liked to sow, different answers came up as for why they like it and why they do not like it. Sixty-four percent of students answered in an affirmative way (as if they enjoyed the activity) and 20% in a negative way, while 16% said they did not know how to do it or they did not answer. This question contrasts with the last one (63% claimed they did not like sowing, whereas 64% reacted to the activity in a positive way). This shows that children regard sowing as a positive activity, but they might not be very familiar with it or have stronger interests elsewhere, or they might not feel proud of it (maybe because it has lost its value for them). However, we do not know the cause for this difference.

## 12.6 CONCLUSIONS

The properties of agricultural and forest soils were significantly different in all cases except for the interchangeable bases. Statistically speaking, there were no significant differences between the agricultural management systems regarding their soil and vegetation qualities. However, there was a downward trend in the values according to the level of intensity of the management system. As the system intensifies, a decrease in porosity and moisture is observed, as well as an increase in bulk density. There is a tendency to a higher organic matter ratio in traditional maize fields, and higher electrical conductivity in intensive ones, possibly due to the addition of agrochemicals. Traditional and medium management show more similar soil and noncultivated plant characteristics between them, compared with the intensive system.

Slight grouping is observed in the vegetation composition of the treatments. The Canonical correspondence analysis highlighted the correlation between electrical conductivity, pH, and vegetation composition. The diversity ratio (monocotyledons/dicotyledons) is relatively similar between treatments, with low values that show that vegetation in the maize fields studied fosters the presence of pollinators. Richness, diversity, and heterogeneity were higher in the maize fields with traditional management, and lower in the intensive ones, which present higher dominance values. This was corroborated in the field. The dominance and tolerance of certain species in the maize fields have represented higher expenses in herbicides for the farmers. Moreover, they have reduced plant diversity, along with the ecosystem services that these plots provide.

When comparing the natural forest soil with soil that received agricultural management, we can say that the agricultural soil in Tlazala has slightly deteriorated. We can recognize that if the intensive management continues, it will lead to a more severe deterioration process.

Most noncultivated plants found in the maize fields could be used by humans. Not all inhabitants are aware of their uses, perhaps because of the loss of traditional practices. The main use for noncultivated plants is as forage, followed by medicinal and food uses.

Traditional management is the most common agricultural management in Tlazala, as the products are for household consumption and the farmers want to invest as little as possible in chemical inputs. These are frequently substituted with traditional knowledge about land fertility and its management.

People in Tlazala feel identified with their maize fields and they claim that they provide them with food (provisioning services), and they bring them satisfaction, safety, and well-being (cultural services). However, they do not recognize supporting or regulating services in their lands (habitats, maintaining genetic diversity, soil quality, air treatment, water, pollination, etc.). The impact their plot management has at a larger scale is not perceived by them, and neither is the fact that local decisions are related to environmental health.

The inhabitants of Tlazala have noticed changes in the environment for 30 years approximately. All the interviewees have noticed a decay in the quality and quantity of the watercourses, and most of them mention changes in the rainfall patterns and loss of biodiversity.

The farmers in Tlazala claim that they like sowing. However, they no longer want to enter the market, as the earnings are low for the agricultural sector compared with other activities.

Agricultural activities in this town are carried out by older people. They claim that younger people are not interested in continuing the maize field tradition, as they are more inclined toward jobs that involve less physical effort and that are related to the city.

Land misuse, lack of information about adequate crops, abandonment of lands, and loss of agricultural activity are some of the problems perceived by the farmers of Tlazala. The economic aspect is the main problem, as they consider that the government aids are scarce and inadequate, and that the market is unfavorable.

## 12.7 PERSPECTIVES

The problems perceived by the farmers of Tlazala could be mitigated or even avoided if they received more information and training regarding agricultural management, and if they were given adequate alternatives and aids regardless of the size of their production.

The traditional management used in Tlazala complies with the characteristics of organic agriculture. This type of agriculture has not been officially recognized by the Mexican government. Therefore, no active regulations exist, government aids are limited and disjointed, and there is no specific institute that can support research on this type of agricultural management, train interested farmers or articulate the results of research with the practical development of management strategies.[31]

At an institutional level, it is essential to spread information about the use of pesticides and their effects on people's health and the environment, so that farmers can broaden their knowledge about which ones to use and how. This will help to avoid bad agricultural practices that end up in occupational exposure, pollution, and environmental decay. Moreover, the species richness and diversity will benefit from a low-intensity management that integrates patrimonial and landscape goals.[16] In contrast to systems that are not diverse, biodiversified agricultural systems promote regulation of weeds, sickness, and plagues, and they increase pollination.[7]

As for environmental value recognition and the relationship between human beings and the environment, it is important to promote values in order to respect it, take care of it, and value it. It is also essential to encourage knowledge sharing regarding its current state, the impact certain actions have on it, and the benefits of a healthy environment.

It is necessary to encourage a rational management that integrates scientific and popular knowledge, and that recognizes both as valuable, as this encourages a sense of identity and allows a deeper, more significant relationship between humans and their environment. This will allow the inhabitants of Tlazala to be empowered in their land and culture.

The way in which a community or an individual perceives the relationship they have with the environment and the impact their decisions have on it is a fundamental field of study if we want to reach environmental conservation and health objectives. This perception is built through education and information, and it is strengthened by making the relationship between human beings and their environment closer, allowing people to value the environmental, cultural, aesthetic, and spiritual benefits provided by nature.

## KEYWORDS

- **modern agriculture**
- **traditional agriculture**
- **land use changes**
- **soil quality**
- **plant diversity**

## REFERENCES

1. Acton, D. F.; Gregorich, L. J. *The Health of Our Soils : Toward Sustainable Agriculture in Canada*; Centre for Land and Biological Resources Research, Research Branch, Agriculture and Agri-Food: Ottawa, 1995.
2. Altieri, M. A. The Impact, Uses, and Ecological Role of Weeds in Agroecosystems. In: *Weed Management in Agroecosystems: Ecological Approaches*; CRC Press: Florida, 1988; pp 2–6.

3. Araujo, C. *Non-Agricultural Employment Growth and Rural Poverty Reduction in Mexico during the 90s*; University of California: Berkeley, 2003.
4. Ávila-Akerberg, V.; González Martínez, T. Participación social y educación ambiental para la conservación. un estudio de caso con niños y jóvenes de una zona rural periurbana. *Teoría y Prax.* **2016,** *12* (19), 119–136. DOI: 10.22403/uqroomx/typ19/06.
5. Ávila-Akerberg, V. *Forest Quality in Te Southwest of Mexico City. Assessment Towards Ecological Restoration of Ecosystem Services*; Institut für Landespflegerg, Schriftenreihe des Instituts für Landespflege der Albert-Ludwigs-Universität Freibu: Freiburg, 2010.
6. Balvanera, P.; Castillo, A.; MartÍnez-Harms, M. J. Ecosystem Services in Seasonally Dry Tropical Forests. In: *Seasonally Dry Tropical Forests*; Island Press: Washington, DC, 2011; pp 259–277. DOI: 10.5822/978-1-61091-021-7_15.
7. Ávila-Akerberg, V. and González-Martínez, T. (2023). Del Monte Baja la Vida. Historia biocultural de la Cuenca Presa de Guadalupe. Editorial Instituto de Ciencias Agropecuarias y Rurales, UAEMéx, 364 pp.
8. Balvanera, P.; Siddique, I.; Dee, L.; Paquette, A.; Isbell, F.; Gonzalez, A.; Byrnes, J.; O'Connor, M. I.; Hungate, B. A.; Griffin, J. N. Linking Biodiversity and Ecosystem Services: Current Uncertainties and the Necessary Next Steps. *Bioscience* **2014,** *64* (1), 49–57. DOI: 10.1093/biosci/bit003.
9. Bennett, E. M.; Balvanera, P. The Future of Production Systems in a Globalized World. *Front. Ecol. Environ.* **2007,** *5* (4), 191–198. DOI: 10.1890/1540-9295(2007)5[191:TFOPSI]2.0.CO;2.
10. Bezdicek, D. F.; Papendick, R. I.; Lal, R. *Methods for Assessing Soil Quality*; Doran, J. W., Jones, A. J., Eds.; SSSA Special Publications; Soil Science Society of America: Madison, WI, USA, 1997. DOI: 10.2136/sssaspecpub49.
11. Bini, D.; Santos, C. A. dos; Carmo, K. B. do; Kishino, N.; Andrade, G.; Zangaro, W.; Nogueira, M. A. Effects of Land Use on Soil Organic Carbon and Microbial Processes Associated with Soil Health in Southern Brazil. *Eur. J. Soil Biol.* **2013,** *55*, 117–123. DOI: 10.1016/j.ejsobi.2012.12.010.
12. Blaikie, P.; Brookfield, H. Defining and Debating the Problem. In *Land Degradation and Society*; Routledge: New York, 2015; pp 27–52.
13. Brussaard, L.; Kuyper, T. W.; Didden, W. A. M.; Goede, R. G. M. de; Bloem, J. Biological Soil Quality from Biomass to Biodiversity—Importance and Resilience to Management Stress and Disturbance. In: *Managing Soil Quality: Challenges in Modern Agriculture*; CAB International: Wallingford, Oxfordshire, 2004; pp 139–161. DOI: 10.1079/9780851996714.0139.
14. Caamal Maldonado, J.A. Castillo Caamal, J. B. Muestreo de Arvenses. In: *Técnicas de muestreo para manejadores de recursos naturlaes*; Bautista Zúñiga, F., Delfín González, H., Palacio Prieto, J. L., Eds.; Universidad Nacional Autónoma de México: Mexico, 2011.
15. Cardinale, B. J.; Duffy, J. E.; Gonzalez, A.; Hooper, D. U.; Perrings, C.; Venail, P.; Narwani, A.; Mace, G. M.; Tilman, D.; Wardle, D. A.; Kinzig, A. P.; Daily, G. C.; Loreau, M.; Grace, J. B.; Larigauderie, A.; Srivastava, D. S.; Naeem, S. Biodiversity Loss and Its Impact on Humanity. *Nature* **2012,** *486* (7401), 59–67. DOI: 10.1038/nature11148.
16. Chacón, J. C.; Gliessman, S. R. Use of the "Non-Weed" Concept in Traditional Tropical Agroecosystems of South-Eastern Mexico. *Agro-Ecosystems* **1982,** *8* (1), 1–11. DOI: 10.1016/0304-3746(82)90010-5.

17. Cohen, M.; Bilodeau, C.; Alexandre, F.; Godron, M.; Andrieu, J.; Grésillon, E.; Garlatti, F.; Morganti, A. What Is the Plant Biodiversity in a Cultural Landscape? A Comparative, Multi-Scale and Interdisciplinary Study in Olive Groves and Vineyards (Mediterranean France). *Agric. Ecosyst. Environ.* **2015,** *212*, 175–186. DOI: 10.1016/j.agee.2015.06.023.
18. Collins, W. W.; Calvin O, Q. *Biodiversity in Agroecosystems*; CRC Press: Florida, 2000. DOI: 10.1201/9781003040668.
19. Comisión Nacional para el conocimiento y uso de la Biodiversidad. Ficha Informativa. Oxalis pes-caprae L. http://www.conabio.gob.mx/malezasdemexico/oxalidaceae/oxalis-pes-caprae/fichas/ficha.htm. (accessed 2017 -04 -05).
20. Cram, S.; Sommer, I.; Fernández, P.; Galicia, L.; Ríos, C.; Barois, I. Soil Natural Capital Modification through Landuse and Cover Change in a Tropical Forest Landscape: Implications for Management. *J. Trop. For. Sci.* **2015,** *27* (2), 189–201.
21. Dominati, E.; Patterson, M.; Mackay, A. A Framework for Classifying and Quantifying the Natural Capital and Ecosystem Services of Soils. *Ecol. Econ.* **2010,** *69* (9), 1858–1868. DOI: 10.1016/j.ecolecon.2010.05.002.
22. Doran, J. W. Soil Health and Global Sustainability: Translating Science into Practice. *Agric. Ecosyst. Environ.* **2002,** *88* (2), 119–127. DOI: 10.1016/S0167-8809(01)00246-8.
23. Doran, J. W.; Parkin, T. B. Defining and Assessing Soil Quality. In Defining Soil Quality for a Sustainable Environment, Volume 35. In: *Soil Science Society of America and American Society of* Agronomy; Doran, J. W., Coleman, D. C., Bezdicek, D. F., Stewart, B. A., Eds.; SSSA Special Publications: Madison, 1994; pp 1–21. DOI: 10.2136/sssaspecpub35.
24. Doran, J. W.; Sarrantonio, M.; Liebig, M. A. Soil Health and Sustainability. *Adv. Agron.* **1996,** 56 (C), 1–54. DOI: 10.1016/S0065-2113(08)60178-9.
25. Escalante, R.; Catalán, H. Situación Actual Del Sector Agropecuario En México: Perspectivas y Retos. *Econ. Inf.* **2008,** *350*, 7–25.
26. Escalante, R.; Catalán, H.; Galindo, L. M.; Reyes, O. Desagrarización En México: Tendencias Actuales y Retos Hacia El Futuro. *Cuad. Desarro. Rural* **2007,** *4* (59), 87–116.
27. Escalante, R.; Galindo, L. M.; Catalán, H. La evolución del producto del sector agropecuario mexicano, 1960–2002: algunas regularidades empíricas. *Cuad. Desarro. Rural* **2005,** *54*, 87–112.
28. Escuela Nacional de Estudios Superiores, U. La obesidad y la sostenibilidad en México: parte 1. Guanajuato, México. http://enes.unam.mx/?lang=es_MX&cat=sostenibilidad&pl=la-obesidad-y-la-sostenibilidad-en-mexico-parte-1 [Online] (accessed 2018-06-02).
29. Fagotti, D. S. L.; Miyauchi, M. Y. H.; Oliveira, A. G.; Santinoni, I. A.; Eberhardt, D. N.; Nimtz, A.; Ribeiro, R. A.; Paula, A. M.; Queiroz, C. A. S.; Andrade, G.; Zangaro, W.; Nogueira, M. A. Gradients in N-Cycling Attributes along Forestry and Agricultural Land-Use Systems Are Indicative of Soil Capacity for N Supply. *Soil Use Manag.* **2012,** *28* (3), 292–298. DOI: 10.1111/j.1475-2743.2012.00418.x.
30. Foley, J. A.; Ramankutty, N.; Brauman, K. A.; Cassidy, E. S.; Gerber, J. S.; Johnston, M.; Mueller, N. D.; O'Connell, C.; Ray, D. K.; West, P. C.; Balzer, C.; Bennett, E. M.; Carpenter, S. R.; Hill, J.; Monfreda, C.; Polasky, S.; Rockström, J.; Sheehan, J.; Siebert, S.; Tilman, D.; Zaks, D. P. M. Solutions for a Cultivated Planet. *Nature* **2011,** *478* (7369), 337–342. DOI: 10.1038/nature10452.

31. Giménez, A. M.; Hernández, P.; Gerez, R.; Ríos, N. A. Diversidad vegetal en siete unidades demostrativas del chaco semiárido Argentino. *Madera y Bosques* **2007,** *13* (1), 61–78. DOI: 10.21829/myb.2007.1311236.
32. Gómez Tovar, L. Propuesta de política de apoyo para la agricultura orgánica de México. *Rev. Vinculando* **2004**.
33. González-Esquivel, C. E.; Gavito, M. E.; Astier, M.; Cadena-Salgado, M.; Del-Val, E.; Villamil-Echeverri, L.; Merlín-Uribe, Y.; Balvanera, P. Ecosystem Service Trade-Offs, Perceived Drivers, and Sustainability in Contrasting Agroecosystems in Central Mexico. *Ecol. Soc.* **2015,** *20* (1). DOI: 10.5751/ES-06875-200138.
34. Gordon, L. J.; Peterson, G. D.; Bennett, E. M. Agricultural Modifications of Hydrological Flows Create Ecological Surprises. *Trends Ecol. Evol.* **2008,** *23* (4), 211–219. DOI: 10.1016/j.tree.2007.11.011.
35. Gray, C.; Baird, D. J.; Baumgartner, S.; Jacob, U.; Jenkins, G. B.; O'Gorman, E. J.; Lu, X.; Ma, A.; Pocock, M. J. O.; Schuwirth, N.; Thompson, M.; Woodward, G. Ecological Networks: The Missing Links in Biomonitoring Science. *J. Appl. Ecol.* **2014,** *51* (5), 1444–1449. DOI: 10.1111/1365-2664.12300.
36. Gregory, R. Using Stakeholder Values to Make Smarter Environmental Decisions. *Environment* **2000,** *42* (5), 34–44. DOI: 10.1080/00139150009604888.
37. Hart, R. *Conceptos Básicos Sobre Agroecosistemas*; Centro Agronómico Tropical de Investigación y Enseñanza: Turrialba, Costa Rica, 1997.
38. *Informe Perspectivas Del Medio Ambiente En México: GEO México 2004*; Cárdenas, A.; Ezcurra E.; Sánchez R.; Zahedi, K.; Arreola, M. E., Eds.; Secretaría de Medio Ambiente y Recursos Naturales, Programa de Naciones Unidas para el Medio Ambiente, Instituto Nacional de Ecología: Mexico, 2004.
39. Jackson, L. E.; Pascual, U.; Hodgkin, T. Utilizing and Conserving Agrobiodiversity in Agricultural Landscapes. *Agric. Ecosyst. Environ.* **2007,** *121* (3), 196–210. DOI: 10.1016/j.agee.2006.12.017.
40. Karlen, D. L.; Mausbach, M. J.; Doran, J. W.; Cline, R. G.; Harris, R. F.; Schuman, G. E. Soil Quality: A Concept, Definition, and Framework for Evaluation (A Guest Editorial). *Soil Sci. Soc. Am. J.* **1997,** *61* (1), 4–10. DOI: 10.2136/sssaj1997.03615995006100010001x.
41. Korkanc, S. Y.; Ozyuvaci, N.; Hizal, A. Impacts of Land Use Conversion on Soil Properties and Soil Erodibility. *J. Environ. Biol.* **2008,** *29* (3), 363–370.
42. Liquete, C.; Cid, N.; Lanzanova, D.; Grizzetti, B.; Reynaud, A. Perspectives on the Link between Ecosystem Services and Biodiversity: The Assessment of the Nursery Function. *Ecol. Indic.* **2016,** *63*, 249–257. DOI: 10.1016/j.ecolind.2015.11.058.
43. Lorencová, E. Š.; Frélichová, J.; Nelson, E.; Vačkář, D. Past and Future Impacts of Land Use and Climate Change on Agricultural Ecosystem Services in the Czech Republic. *Land use policy* **2013,** *33*, 183–194. DOI: 10.1016/j.landusepol.2012.12.012.
44. Maass, J. M.; Balvanera, P.; Castillo, A.; Daily, G. C.; Mooney, H. A.; Ehrlich, P.; Quesada, M.; Miranda, A.; Jaramillo, V. J.; García-Oliva, F.; Martínez-Yrizar, A.; Cotler, H.; López-Blanco, J.; Pérez-Jiménez, A.; Búrquez, A.; Tinoco, C.; Ceballos, G.; Barraza, L.; Ayala, R.; Sarukhán, J. Ecosystem Services of Tropical Dry Forests: Insights from Long-Term Ecological and Social Research on the Pacific Coast of Mexico. *Ecol. Soc.* **2005,** *10* (1), art17. DOI: 10.5751/ES-01219-100117.

45. Martínez Alva, G. Propiedades bioquímicas en suelos agrícolas del Nevado de Toluca: Efectos del laboreo. PhD Dissertation, Universidad Nacional Autónoma de México, México, 2015.
46. Martín-López, B.; Iniesta-Arandia, I.; García-Llorente, M.; Palomo, I.; Casado-Arzuaga, I.; Del Amo, D. G.; Gómez-Baggethun, E.; Oteros-Rozas, E.; Palacios-Agundez, I.; Willaarts, B.; González, J. A.; Santos-Martín, F.; Onaindia, M.; López-Santiago, C.; Montes, C. Uncovering Ecosystem Service Bundles through Social Preferences. *PLoS One* **2012,** *7* (6). DOI: 10.1371/journal.pone.0038970.
47. Mas, J. F.; Velázquez, A.; Díaz-Gallegos, J. R.; Mayorga-Saucedo, R.; Alcántara, C.; Bocco, G.; Castro, R.; Fernández, T.; Pérez-Vega, A. Assessing Land Use/Cover Changes: A Nationwide Multidate Spatial Database for Mexico. *Int. J. Appl. Earth Obs. Geoinf.* **2004,** *5* (4), 249–261. DOI: 10.1016/j.jag.2004.06.002.
48. Mas, J.-F.; Velázquez, A.; Couturier, S. La Evaluación de Los Cambios de Cobertura / Uso Del Suelo En La República Mexicana. *Investig. Ambient.* **2009,** *1* (1), 23–39.
49. Nortcliff, S. Standardisation of Soil Quality Attributes. *Agric. Ecosyst. Environ.* **2002,** *88* (2), 161–168. DOI: 10.1016/S0167-8809(01)00253-5.
50. Osawa, T.; Kohyama, K.; Mitsuhashi, H. Trade-off Relationship between Modern Agriculture and Biodiversity: Heavy Consolidation Work Has a Long-Term Negative Impact on Plant Species Diversity. *Land use policy* **2016,** *54*, 78–84. DOI: 10.1016/j.landusepol.2016.02.001.
51. Pierzynski, G. M.; Vance, G. F.; Sims, J. T. *Soils and Environmental Quality*; CRC Press: Florida, 1994. DOI: 10.1016/s0016-7061(01)00135-5.
52. Porter, J.; Costanza, R.; Sandhu, H.; Sigsgaard, L.; Wratten, S. The Value of Producing Food, Energy, and Ecosystem Services within an Agro-Ecosystem. *Ambio* **2009,** *38* (4), 186–193. DOI: 10.1579/0044-7447-38.4.186.
53. Raudsepp-Hearne, C.; Peterson, G. D.; Bennett, E. M. Ecosystem Service Bundles for Analyzing Tradeoffs in Diverse Landscapes. *Proc. Natl. Acad. Sci. U. S. A.* **2010,** *107* (11), 5242–5247. DOI: 10.1073/pnas.0907284107.
54. Rodríguez, G.; Taylor, J. E.; Yúnez-Naude, A. The Impacts of Economic Reforms on an Ejido Community: A Quantitative Analysis. In *The transformation of Rural Mexico: Reforming Rural Mexico*; Cornelius, W.; Myhre, D., Eds.; Center for US-Mexican Studies, University of California, 1998; pp 333–354.
55. Sánchez-Colón, S.; Flores Martínez, A.; Cruz-Leyva, I. A.; Velázquez, A. *Estado y Transformación de Los Ecosistemas Terrestres Por Causas Humanas. II Estudio de País*; CONABIO: Mexico, 2008.
56. Sandhu, H. S.; Crossman, N. D.; Smith, F. P. Ecosystem Services and Australian Agricultural Enterprises. *Ecol. Econ.* **2012,** *74*, 19–26. DOI: 10.1016/j.ecolecon.2011.12.001.
57. Sandhu, H.; Wratten, S. Ecosystem Services in Farmland and Cities. In *Ecosystem Services in Agricultural and Urban Landscapes*; Wiley Blackwell: Oxford, 2013; pp 1–15. DOI: 10.1002/9781118506271.ch1.
58. Secretaria de Medio Ambiente y Recursos Naturales. NORMA Oficial Mexicana NOM-021-RECNAT-**2000,** que establece las especificaciones de fertilidad, salinidad y clasificación de suelos. estudios, muestreo y análisis. *D. Of. la Fed.* **2002,** 73.
59. Taylor, J. E.; Mora, J.; Adams, R. H.; Lopez-Feldman, A. *Remittances, Inequality and Poverty: Evidence from Rural Mexico*; 05-003; 2005.

60. Ter Braak, C. J. F. Canonical Correspondence Analysis: A New Eigenvector Technique for Multivariate Direct Gradient Analysis. *Ecology* **1986,** *67* (5), 1167–1179. DOI: 10.2307/1938672.
61. Tscharntke, T.; Klein, A. M.; Kruess, A.; Steffan-Dewenter, I.; Thies, C. Landscape Perspectives on Agricultural Intensification and Biodiversity - Ecosystem Service Management. *Ecol. Lett.* **2005,** *8* (8), 857–874. DOI: 10.1111/j.1461-0248.2005.00782.x.
62. United States Department of Agriculture. Soil Quality Indicators : Biological Indicators and Soil Functions. *USDA Nat. Resour. Conserv. Serv.* **2015,** No. April, pp 1–4.
63. Villegas Martínez, D. Turismo rural como estrategia de desarrollo local en Isidro Fabela y la region Monte Alto. Master Dissertation, Universidad Autónoma del Estado de México, Estado de México, 2015.
64. Worm, B.; Barbier, E. B.; Beaumont, N.; Duffy, J. E.; Folke, C.; Halpern, B. S.; Jackson, J. B. C.; Lotze, H. K.; Micheli, F.; Palumbi, S. R.; Sala, E.; Selkoe, K. A.; Stachowicz, J. J.; Watson, R. Impacts of Biodiversity Loss on Ocean Ecosystem Services. *Science* **2006,** *314* (5800), 787–790. DOI: 10.1126/science.1132294.
65. Yaalon, D. H. Human-Induced Ecosystem and Landscape Processes Always Involve Soil Change. *Bioscience* **2007,** *57* (11), 918–919. DOI: 10.1641/B571102.
66. Yussefi, M.; Willer, H. Organic Farming Worldwide 2007: Overview & Main Statistics. *World Org. Agric. 2007. Stat. Emerg. Trends* **2007,** 9–16.
67. Zhang, W.; Ricketts, T. H.; Kremen, C.; Carney, K.; Swinton, S. M. Ecosystem Services and Dis-Services to Agriculture. *Ecol. Econ.* **2007,** *64* (2), 253–260. DOI: 10.1016/j.ecolecon.2007.02.024.

# CHAPTER 13

# Oregano and Laurel Essential Oils: Trends in Eco-Friendly Extraction Processes

CRISTIAN MARTÍNEZ-ÁVILA[1], PEDRO AGUILAR-ZARATE[2], JULIO C. TAFOLLA-ARELLANO[3], and ROMEO ROJAS[1*]

[1]*School of Agronomy, Universidad Autónoma de Nuevo León, General Escobedo, Nuevo León, México*

[2]*Tecnológico Nacional de México Campus Ciudad Valles, Departamento de Ingenierías, Vista Hermosa. Cd. Valles, S.L.P., México*

[3]*Universidad Autonoma Agraria Antonio Narro. Departamento de Ciencias Básicas, Laboratorio de Biotecnología y Biología Molecular, Buenavista, Saltillo, Coahuila, México*

## ABSTRACT

Oregano and laurel are some of the most consumed spices worldwide due to their high content of essential oil that adds flavor to traditional foods. Mexican oregano is known as an aromatic plant of three different species: *L. graveolens* (or *L. berlandieri*), *P. longiflora*, and *Monarda fistulosa* var. *menthifolia*. In Mexico, *L. graveolens* is the most commercialized species. Laurel (*Laurus nobilis*) is also used as a condiment in world cuisine. The traditional/conventional extraction is carried out by hydrodistillation; however, this technique is very time consuming, generally leads to the degradation of the more thermolabile molecules, requires large amounts of solvents such as methanol, ethanol, n-hexane, petroleum ether, toluene, chloroform, benzene, diethyl ether, etc. In this way, in recent years, they have proposed novel techniques to extract essential oils. Some of these

Sustainable Agricultural Production Systems: Traditional and Modern Perspectives.
Romeo Rojas, Guillermo Cristian Guadalupe Martínez-Ávila, Julia Mariana Márquez-Reyes, and Ma del Carmen Ojeda-Zacarías (Eds.)

techniques are: ohmic heating, microwave-assisted extraction, ultrasonic-assisted extraction, and supercritical fluid extraction, which offer economic, environmental, time saving, and high-yield advantages. Today, there is no recent information indicating an overview of green trends for the oregano and laurel essential oil extraction process. This document includes relevant information related to the most recent, efficient, and emerging techniques used in recent years for the extraction of essential oil of oregano and laurel. In addition, it includes information based on the methodologies and the best extraction conditions for each one.

## 13.1 INTRODUCTION

Population growth, as well as the changing lifestyle of people, has caused health problems. Therefore, essential oils of various species have become important in recent years. Essential oils are aromatic (oily) liquids that are extracted from different parts of plants such as seeds, roots, and leaves mainly. You can be by hydrodistillation or with solvents. Likewise, oils have biological activities, among which antibacterial, antiviral, and antioxidant activities stand out, among others.[1] These essential oils have been used in the pharmaceutical, cosmetic, food, and chemical industries. However, the extraction process par excellence is hydrodistillation and hence the area of opportunity for the use of emerging and eco-friendly technologies to obtain essential oils. Two of the main species for obtaining oils are oregano and laurel.

The name of the genus *Origanum* is derived from the Greek "oros" (mountain) and ganos (joy), in reference to the beauty of this plant in regions where it is grown.[2] It includes different plant species that have a characteristic smell and taste. Four groups of oregano have been named according to their use and economic importance: Mexican oregano (*Lippia graveolens* Kunth or *Lippia berlandieri*), Greek oregano (*Origanum vulgare* subsp hirtus (Link) letswaart), Turkish oregano (*Origanum onites* L), and Spanish oregano. (*Coridohymus capitatu* (L) Hoffmanns and Links).[3] The Origanum vulgare is one of the most consumed spices in the world due to its high essential oil content that provides the desired flavors in traditional foods. The main components of oregano essential oil are carvacrol, thymol, β-caryophyllene, α-humulene, and caryophyllene oxide[4] that have strong biological activities, including anti-inflammatory, antibacterial, and antioxidant.[5] The main components are thymol and carvacrol.[1]

The laurel (*Laurus nobilis* L.) is an aromatic evergreen tree, native to the Mediterranean area. For its flavor, the Greeks and Romans used bay leaves as an aromatic herb for seasoning to add a distinctive aroma to food.[6] In Mexico, it is called *Litsea glaucescens* Humb., Bonpl. & Kunth var. *glaucescens*. It is a shrub up to 6 m tall that grows from Tamaulipas to Veracruz, Chiapas, and Tepic, including Central America. It is commonly called laurel (Veracruz, Mexico, Michoacán, Guerrero, Oaxaca, Chiapas, Baja Verapaz, Chimaltenango, Quezeltenango, and Guatemala), laurel (Zacatecas, Nayarit, and Jalisco), laurel (Michoacán), laurel of the Sierra (Sinaloa), aromatic laurel (Petén), species laurel (El Salvador); suficaya (Veracruz) and ziz-uch (Chiapas). It is used as a condiment and in traditional medicine as an infusion to treat colic and sore throat.[7] The main volatile compounds are commonly 1,8-cineole, methyl eugenol, and α-terpinyl acetate, α-pinene, β-pinene, sabinene, and linalool, being 1,8-cineol (oxygenated monoterpene) one of the main components of the essential oil of leaves and berries obtained from Laurus nobilis.[8]

In recent years, emerging technologies (ohmic heating, microwaves, ultrasound, supercritical fluids, and biotechnological extraction) for the extraction of compounds (phenolics and essential oils) are the most efficient methods for extracting bioactive compounds, because they allow the use of more suitable solvents, in smaller proportions, as well as the reduction of processing times.[9] Therefore, these methodologies have been accepted as alternatives to conventional extraction methods. This essay focuses on the use of emerging technologies for the extraction of essential oils from laurel and oregano, as well as the best extraction conditions for each method and the partial characterization of the components of the oils.

## 13.2 CONVENTIONAL EXTRACTION (CE)

The CE is a hydrodistillation system that consists of immersing the plant (sample) in water, heating it, and generating vapors that contain the aromatic compounds. It is then condensed to recover the oily compounds.[10] It is considered the most widely used and versatile method for extracting essential oils. Hydrodistillation is used for the extraction of oregano and laurel oils and also as a benchmark for the development and comparison of alternative technologies for oil extraction. Because oils are a complex mixture of secondary metabolites, including terpenes, phenols, and alcohols, the methods focus on the main compounds or those of interest to the industry. In the case of oregano, thymol and carvacrol are sought[11,12] and for laurel eucalyptol, 1-8-cineole, terpinyl acetate, and sabinene *trans*-hydrate[13–15] (Table 13.1).

**TABLE 13.1** Extraction of Essential Oils by Conventional Technologies.

| Species | Conditions | Yields | Characterization | References |
|---|---|---|---|---|
| *Laurus nobilis* L. | Ratio 1:10 g/mL of sample: $H_2O$. Hydrodistillation for 2.5 h | 1.40% | Eucalyptol (49.73%). α-terpinenyl acetate (16.92%). 0.84 ± 0.3 μg/mL are needed to inhibit the $DPPH^{•+}$ radical by 50%. | [13] |
| | Ratio 1:33 g/mL of sample: $H_2O$. Hydrodistillation with Clevenger for 3 h. | NR | NR | [16] |
| | Ratio 1:21.5 g/mL of sample: $H_2O$. Hydrodistillation for 1 h | 0.784% | 1-8-cineole (35.7 ± 2.6%), trans-Sabinene hydrate (9.7 ± 0.3%), α-terpinyl acetate (9.3 ± 0.6%). | [17] |
| | Ratio 1:6 or 1:10 g/mL of sample: $H_2O$. Hydrodistillation for 2 h. | 1.3% | 1-8-cineole (42.8%), Sabinene (8.7%), α-terpinyl acetate (7.4%). | [14] |
| *Origanum heracleoticum* | Hydrodistillation in a Clevenger apparatus for 4 h. | 1.34% | Carvacrol (74.65%). 30 mg/mL to inhibit DPPH•+ radicals at 60.83%. 0.21% of total flavonoids and 28.4 mg GA/g of total phenols. | [18] |
| *Origanum vulgare* subsp. *viride* | Ratio 1:25 g/mL of sample: $H_2O$. Hydrodistillation for 95 min in a Clevenger apparatus. | 9.5 ± 0.8% | Thymol (28.1 ± 0.5%). Carvacrol (4.7 0.3%). 33.4 μg/mL to inhibit $DPPH^{•+}$ radicals ($IC_{50}$). 6.25 μg/mL is needed to inhibit *S. aureus* and *B. cereus* by 50%. | [19] |
| *Origanum vulgare* L. | Ratio 1:10 g/mL of sample: $H_2O$. Hydrodistillation for 3 h. | 4.8% | Thymol (70.20%). Carvacrol (4.46%). ρ-Cymene (6.77%). | [20] |
| | Ratio 1:25 g/mL of sample: $H_2O$. Hydrodistillation for 3 h. | 1.76% | Carvacrol (46.86 ± 2.47%), Thymol (3.69 ± 0.53%), γ-Terpinene (14.16 ± 2.77). | [21] |

*NR*: Non-reported; $DPPH^{•+}$: 2,2–diphenyl–1–picrylhydrazyl

However, its operation is easy and inexpensive,[22] but it requires a lot of processing time, solvent residues are produced, low yields, partial degradation of thermolabile compounds occurs, resulting in the loss of volatile compounds.[23] For this reason, the heating system must be taken care of since it is a very obsolete technology.

## 13.3 EXTRACTION METHODS BY EMERGING TECHNOLOGIES

### *13.3.1 OHMIC HEATING (OH)*

OH has a lot of applications, like scalding,[24,25] evaporation,[26,27] dehydration,[28] extraction,[29,30] and sterilization.[31] It is an alternating current that passes through the sample and therefore heats up quickly. Due to the inherent resistance, heat generation occurs within the medium.[32] Electroporation of cell membranes could be induced by CO and is assumed to dominate in the experiments, when the biological tissue is crossed by the electric current; this increases the temperature and damages the membrane, causing the diffusion of solutes into it of the cell structure.[33,34] However, it has been little used for the extraction of essential oils from oregano and laurel (Table 13.2), which opens up an opportunity for the development of research in this area.

**TABLE 13.2** Oil Extraction by Ohmic Heating

| Species | Conditions | Yields | Characterization | References |
|---|---|---|---|---|
| *Laurus nobilis* L. | Ratio 1:10 g/mL of sample:NaCl solution (0.3%). 120V/45min | 0.83% | Eucalyptol (34.4–50.0%), α-terpinenyl acetate (14.9–18.8%), terpinene-4-ol (4.7–6.0%), and sabinene (4.9–5.9%). 89.7 ± 1.6% inhibition of DPPH$^{\bullet+}$ at 10 μL/mL. SEM analysis. | [13] |
| *Origanum vulgare* subsp. viride | Ratio 1:25 g/mL of sample:NaCl solution (2.85%). 200V/31 min. | 10.6 ± 0.4% | Thymol (32%), 4-terpineol (12.1%) and α terpinene (10%) were the predominant compounds identified. DPPH$^{\bullet+}$ IC50 of 31.2 μg/mL. The MIC of the essential oil was 12.5, 12.5, 25, 6.25, and 6.25 μg/mL for S. *Typhimurium*, *E. coli* O157:H7, P. aeruginosa, *S. aureus*, and *B. cereus*, respectively. | [19] |

*DPPH*$^{\bullet+}$: 2,2–diphenyl–1–picrylhydrazyl; V: Voltio; SEM: scanning electron microscopy; $IC_{50}$: number of compounds needed to inhibit a reaction 50%; MIC: Minimum inhibition concentration.

For the extraction of essential oils of oregano and laurel there are reports in which 120–200 V are used for a period of 30–45 min. For oregano,[19] the main compounds identified are thymol (32%) and 4-terpineol (12.1%) and for laurel[13] eucalyptol (34.4–50,0%), α-terpinenyl acetate (14.9–18.8%). It is worth mentioning that the process consists of substituting conventional heating for ohmic heating and then distilling; in addition, NaCl is used to promote increasing the electrical conductivity of the system. Using CO is a much more selective technology for the extraction of oxygenated monoterpenes, whereas hydrodistillation promotes the extraction of sesquiterpenes, which is probably attributed to the greater dipole moment and the decrease in thermal effects.[17,35] There is no significant difference in yields, but extraction times are reduced from 150 to 45 min, thus reducing the risk of degradation of thermolabile compounds.[17]

### 13.3.2 MICROWAVE-ASSISTED EXTRACTION (MAE)

Microwaves are electromagnetic energy (300 MHz to 300 GHz). It is transmitted as waves, penetrates biomaterials, interacts with free water molecules (generating heat), and thus generates focused heating. Because of this, the temperature rises rapidly to almost or above the boiling point of water. Likewise, it breaks the walls due to the rapid expansion.[36] The low specific heat of lipids makes them susceptible to this radiation. This allows permanent pores to be generated in the plant material and therefore yield is increased.[13]

In recent years, this technology is used to extract essential oils from various sources such as Myristicae arillus,[37] *Pongamia pinnata* seeds,[36] beaver seeds,[38] *Laurus nobilis*,[39] *Hura crepitans* seeds,[40] and others. For the MAE of laurel and oregano oils, powers of 300–1000 W are used for periods of 5–90 min. The main compounds are identified for oregano,[20,41] they are thymol (0.27–73.22%), carvacrol (5.17–22.8%), and linalool (26.66%); and for laurel[13,17,39] eucalyptol (34.4–50.0%), 1-8-cineole (31.4%), and α-terpinenyl acetate (11.4–18.8%), previously reported by other authors[42,43] (Table 13.3). The use of microwaves allows the recovery of oxygenated monoterpenes and phenylpropanides attributed to the higher dipole moment and the decrease in thermal effects,[17] as well as an increase in the yield of up to 30% of eucalyptol.[14] In this way it is shown that a longer processing time can deteriorate some compounds. In addition, the use of EAM allows a destructive effect on secretory cells, while conventional hydrodistillation does not damage many cells in the leaves.[13]

### 13.3.3 ULTRASOUND-ASSISTED EXTRACTION (UAE)

The UAE is a technique used to extract compounds from plants since it reduces the extraction time, use of solvents and manages to increase

**TABLE 13.3** Microwave-Assisted Extraction of Essential Oils.

| Species | Conditions | Yields | Characterization | References |
|---|---|---|---|---|
| *Laurus nobilis* L. | Ratio 1:1 g/mL of sample: H2O. 540 W at 400 rpm | 1.08 ± 0.04% | The authors only focused on performance optimization. | [39] |
| | Ratio 1:10 g/mL of sample:NaCl solution (0.3%). 400 W/45 min. | 1.0% | Eucalyptol (34.4–50.0%), α-terpinenyl acetate (14.9–18.8%), terpinene-4-ol (4.7–6.0%), and sabinene (4.9–5.9%). 83.7 ± 7.0% inhibition of DPPH$^{\bullet+}$ at 10 μL/mL. Scanning electron microscopy. | [13] |
| | Ratio 1:21.5 g/mL of sample: H2O. 300 W/1 h. | 1.132% | 1-8-cineole (31.4 ± 1.0%), trans-Sabinene hydrate (9.8 ± 0.5%), α-terpinenyl acetate (11.4 ± 0.3%). | [17] |
| *Origanum vulgare* L | Ratio 1:14 g/mL of sample: H2O. 622 W/5 min | 5.4% | Thymol (73.22%). Carvacrol (5.17%). | [20] |
| *Origanum vulgare* L. ssp. hirtum (Link) Ietswaart | Ratio 1:6 g/mL of sample: H2O. hydro-distillation for 90 min. 1000 W/20 min | 0.73% | Carvacrol (22.80%), Thymol (0.27%), Linalool (26.66%). | [41] |

DPPH$^{\bullet+}$, 2,2–diphenyl–1–picrylhydrazyl; W, Watt.

yields since it uses a simpler methodology than conventional ones. It is based on cavitation (implosion of microbubbles) which breaks the cell wall of plant tissue. The damage increases the turbulence and the degree of penetration of the solvent in the plant matrix, releasing the intracellular content. Likewise, increase the mass transfer rate of the solvent to the internal area of the matrix and of the soluble components to the solvent. Also, cavitation causes disturbance of the cell wall structure, caused by micro-jets of solvent, which also contributes to the increase in mass transfer.[44] Using EUA to extract laurel essential oil[15] it is possible to recover eucalyptol (16.32%), terpinenyl acetate (9.37%), and α-terpinene (4.35%) mainly and for oregano[15] cymene (9.73%), carvacrol (0.82%), and thymol (0.8%), and allows the extraction of phenolic compounds[45,46] (Table 13.4).

**TABLE 13.4** Ultrasound-Assisted Extraction of Essential Oils.

| Species | Conditions | Yields | Characterization | References |
|---|---|---|---|---|
| *Laurus nobilis* L. | Ratio 1:25 g/mL of sample: EtOH for 10 min at 450 W. It was distilled for 3 h | NR | Eucalyptol (16.32%), Terpinenyl Acetate (9.37%), α-Terpinene (4.35%) | [15] |
| *Oreganum majorana* | Ratio 1:25 g/mL of sample: EtOH for 10 min at 450 W. It was distilled for 3 h | NR | Cymene (9.73%), carvacrol (0.82%), thymol (0.80%). | [15] |
| *Laurus nobilis* L. | Ratio 1:12 g/mL of sample:EtOH 35%. 40 min at 40 kHz | 1.73 ± 0.15% | It inhibits the $DPPH^{\bullet+}$ radical by 94.73 ± 0.49% and lipid oxidation by 73.55 ± 1.91%. | [47] |

NR, Non-reported; EtOH, ethanol; $DPPH^{\bullet+}$, 2,2–diphenyl–1–picrylhydrazyl; W, Watt; kHz, Kilohertz.

Additionally, it is used to extract essential oils from a wide range of conventional and unconventional sources such as radish sedes,[44] virgin olive oil,[48] linseed,[49] papaya seed,[50] sweet passion fruit seeds,[51] and others. The quality, as well as the quantity (yield) of the oil and its components, is due to the diminution of extraction times and the loss of compounds when extracted by hydrodistillation. Therefore, the EAU is less expensive, faster, profitable, and efficient due to the rupture of the cell wall of the plant tissue caused by the implosion of the bubbles that allows the release of the compounds.[15] In addition, it is a more option for the extraction of essential oils and/or compounds of high interest for several industries like food, pharmaceutical, and cosmetics.

### *13.3.4 SUPERCRITICAL FLUID EXTRACTION (SFE)*

SFE, primarily using supercritical carbon dioxide (SC-$CO_2$), can be used to recover essential oils from natural products because it does not cause great thermal degradation and also does not use organic solvents. SC-$CO_2$ is a extensively used solvent since it has various advantages, such as cheap, readily available, nontoxic, lower critical points, high diffusivity, environment-friendly, recyclable, etc.[52] Given the quality of the oils that are recovered using SFE technology and the benefits that it presents, several studies have been carried out for the recovery of oils from laurel and oregano, as well as for other aromatic species. $CO_2$ is regularly used at pressures of 8–68 MPa for times from 10 min to 2.5 h (Table 13.5). In oregano oil[53] mainly thymol (6.08–83.64%) and carvacrol (74.22–79.86%) have been identified and in the laurel[54] 1-8-cineol (24.84–43%) and linanol (14.46%) (Table 13.5).

**TABLE 13.5** Extraction of Essential Oils by Supercritical Fluids.

| Species | Conditions | Yields | Characterization | References |
|---|---|---|---|---|
| *Origanum heracleoticum* | 45 g sample. CO2 at 10 MPa/40°C and 30 MPa/100°C. CO2 flow 0.5 kg/h and 0.3 kg/h, respectively. | 1.02% (10 MPa, 40°C). 0.74% (30 MPa, 40°C). 1.5% (30 MPa, 100°C). | Carvacrol (79.86%)–(10 MPa, 40°C) with 342.1 AG/g. 30 MPa, 100°C. $DPPH^{\bullet+}$ radical reduction (100%) at 7.5 mg/mL and 0.67% of total flavonoids. | [18] |
| *Lippia thymoides* Mart. & Schauer (Verbenaceae) | 10 g of sample at 30 MPa of CO2 with a density of 879.43 kg/m$^3$ at 50 °C for 2.5 h. | 1.63 ± 0.01% | Thymol (83.64 ± 0.29%), (E)-caryophyllene (4.42 ± 0.49%), thymol acetate (1.67 ± 0.26). | [53] |
| *Lippia graveolens* | 15 g of sample at 35 MPa of CO2 with a density of 934.9 kg/m$^3$ at 60°C for 10 min. | 2.7 ± 0.46% | Carvacrol (74.22%). Thymol (6.08%), hexadecanoic acid (7.53%). | [55] |
| | 2 g of sample at 68 MPa of CO2 at 75°C for 1 h. | 26.95% | Lauric acid (43.1 ± 0.9%), oleic acid (37.2 ± 0.6%), linoleic acid (14.7 ± 1.4%). | [56] |
| *Laurus nobilis* L. | 2 g of sample at 8 MPa of CO2 at 80°C for 20 min. | 1.3% | 1-8-cineole (43%), α-Terpinyl acetate (10.8%), trans-Sabinene hydrate (10.3%). | [57] |
| | 3.8 kg of sample in 9 L extractor vessels at 11 MPa of CO2 at 40°C | 0.6% | 1-8-cineole (24.84%), linalool (14.46%), Terpineol acetate (12.36%). | [54] |

MPa, Megapascals; $DPPH^{\bullet+}$, 2,2–diphenyl–1–picrylhydrazyl.

It is very essential to say that the yields and concentration of oil components depend on various factors, from the physiological state of the plant, location, etc., but also to the density of the fluid, pressure, temperature, flow, etc. However, EFS promotes the recovery of oxygenated monoterpenes, sesquiterpenes, and oxygenated sesquiterpenes, and monoterpenes to a lesser extent.[53] The integrity of the compounds extracted by this methodology is mainly due to the processing temperatures (40–60°C) avoiding the thermal degradation of the bioactive compounds; in addition to that it can be used to carry out fractional separations of compounds.[58] Additionally, this technology points to be very promising due to the use of eco-friendly and totally clean solvents, as well as its high selectivity to obtain high purity oils. However, the critical problem lies in the amount of energy used for the compression and decompression stage.[11]

## 13.4 CONCLUSION

The significance of the oils (oregano and laurel) is not just due to their culinary value, but also to their application in the pharmaceutical and cosmetic industries. They can microporate the cell wall of microorganisms, affect the metabolites and enzymes (from the tricarboxylic acid cycle), and inhibit microorganisms like S. aureus, limiting the production of extracellular toxins.[1] Therefore, the use of emerging technologies allows obtaining essential oils of high quality and purity. Technologies such as microwaves, supercritical fluids, ohmic heating, and ultrasound allow to reduce production costs, speed up process times, avoid the degradation of compounds with high added value, use eco-friendly solvents, and increase yields, unlike hydrodistillation. However, it is of highest importance to create tailor-made processes for each species to be evaluated, as well as to review in detail the advantages and disadvantages that each extraction methodology may have. In Mexico, through CONAFOR, projects are developed to improve these essential oil extraction systems, such is the case of the project entitled, "Standardization of the extraction process of essential oils from aromatic species: design and construction of micro-industrial equipment" by the Autonomous University of Nuevo León to benefit producers.

## ACKNOWLEDGMENTS

The authors thank the projects CONACYT-CONAFOR 2018-1-B-S-65769 "Estandarización de proceso de extracción de aceites esenciales de especies aromáticas: diseño y construcción de equipo microindustrial" and 2018-2-B-S-13-1466 "Diseño y construcción de equipo semiautomático para la extracción de cera de candelilla orgánica."

## KEYWORDS

- **emerging technologies**
- **supercritical fluids**
- **ohmic heating**
- **ultrasound**
- **microwave**

## REFERENCES


1. Cui, H.; Zhang, C.; Li, C.; Lin, L. Antibacterial Mechanism of Oregano Essential Oil. *Ind. Crops Products* **2019,** *139*, 111498. DOI: 10.1016/j.indcrop.2019.111498.
2. Cid-Pérez, T. S.; Nevárez-Moorillón, G. V.; Torres-Muñoz, J. V.; Palou, E.; López-Malo, A. Chapter 63—Mexican Oregano (Lippia Berlandieri and Poliomintha Longiflora) Oils. In: *Preedy Flavor and Safety* ; Academic Press: San Diego, 2016; pp 551–560. DOI: 10.1016/B978-0-12-416641-7.00063-8.
3. Arcila-Lozano, C. C.; Loarca-Piña, G.; Lecona-Uribe, S.; González de Mejía, E. *El Orégano: Propiedades, Composición y Actividad Biológica de Sus Componentes*; Archivos Latinoamericanos de Nutrición: scielon, 2004; pp 100–111.
4. Barbieri, N.; Sanchez-Contreras, A.; Canto, A.; Cauich-Rodriguez, J. V.; Vargas-Coronado, R.; Calvo-Irabien, L. M. Effect of Cyclodextrins and Mexican Oregano (Lippia Graveolens Kunth) Chemotypes on the Microencapsulation of Essential Oil. *Ind. Crops Products* **2018,** *121*, 114–123. DOI: 10.1016/j.indcrop.2018.04.081.
5. Popović-Djordjević, J.; Cengiz, M.; Ozer, M. S.; Sarikurkcu, C. Calamintha Incana: Essential Oil Composition and Biological Activity. *Ind. Crops Products* **2019,** *128*, 162–166. DOI: 10.1016/j.indcrop.2018.11.003.
6. Chaumun, M.; Goëlo, V.; Ribeiro, A. M.; Rocha, F.; Estevinho, B. N. In Vitro Evaluation of Microparticles with Laurus Nobilis L. Extract Prepared by Spray-Drying for Application in Food and Pharmaceutical Products. *Food Bioprod. Process.* **2020,** *122*, 124–135. DOI: 10.1016/j.fbp.2020.04.011.
7. Tucker, A. O.; Maciarello, M. J.; Hill, M. Litsea Glaucescens Humb., Bonpl. & Kunth Var.Glaucescens (Lauraceae): A Mexican Bay. *Econ. Bot.* **1992,** *46* (1), 21–24. https://doi.org/10.1007/BF02985250.
8. Kasabri, V.; Abu-Dahab, R.; Afifi, F. "'Evaluation of the Volatile Oil Composition and Antiproliferative Activity of Laurus Nobilis L. (Lauraceae) on Breast Cancer Cell Line Models.'" Records of Natural Products, 2013.
9. Wong-Paz, J. E.; Muñiz-Márquez, D. B.; Aguilar-Zárate, P.; Ascacio-Valdés, J. A.; Cruz, K.; Reyes-Luna, C.; Rodríguez, R.; Aguilar, C. N. Chapter 5—Extraction of Bioactive Phenolic Compounds by Alternative Technologies. In Handbook of Food Bioengineering; Grumezescu, A. M., et al., Eds.; Academic Press, 2017; pp 229–252. DOI: 10.1016/B978-0-12-811521-3.00005-3.
10. Perović, A.; Stanković, M. Z.; Veljković, V. B.; Kostić, M. D.; Stamenković, O. S. A Further Study of the Kinetics and Optimization of the Essential Oil Hydrodistillation from Lavender Flowers. *Chinese J. Chem. Eng.* 2020. DOI: 10.1016/j.cjche.2020.06.028.
11. Moncada, J.; Tamayo, J. A.; Cardona, C. A. Techno-Economic and Environmental Assessment of Essential Oil Extraction from Oregano (Origanum Vulgare) and Rosemary (Rosmarinus Officinalis) in Colombia. *J. Clean. Product.* **2016,** *112*, 172–181. DOI: 10.1016/j.jclepro.2015.09.067.
12. Dutra, T. V.; Castro, J. C.; Menezes, J. L.; Ramos, T. R.; do Prado, I. N.; Machinski, M.; Mikcha, J. M. G.; Filho, B. A. de A. Bioactivity of Oregano (Origanum Vulgare) Essential Oil against Alicyclobacillus Spp. *Ind. Crops Products* **2019,** *129*, 345–349. DOI: 10.1016/j.indcrop.2018.12.025.
13. Taban, A.; Saharkhiz, M. J.; Niakousari, M. Sweet Bay (Laurus Nobilis L.) Essential Oil and Its Chemical Composition, Antioxidant Activity and Leaf Micromorphology under

Different Extraction Methods. *Sustain. Chem. Pharm.* **2018,** *9*, 12–18. DOI: 10.1016/j.scp.2018.05.001.

14. Di Leo Lira, P.; Retta, D.; Tkacik, E.; Ringuelet, J.; Coussio, J. D.; van Baren, C.; Bandoni, A. L. Essential Oil and By-Products of Distillation of Bay Leaves (Laurus Nobilis L.) from Argentina. *Ind. Crops Products* **2009,** *30* (2), 259–264. DOI: 10.1016/j.indcrop.2009.04.005.
15. Roldán-Gutiérrez, J. M.; Ruiz-Jiménez, J.; Luque de Castro, M. D. Ultrasound-Assisted Dynamic Extraction of Valuable Compounds from Aromatic Plants and Flowers as Compared with Steam Distillation and Superheated Liquid Extraction. *Talanta* **2008,** *75* (5), 1369–1375. DOI: 10.1016/j.talanta.2008.01.057.
16. Reis, P. M. C. L.; Mezzomo, N.; Aguiar, G. P. S.; Senna, E. M. T. L.; Hense, H.; Ferreira, S. R. S. Ultrasound-Assisted Emulsion of Laurel Leaves Essential Oil (Laurus Nobilis L.) Encapsulated by SFEE. *J. Supercritical Fluids* **2019,** *147*, 284–292. DOI: 10.1016/j.supflu.2018.11.018.
17. Flamini, G.; Tebano, M.; Cioni, P. L.; Ceccarini, L.; Ricci, A. S.; Longo, I. Comparison between the Conventional Method of Extraction of Essential Oil of Laurus Nobilis L. and a Novel Method Which Uses Microwaves Applied in Situ, without Resorting to an Oven. *J. Chromatogr. A* **2007,** *1143* (1), 36–40. DOI: 10.1016/j.chroma.2007.01.031.
18. Stamenic, M.; Vulic, J.; Djilas, S.; Misic, D.; Tadic, V.; Petrovic, S.; Zizovic, I. Free-Radical Scavenging Activity and Antibacterial Impact of Greek Oregano Isolates Obtained by SFE. *Food Chem.* **2014,** *165*, 307–315. DOI: 10.1016/j.foodchem.2014.05.091.
19. Hashemi, S. M. B.; Nikmaram, N.; Esteghlal, S.; Mousavi Khaneghah, A.; Niakousari, M.; Barba, F. J.; Roohinejad, S.; Koubaa, M. Efficiency of Ohmic Assisted Hydrodistillation for the Extraction of Essential Oil from Oregano (Origanum Vulgare Subsp. Viride) Spices. *Innov. Food Sci. Emerg. Technol.* **2017,** *41*, 172–178. DOI: 10.1016/j.ifset.2017.03.003.
20. Bayramoglu, B.; Sahin, S.; Sumnu, G. Solvent-Free Microwave Extraction of Essential Oil from Oregano. *J. Food Eng.* **2008,** *88* (4), 535–540. DOI: 10.1016/j.jfoodeng.2008.03.015.
21. Morshedloo, M. R.; Salami, S. A.; Nazeri, V.; Maggi, F.; Craker, L. Essential Oil Profile of Oregano (Origanum Vulgare L.) Populations Grown Under Similar Soil and Climate Conditions. *Ind. Crops Products* **2018,** *119*, 183–190. DOI: 10.1016/j.indcrop.2018.03.049.
22. Chen, G.; Sun, F.; Wang, S.; Wang, W.; Dong, J.; Gao, F. Enhanced Extraction of Essential Oil from Cinnamomum Cassia Bark by Ultrasound Assisted Hydrodistillation. *Chinese J. Chem. Eng.* **2020**. DOI: 10.1016/j.cjche.2020.08.007.
23. Binello, A.; Orio, L.; Pignata, G.; Nicola, S.; Chemat, F.; Cravotto, G. Effect of Microwaves on the in Situ Hydrodistillation of Four Different Lamiaceae. *Comptes Rendus Chimie* **2014,** *17* (3), 181–186. DOI: 10.1016/j.crci.2013.11.007.
24. Gomes, C. F.; Sarkis, J. R.; Marczak, L. D. F. Ohmic Blanching of Tetsukabuto Pumpkin: Effects on Peroxidase Inactivation Kinetics and Color Changes. *J. Food Eng.* **2018,** *233*, 74–80. DOI: 10.1016/j.jfoodeng.2018.04.001.
25. Guida, V.; Ferrari, G.; Pataro, G.; Chambery, A.; Di Maro, A.; Parente, A. The Effects of Ohmic and Conventional Blanching on the Nutritional, Bioactive Compounds and Quality Parameters of Artichoke Heads. *LWT—Food Sci. Technol.* **2013,** *53* (2), 569–579. DOI: 10.1016/j.lwt.2013.04.006.

26. Cokgezme, O. F.; Sabanci, S.; Cevik, M.; Yildiz, H.; Icier, F. Performance Analyses for Evaporation of Pomegranate Juice in Ohmic Heating Assisted Vacuum System. *J. Food Eng.* **2017,** *207*, 1–9. DOI: 10.1016/j.jfoodeng.2017.03.015.
27. Icier, F.; Yildiz, H.; Sabanci, S.; Cevik, M.; Cokgezme, O. F. Ohmic Heating Assisted Vacuum Evaporation of Pomegranate Juice: Electrical Conductivity Changes. *Innov. Food Sci. Emerg. Technol.* **2017,** *39*, 241–246. DOI: 10.1016/j.ifset.2016.12.014.
28. Moreno, J.; Gonzales, M.; Zúñiga, P.; Petzold, G.; Mella, K.; Muñoz, O. Ohmic Heating and Pulsed Vacuum Effect on Dehydration Processes and Polyphenol Component Retention of Osmodehydrated Blueberries (Cv. Tifblue). *Innov. Food Sci. Emerg. Technol.* **2016,** *36*, 112–119. DOI: 10.1016/j.ifset.2016.06.005.
29. Pereira, R. N.; Rodrigues, R. M.; Genisheva, Z.; Oliveira, H.; de Freitas, V.; Teixeira, J. A.; Vicente, A. A. Effects of Ohmic Heating on Extraction of Food-Grade Phytochemicals from Colored Potato. *LWT*, **2016,** *74*, 493–503. DOI: 10.1016/j.lwt.2016.07.074.
30. Termrittikul, P.; Jittanit, W.; Sirisansaneeyakul, S. The Application of Ohmic Heating for Inulin Extraction from the Wet-Milled and Dry-Milled Powders of Jerusalem Artichoke (Helianthus Tuberosus L.) Tuber. *Innov. Food Sci. Emerg. Technol.* **2018,** *48*, 99–110. DOI: 10.1016/j.ifset.2018.05.022.
31. Mesías, M.; Wagner, M.; George, S.; Morales, F. J. Impact of Conventional Sterilization and Ohmic Heating on the Amino Acid Profile in Vegetable Baby Foods. *Innov. Food Sci. Emerg. Technol.* **2016,** *34*, 24–28. DOI: 10.1016/j.ifset.2015.12.031.
32. Jha, S. N.; Narsaiah, K.; Basediya, A. L.; Sharma, R.; Jaiswal, P.; Kumar, R.; Bhardwaj, R. Measurement Techniques and Application of Electrical Properties for Nondestructive Quality Evaluation of Foods—A Review. *J. Food Sci. Technol.* **2011,** *48* (4), 387–411. https://doi.org/10.1007/s13197-011-0263-x.
33. Aamir, M.; Jittanit, W. Ohmic Heating Treatment for Gac Aril Oil Extraction: Effects on Extraction Efficiency, Physical Properties and Some Bioactive Compounds. *Innov. Food Sci. Emerg. Technol.* **2017,** *41*, 224–234. DOI: 10.1016/j.ifset.2017.03.013.
34. Lebovka, N.; Vorobiev, E. *Electrotechnologies for Extraction from Food Plants and Biomaterials*, 1st ed.; Springer: New York, 2009. DOI: 10.1007/978-0-387-79374-0.
35. Seidi Damyeh, M.; Niakousari, M.; Golmakani, M. T.; Saharkhiz, M. J. Microwave and Ohmic Heating Impact on the In Situ Hydrodistillation and Selective Extraction of Satureja Macrosiphonia Essential Oil. *J. Food Process. Preserv.* **2016,** *40* (4), 647–656. DOI: 10.1111/jfpp.12644.
36. Kumar, R. C.; Benal, M. M.; Prasad, B. D.; Krupashankara, M. S.; Kulkarni, R. S.; Siddaligaswamy, N. H. Microwave Assisted Extraction of Oil from Pongamia Pinnata Seeds. *Mater. Today: Proc.* **2018,** *5* (1, Part 3), 2960–2964. DOI: 10.1016/j.matpr.2018.01.094.
37. Megawati; Fardhyanti, D. S.; Sediawan, W. B.; Hisyam, A. Kinetics of Mace (Myristicae Arillus) Essential Oil Extraction Using Microwave Assisted Hydrodistillation: Effect of Microwave Power. *Ind. Crops Products* **2019,** *131*, 315–322. DOI: 10.1016/j.indcrop.2019.01.067.
38. Ibrahim, N. A.; Zaini, M. A. A. Microwave-Assisted Solvent Extraction of Castor Oil from Castor Seeds. *Chinese J. Chem. Eng.* **2018,** *26* (12), 2516–2522. DOI: 10.1016/j.cjche.2018.07.009.
39. Franco-Vega, A.; Ramírez-Corona, N.; López-Malo, A.; Palou, E. Studying Microwave Assisted Extraction of Laurus Nobilis Essential Oil: Static and Dynamic Modeling. *J. Food Eng*, **2019,** *247*, 1–8. DOI: 10.1016/j.jfoodeng.2018.11.007.

40. Ibrahim, A. P.; Omilakin, R. O.; Betiku, E. Optimization of Microwave-Assisted Solvent Extraction of Non-Edible Sandbox (Hura Crepitans) Seed Oil: A Potential Biodiesel Feedstock. *Renew. Energy* **2019,** *141*, 349–358. DOI: 10.1016/j.renene.2019.04.010.
41. Binello, A.; Orio, L.; Pignata, G.; Nicola, S.; Chemat, F.; Cravotto, G. Effect of Microwaves on the in Situ Hydrodistillation of Four Different Lamiaceae. *Comptes Rendus Chimie* **2014,** *17* (3), 181–186. DOI: 10.1016/j.crci.2013.11.007.
42. Pilar Santamarina, M.; Roselló, J.; Giménez, S.; Amparo Blázquez, M. Commercial Laurus Nobilis L. and Syzygium Aromaticum L. Merr. & Perry Essential Oils against Post-Harvest Phytopathogenic Fungi on Rice. *LWT—Food Sci. Technol.* **2016,** *65*, 325–332. DOI: 10.1016/j.lwt.2015.08.040.
43. Sellami, I. H.; Wannes, W. A.; Bettaieb, I.; Berrima, S.; Chahed, T.; Marzouk, B.; Limam, F. Qualitative and Quantitative Changes in the Essential Oil of Laurus Nobilis L. Leaves as Affected by Different Drying Methods. *Food Chem.* **2011,** *126* (2), 691–697. DOI: 10.1016/j.foodchem.2010.11.022.
44. Stevanato, N.; da Silva, C. Radish Seed Oil: Ultrasound-Assisted Extraction Using Ethanol as Solvent and Assessment of Its Potential for Ester Production. *Ind. Crops Products* **2019,** *132*, 283–291. DOI: 10.1016/j.indcrop.2019.02.032.
45. Castro-López, C.; Ventura-Sobrevilla, J. M.; González-Hernández, M. D.; Rojas, R.; Ascacio-Valdés, J. A.; Aguilar, C. N.; Martínez-Ávila, G. C. G. Impact of Extraction Techniques on Antioxidant Capacities and Phytochemical Composition of Polyphenol-Rich Extracts. *Food Chem.* **2017,** 237. https://doi.org/10.1016/j.foodchem.2017.06.032.
46. Muñiz-Márquez, D. B.; Martínez-Ávila, G. C.; Wong-Paz, J. E.; Belmares-Cerda, R.; Rodríguez-Herrera, R.; Aguilar, C. N. Ultrasound-Assisted Extraction of Phenolic Compounds from Laurus Nobilis L. and Their Antioxidant Activity. *Ultrasonics Sonochem.* **2013,** *20* (5), 1149–1154. https://doi.org/http://dx.doi.org/10.1016/j.ultsonch.2013.02.008.
47. Muñiz-Márquez, D. B.; Martínez-Ávila, G. C.; Wong-Paz, J. E.; Belmares-Cerda, R.; Rodríguez-Herrera, R.; Aguilar, C. N. Ultrasound-Assisted Extraction of Phenolic Compounds from Laurus Nobilis L. and Their Antioxidant Activity. *Ultrasonics Sonochem.* **2013,** *20* (5), 1149–1154. https://doi.org/http://dx.doi.org/10.1016/j.ultsonch.2013.02.008.
48. Taticchi, A.; Selvaggini, R.; Esposto, S.; Sordini, B.; Veneziani, G.; Servili, M. Physicochemical Characterization of Virgin Olive Oil Obtained Using an Ultrasound-Assisted Extraction at an Industrial Scale: Influence of Olive Maturity Index and Malaxation Time. *Food Chem.* **2019,** *289*, 7–15. DOI: 10.1016/j.foodchem.2019.03.041.
49. Zhang, Z.-S.; Wang, L.-J.; Li, D.; Jiao, S.-S.; Chen, X. D.; Mao, Z.-H. Ultrasound-Assisted Extraction of Oil from Flaxseed. *Sep. Purif. Technol.* **2008,** *62* (1), 192–198. DOI: 10.1016/j.seppur.2008.01.014.
50. Samaram, S.; Mirhosseini, H.; Tan, C. P.; Ghazali, H. M.; Bordbar, S.; Serjouie, A. Optimisation of Ultrasound-Assisted Extraction of Oil from Papaya Seed by Response Surface Methodology: Oil Recovery, Radical Scavenging Antioxidant Activity, and Oxidation Stability. *Food Chem.*, **2015,** *172*, 7–17. DOI: 10.1016/j.foodchem.2014.08.068.
51. Pereira, M. G.; Hamerski, F.; Andrade, E. F.; Scheer, A. de P.; Corazza, M. L. Assessment of Subcritical Propane, Ultrasound-Assisted and Soxhlet Extraction of Oil from Sweet Passion Fruit (Passiflora Alata Curtis) Seeds. *J. Supercrit. Fluids* **2017,** *128*, 338–348. DOI: 10.1016/j.supflu.2017.03.021.

52. Priyanka; Khanam, S. Influence of Operating Parameters on Supercritical Fluid Extraction of Essential Oil from Turmeric Root. *J. Clean. Product.* **2018,** *188*, 816–824. DOI: 10.1016/j.jclepro.2018.04.052.
53. Silva, S. G.; de Oliveira, M. S.; Cruz, J. N.; da Costa, W. A.; da Silva, S. H. M.; Barreto Maia, A. A.; de Sousa, R. L.; Carvalho Junior, R. N.; de Aguiar Andrade, E. H. Supercritical $CO_2$ Extraction to Obtain Lippia Thymoides Mart. & Schauer (Verbenaceae) Essential Oil Rich in Thymol and Evaluation of Its Antimicrobial Activity. *J. Supercrit. Fluids* **2021,** *168*, 105064. DOI: 10.1016/j.supflu.2020.105064.
54. De Corato, U.; Maccioni, O.; Trupo, M.; Di Sanzo, G. Use of Essential Oil of Laurus Nobilis Obtained by Means of a Supercritical Carbon Dioxide Technique Against Post Harvest Spoilage Fungi. *Crop Protect.* **2010,** *29* (2), 142–147. DOI: 10.1016/j.cropro.2009.10.012.
55. Soto-Armenta, L. C.; Sacramento-Rivero, J. C.; Ruiz-Mercado, C. A.; Lope-Navarrete, M. C.; Rocha-Uribe, J. A. Extraction Yield and Kinetic Study of Lippia Graveolens with Supercritical $CO_2$. *J. Supercrit. Fluids*, **2019,** *145*, 205–210. DOI: 10.1016/j.supflu.2018.12.018.
56. Beis, S. H.; Dunford, N. T. Supercritical Fluid Extraction of Daphne (Laurus Nobilis L.) Seed Oil. *J. Am. Oil Chem. Soc.* **2006,** *83* (11), 953–957. https://doi.org/10.1007/s11746-006-5052-8.
57. Ozek, T.; Bozan, B.; Baser, K. H. C. Supercritical $CO_2$ Extraction of Volatile Components from Leaves OfLaurus Nobilis L. *Chem. Nat. Compd.* **1998,** *34* (6), 668–671. https://doi.org/10.1007/BF02336090.
58. Fornari, T.; Ruiz-Rodriguez, A.; Vicente, G.; Vázquez, E.; García-Risco, M. R.; Reglero, G. Kinetic Study of the Supercritical $CO_2$ Extraction of Different Plants from Lamiaceae Family. *J. Supercrit. Fluids* **2012,** *64*, 1–8. DOI: 10.1016/j.supflu.2012.01.006.

# Index

# Phonetica

**International Journal of Phonetic Science**

Founded 1957 by E. Zwirner.
Edited by E. Zwirner 1957–1969, B. Malmberg 1962–1969,
H. Pilch 1962–1978, W. Bethge 1963–1969, G. Ungeheuer 1970–1978.
Redactor 1970–1978 W. Bethge. Edited by K.J. Kohler 1979–2013.





KARGER

Printed in Germany on
acid-free and non-aging paper (ISO 9706)
by Stückle Druck, Ettenheim

**Phonetica**

**Guidelines for Authors**
We strongly encourage authors to read the Guidelines for Authors at www.karger.com/pho_guidelines prior to submitting an article.

**Journal Contact**
For questions or comments, please contact the persons responsible who can be found at http://www.karger.com/Journal/Contact/224275.

## Journal Information

### Aims and Scope

Contemporary interdisciplinary research on phonetics employs a wide range of approaches, from instrumental measures to perceptual and neurocognitive procedures, to computational modelling, for investigating the properties and principles of phonetics in communicative settings across the world's languages. It also ranges across styles, types of language users, and communicative modalities (speech, sign, song). *Phonetica* is an international forum for phonetic science that covers all aspects of the subject matter, from phonetic and phonological descriptions, to articulatory and signal analytic measures of production, to perception, acquisition, and phonetic variation and change. *Phonetica* thus provides a platform for a comprehensive understanding of producer-perceiver interaction across languages and dialects, and of learning throughout the lifespan and across contexts. Papers published in this journal report expert original work that deals both with theoretical issues, new empirical data, and innovative methods and applications that help to advance the field.

**ISSN Print Edition:** 0031-8388
**ISSN Online Edition:** 1423-0321

**Journal Homepage:** www.karger.com/pho
**Bibliographic Indices:** This journal is regularly listed in bibliographic services, including Current Contents® and PubMed/MEDLINE.

**Publication Data:** *Phonetica* is published 6 times a year. Volume 76 with 6 issues appears in 2019.



**Subscription Orders:**
Orders can be placed at agencies, bookstores, or directly with the Publisher.

**S. Karger AG**
Medical and Scientific Publishers
Allschwilerstrasse 10
CH-4009 Basel
Switzerland

t: +41 61 306 11 11
f: +41 61 306 12 34
e: karger@karger.com
w: www.karger.com

(for courier services only:
Allschwilerstrasse 10
CH-4055 Basel)

**Change of Address:**
Both old and new addresses should be sent to the subscription source.
**Subscription Rates:** Subscriptions run for a full calendar year. Prices are given per year.

**Personal subscription:**

| *Print or Online* | *Print+Online combined* |
|---|---|
| CHF 303.00 | CHF 356.00 |
| EUR 273.00 | EUR 316.00 |
| USD 329.00 | USD 382.00 |

***postage and handling***
(added to print and print+online)
CHF 50.40 Europe, CHF 72.00 Overseas
EUR 45.60
USD 69.60

**Institutional subscription:**

| *Print or Online* | *Print+Online combined* |
|---|---|
| CHF 1514.00 | CHF 1741.00 |
| EUR 1364.00 | EUR 1569.00 |
| USD 1646.00 | USD 1893.00 |

***postage and handling***
(added to print and print+online)
CHF 63.00 Europe, CHF 90.00 Overseas
EUR 57.00
USD 87.00

**Back Volumes and Single Issues:** Information on availability and prices of single print issues and print or electronic back volumes can be obtained from Customer Service at *service@karger.com.*

KARGER


www.karger.com
karger@karger.com